Hatim Machrafi

Extended Non-Equilibrium Thermodynamics

Hatim Machrafi

Extended Non-Equilibrium Thermodynamics

From Principles to Applications in Nanosystems

CRC Press
Taylor & Francis Group
Boca Raton London New York

CRC Press is an imprint of the
Taylor & Francis Group, an **informa** business

CRC Press
Boca Raton and London

CRC Press
Taylor & Francis Group
6000 Broken Sound Parkway NW, Suite 300
Boca Raton, FL 33487-2742

CRC Press is an imprint of Taylor & Francis Group, an Informa business

Printed on acid-free paper

International Standard Book Number-13: 978-1-138-49639-2 (Hardback)

Visit the Taylor & Francis Web site at
http://www.taylorandfrancis.com

and the CRC Press Web site at
http://www.crcpress.com

MIX
Paper from responsible sources
FSC www.fsc.org FSC™ C013985

Printed in the United Kingdom
by Henry Ling Limited

Contents

Preface

Science and technology at nanoscale is occupying a larger space in academic and industrial activities. Systems that operate at nanoscales show behaviours that can be completely different from macroscale ones. Therefore, classical theories are unable to describe such systems. Several theories exist that describe systems at nanoscale, but often involve either serious assumptions or complicated formulations that are not "user-friendly," are in need of large computing times or are strongly limited to specific cases. Extended non-equilibrium thermodynamics provides powerful tools departing not from empirical or statistical considerations but from fundamental thermodynamic laws, proposing final solutions that are readily usable and recognizable for students, researchers and industry. Furthermore, the book deals with methods that allow combining easily the present theory with other fields of science. This book aims also at providing for a systematic presentation of extended non-equilibrium thermodynamics in nanosystems with a high degree of applicability. The book deals with how physical properties of systems behave as a function of their sizes. Moreover, we provide for a systematic approach to understand the behaviour of thermal, electrical, thermoelectric, photovoltaic and nanofluid properties in nanosystems. Case studies, applying such theories, are explored including nanoporous systems, solar panels, nanomedicine drug permeation and properties of nanoporous scaffolds.

The text in this book is intended to bring attention to how our theory can be applied to real-life applications in nanoscaled environments. Therefore, the objective is based to present an overview of the theory, its fundamental aspects and the development towards formulae, easily applicable to the energy and medical sector. Other non-equilibrium thermodynamics theories, classical (ir)reversible thermodynamics and theoretical considerations on extended thermodynamics have already been discussed in large extent elsewhere (*Understanding non-equilibrium thermodynamics* (Springer 2008), and *Extended irreversible thermodynamics* (Springer 2010)), including many academic discussions with some of the authors (G. Lebon and D. Jou). Fundamental theories, applied to nanosystems, are still quite a challenge, which encourages entering immediately into the subject of this book in order to put the focus on the link between theory and real application. Since basic concepts of this theory, such as temperature and entropy, become difficult to assess beyond the local equilibrium hypothesis, the solutions presented in this work are to be seen as pioneering, inviting for more research in this new domain.

Chapters 1–3 present the development of constitutive equations from the basics of extended non-equilibrium thermodynamics and show a methodology of obtaining such equations for several processes: heat conduction, electric conduction and mass flux are used as representative examples. Furthermore, heat conduction and transfer are employed as typical examples of how our methodology can be applied to nanomaterials and nanocomposites, validated by comparison with experimental data. Among nanocomposites, nanoporous systems are also considered. A set of basic rules and methods is established in these chapters, which will be used for more elaborated and complex processes in the next chapters.

Chapters 4–8 build on the methodology of the preceding ones. More complex processes are studied by extending the models from Chapters 1 to 3, leading to models for thermoelectric phenomena, thermal rectification, viscosity and thermal conduction of nanofluids, nanoporous media and their permeability as well as photovoltaic modelling coupled to thermoelectricity. These chapters go deeper into applicable fields, where case studies (from experimental data) are used to illustrate the link between the theory and the various applications.

Chapters 9–11 focus on real existing systems, modelled by the developments performed in the previous chapters. The three topics studied as examples are solar energy, nanomedicine and self-assembled structures. It is to be noted that modelling such topics, using the theory in this book, is quite recent with new developments. As such, these chapters serve as preliminary studies of how one may address such problems and propose solutions thereof. These solutions also open ways to further developments, inviting for more reflection on the proposed topics and applications to other neighbouring topics.

The applications presented in these chapters are still open to more elaborations. In that light, it is the purpose to present examples of how the tools that are developed in this book can be used for those applications. Moreover, where it is concerned, quantitative examples are presented. Finally, the methods of the theory and the corresponding applications facilitate further developments.

Hatim Machrafi
Liège-Brussels-Paris
July 2018

Author

Hatim Machrafi obtained his PhD from the Pierre and Marie Curie University (Paris 6), Paris, France. He has been working as a senior researcher at the University of Liège (ULiège), Liège, Belgium, and as visiting researcher at the free University of Brussels (ULB), Brussels, Belgium, specializing in non-equilibrium thermodynamics, nanotechnology applied to energetics and nanomedicine, and advanced materials. Recently, he is also active at the Sorbonne University, Paris, France, in the field of microfluidics and renewable energy.

Part I

General Considerations

1

Extended Non-Equilibrium Thermodynamics: Constitutive Equations at Small Length Scales and High Frequencies

1.1 Introduction

The increasing interest in nanotechnology has led to new insights in the study of heat transport. It is well known that heat transfer at micro- and nanoscale behaves differently from that at macroscales [1,2]. At small length scales, transport of heat in complex systems is best quantified by means of the so-called Knudsen number $Kn \equiv \ell/L$ with ℓ denoting the mean free path of the heat carriers, namely, phonons, and L, the characteristic dimension of the system under study. The Knudsen number becomes typically comparable or larger than one for micro- and nanosystems, in which case heat transport is referred to as ballistic, i.e. dominated by phonon collisions with the walls. If the Knudsen number is much smaller than one, heat transport is simply diffusive, i.e. dominated by phonon–phonon collisions inside the system and described by Fourier's law. For small-length-scale systems, as well as for high-frequency processes, Fourier's law is no longer valid. Another drawback associated with Fourier's law is that temperature spreads infinitely fast through the whole body, which is physically unsustainable. These observations justify the need to generalize Fourier's law. This can be achieved by various ways, for instance, via the resolution of Boltzmann's equation [3], making use of the dual-time approach [4] or by computer simulations [5]. Here, we follow a different route based on extended on-equilibrium thermodynamics [6] whose main characteristic is to upgrade the heat flux and higher order heat fluxes to the rank of independent variables. In most applications, for convenience, the analysis is restricted by taking heat flux as single extra variable. In this chapter, we go one step further by selecting an infinite number of higher order fluxes. The use of a large number of flux variables finds its justification in the recent progress of nanotechnology and high-frequency processes. The procedure described in this chapter is by no means limited to heat transfer but can be easily generalized to other transport phenomena, as electrical conduction and matter diffusion. The theoretical model for the general heat transport equation is presented in the next section.

1.2 A General Heat Transport Equation in Terms of High-Order Heat Fluxes

At micro- and nanoscales, heat transport is mostly influenced by non-local effects and high-frequency processes. The classical Fourier law

$$\boldsymbol{q} = -\lambda \nabla T, \tag{1.1}$$

relating the heat flux vector $\boldsymbol{q}$ to the temperature gradient ∇T, with λ denoting the thermal conductivity, is not applicable at short times and small spatial scales. In order to account for high frequencies, Fourier's law has been generalized by Cattaneo [7] under the form

$$\tau \partial_t \boldsymbol{q} + \boldsymbol{q} = -\lambda \nabla T, \tag{1.2}$$

with τ designating the relaxation time of the heat flux and ∂_t the partial time derivative. Cattaneo's relation is easily derivable by assuming that the entropy $\mathcal{s}(e,\boldsymbol{q})$ is depending not only on the internal energy e but also on the heat flux vector $\boldsymbol{q}$. The corresponding Gibbs equation in a rigid conductor at rest can be written as follows:

$$\partial_t \mathcal{s}(e,\boldsymbol{q}) = T^{-1}\partial_t e - \gamma_1 \boldsymbol{q} \cdot \partial_t \boldsymbol{q}, \tag{1.3}$$

where $\mathcal{s}$ and e are measured per unit volume, γ_1 is a phenomenological coefficient assumed to be $\boldsymbol{q}$-independent and a dot stands for the scalar product. It will be identified later on and shown to be related to the relaxation time τ and the thermal conductivity λ. Combining (1.3) with the energy conservation law for heat conductors $\partial_t e = -\nabla \cdot \boldsymbol{q}$ leads to Cattaneo's relation (1.2).

However, Cattaneo's relation is not able to cope with non-local effects which are dominant at small length scales. To take them into account, it is suggested to introduce a hierarchy of fluxes $\boldsymbol{Q}^{(1)}$, $\boldsymbol{Q}^{(2)}, \ldots, \boldsymbol{Q}^{(n)}$ with $\boldsymbol{Q}^{(1)}$ identical to the heat flux vector $\boldsymbol{q}$: $\boldsymbol{Q}^{(2)}$ (a tensor of rank two) is the flux of $\boldsymbol{q}$, $\boldsymbol{Q}^{(3)}$ the flux of $\boldsymbol{Q}^{(2)}$ and so on. From the kinetic theory point of view, the quantities $\boldsymbol{Q}^{(2)}$ and $\boldsymbol{Q}^{(3)}$ represent the higher moments of the velocity distribution. Up to the nth-order flux, the Gibbs equation generalizing relation (1.3) becomes

$$\partial_t \mathcal{s}\left(e, \boldsymbol{q}, \boldsymbol{Q}^{(1)}, \ldots, \boldsymbol{Q}^{(n)}\right) = T^{-1}\partial_t e - \gamma_1 \boldsymbol{q} \cdot \partial_t \boldsymbol{q} - \gamma_2 \boldsymbol{Q}^{(2)} \otimes \partial_t \boldsymbol{Q}^{(2)} - \cdots - \gamma_N \boldsymbol{Q}^{(N)} \otimes \partial_t \boldsymbol{Q}^{(N)}, \tag{1.4}$$

where the symbol $\otimes$ denotes the inner product of the corresponding tensors. Moreover, the time evolution of entropy is governed by a general balance equation which is given in the form:

$$\eta^s = \partial_t \mathcal{s} + \nabla \cdot \boldsymbol{J}^s \geq 0, \tag{1.5}$$

where $\boldsymbol{J}^s$ stands for the entropy flux and η^s for the rate of entropy production per unit volume which is positive definite to fulfil the second law of thermodynamics. To derive an expression for η^s, we need the result (1.4) for $\partial_t \mathcal{s}$ and a constitutive relation for $\boldsymbol{J}^s$ in terms of the set of variables.

It is natural to expect that $\boldsymbol{J}^s$ is not simply given by the classical expression $T^{-1}\boldsymbol{q}$ [8], but it will depend on all the higher order fluxes up to order n, which is given as

$$J^s = T^{-1}\boldsymbol{q} + \beta_1 \boldsymbol{Q}^{(2)} \cdot \boldsymbol{q} + \cdots + \beta_{N-1} \boldsymbol{Q}^{(N)} \otimes \boldsymbol{Q}^{(N-1)}, \tag{1.6}$$

where β_N denotes phenomenological coefficients. It is checked that, after substituting the expressions of $\partial_t \mathcal{s}$ and $\boldsymbol{J}^s$ given by Eqs. (1.4) and (1.6) in Eq. (1.5), respectively, and eliminating the time derivative $\partial_t e$ using the energy conservation law in Eq. (1.4), $\partial_t e = -\nabla \cdot \boldsymbol{q}$, the following is obtained for the entropy production:

$$\begin{aligned}\eta^s = &-\left(-\nabla T^{-1} + \gamma_1 \partial_t \boldsymbol{q} - \beta_1 \nabla \cdot \boldsymbol{Q}^{(2)}\right) \cdot \boldsymbol{q} \ldots \\ &- \sum_{n=2}^{N} \boldsymbol{Q}^{(n)} \otimes \left(\gamma_n \partial_t \boldsymbol{Q}^{(n)} - \beta_n \nabla \cdot \boldsymbol{Q}^{(n+1)} - \beta_{n-1} \nabla \boldsymbol{Q}^{(n-1)}\right) \geq 0\end{aligned} \tag{1.7}$$

The above bilinear expression in fluxes and forces (the quantities between parentheses) suggests the following hierarchy of linear flux–force relations:

$$\nabla T^{-1} - \gamma_1 \partial_t \boldsymbol{q} + \beta_1 \nabla \cdot \boldsymbol{Q}^{(2)} = \mathrm{v}_1 \boldsymbol{q} \tag{1.8}$$

$$\beta_{n-1} \nabla \boldsymbol{Q}^{(n-1)} - \gamma_n \partial_t \boldsymbol{Q}^{(n)} + \beta_n \nabla \cdot \boldsymbol{Q}^{(n+1)} = \mathrm{v}_n \boldsymbol{Q}^{(n)} \quad (n = 2, 3, \ldots N) \tag{1.9}$$

where we state that for a rigid body, $d_t \equiv \partial_t$. Our purpose is to replace the set of relations (1.8)–(1.9) by one single constitutive equation taking into account all the nth-order fluxes. For the sake of clarity, we will make the development up to the fourth order ($n = 4$) and generalize afterwards. The fourth-order flux equation [$n = 4$ in Eq. (1.9)] is given by

$$\beta_3 \nabla \boldsymbol{Q}^{(3)} - \gamma_4 \partial_t \boldsymbol{Q}^{(4)} + \beta_4 \nabla \cdot \boldsymbol{Q}^{(5)} = \mathrm{v}_4 \boldsymbol{Q}^{(4)} \tag{1.10}$$

Taking the divergence ($\nabla \cdot$) of Eq. (1.10), substituting it subsequently in Eq. (1.9) with $n = 3$, and omitting any flux $n > 4$, leads to

$$\beta_2 \nabla \boldsymbol{Q}^{(2)} - \gamma_3 \partial_t \boldsymbol{Q}^{(3)} + \frac{\beta_3}{\mathrm{v}_4} \left(\beta_3 \nabla \cdot \nabla \boldsymbol{Q}^{(3)} - \gamma_4 \nabla \cdot \partial_t \boldsymbol{Q}^{(4)} \right) = \mathrm{v}_3 \boldsymbol{Q}^{(3)} \tag{1.11}$$

Apply again operator divergence on (1.11), substitute it on its turn in Eq. (1.9) but with $n = 2$, and repeat the same operation until arriving at Eq. (1.9) with $n = 1$ [which is now equivalent to Eq. (1.8)]; the final result is

$$\begin{aligned} &\nabla T^{-1} - \gamma_1 \partial_t \boldsymbol{q} + \frac{\beta_1}{\mathrm{v}_2} \left(\beta_1 \nabla \cdot \nabla \boldsymbol{q} - \gamma_2 \nabla \cdot \partial_t \boldsymbol{Q}^{(2)} + \frac{\beta_2}{\mathrm{v}_3} \left(\beta_2 \nabla \cdot \nabla \cdot \nabla \boldsymbol{Q}^{(2)} \right.\right. \\ &\left.\left. - \gamma_3 \nabla \cdot \nabla \cdot \partial_t \boldsymbol{Q}^{(3)} + \frac{\beta_3}{\mathrm{v}_4} \left(\beta_3 \nabla \cdot \nabla \cdot \nabla \cdot \nabla \boldsymbol{Q}^{(3)} - \gamma_4 \nabla \cdot \nabla \cdot \nabla \cdot \partial_t \boldsymbol{Q}^{(4)} \right) \right) \right) = \mathrm{v}_1 \boldsymbol{q} \end{aligned} \tag{1.12}$$

This relation may be viewed as a generalized Cattaneo's relation up to the fourth-order flux $\boldsymbol{Q}^{(4)}$. However, it is not very convenient from a practical point of view because of the presence of the higher order fluxes which are not directly measurable and whose physical meaning is not clear. This is the reason why we propose in the next section a more suitable expression, wherein all the high-order fluxes have been eliminated.

1.3 A Generalized Transport Equation in Terms of the Heat Flux

Our purpose is to express Eq. (1.12) exclusively in terms of the more physical quantities that are the temperature T and the classical heat flux $\boldsymbol{q}$. The higher order fluxes $\boldsymbol{Q}^{(n)}$ are eliminated by making use of the very definition of the high-order flux, i.e.

$$\partial_t \boldsymbol{Q}^{(n)} = -\nabla \cdot \boldsymbol{Q}^{(n+1)} \quad (n = 1, 2, 3, \ldots N) \tag{1.13}$$

In virtue of Eq. (1.13) and after some rearrangements, Eq. (1.12) can be rewritten as

$$\begin{aligned} \frac{\gamma_1}{\mathrm{v}_1} \partial_t \boldsymbol{q} + \boldsymbol{q} = &-\frac{1}{\mathrm{v}_1 T^2} \nabla T + \frac{1}{\mathrm{v}_1} \frac{\beta_1}{\mathrm{v}_2} \beta_1 \nabla^2 \boldsymbol{q} + \frac{1}{\mathrm{v}_1} \frac{\beta_1}{\mathrm{v}_2} \gamma_2 \partial_t \partial_t \boldsymbol{q} - \frac{1}{\mathrm{v}_1} \frac{\beta_1}{\mathrm{v}_2} \frac{\beta_2}{\mathrm{v}_3} \beta_2 \partial_t \nabla^2 \boldsymbol{q} \\ &- \frac{1}{\mathrm{v}_1} \frac{\beta_1}{\mathrm{v}_2} \frac{\beta_2}{\mathrm{v}_3} \gamma_3 \partial_t \partial_t \partial_t \boldsymbol{q} + \frac{1}{\mathrm{v}_1} \frac{\beta_1}{\mathrm{v}_2} \frac{\beta_2}{\mathrm{v}_3} \frac{\beta_3}{\mathrm{v}_4} \beta_3 \partial_t \partial_t \nabla^2 \boldsymbol{q} + \frac{1}{\mathrm{v}_1} \frac{\beta_1}{\mathrm{v}_2} \frac{\beta_2}{\mathrm{v}_3} \frac{\beta_3}{\mathrm{v}_4} \gamma_4 \partial_t \partial_t \partial_t \partial_t \boldsymbol{q} \end{aligned} \tag{1.14}$$

By repeating the exercise for higher order fluxes with $n = N$, we obtain the following general equation governing heat conduction in rigid bodies:

$$\begin{aligned}\frac{\gamma_1}{\nu_1}\partial_t \boldsymbol{q} + \boldsymbol{q} = &-\frac{1}{\nu_1 T^2}\nabla T \\ &+ \frac{1}{\nu_1}\sum_{n=1}^{N}\left(\left(\prod_{i=1}^{n}\frac{\beta_i}{\nu_{i+1}}\right)(-1)^{n+1}\beta_n\frac{\partial^{n-1}}{\partial t^{n-1}}\nabla^2\boldsymbol{q}\right. \\ &+ \left.\left(\prod_{i=1}^{n}\frac{\beta_i}{\nu_{i+1}}\right)(-1)^{n+1}\gamma_{n+1}\frac{\partial^{n+1}}{\partial t^{n+1}}\boldsymbol{q}\right) \\ &(n = 1, 2, 3, \ldots, N),\ \text{with}\ N \in \mathbb{N}^+\end{aligned} \tag{1.15}$$

Note that for $n = 0$, (1.15) reduces to Cattaneo's law by setting $\beta_0 = 0$, i.e. by omitting non-local contributions. It remains to relate the various phenomenological coefficients to physical quantities. Setting $\gamma_1 = \beta_1 = 0$, Eq. (1.8) reduces to Fourier's law from which follows that $\nu_1 = \frac{1}{\lambda T^2}$. Comparing unities in Eq. (1.8) indicates that the dimension of the ratio γ_1/ν_1 is that of a time (t) so that $[\gamma_1] = [t][\nu_1]$. Defining the time unit as the heat flux ($\boldsymbol{q}$) relaxation time τ_1, one has $\gamma_1 = \tau_1\nu_1 = \frac{\tau_1}{\lambda T^2}$. As the heat flux $\boldsymbol{q}$ is expressed in W m^{-2}, the unity of $\boldsymbol{Q}^{(2)}$ is W (m^{-1}*s^{-1}) according to (1.13): $\left[\boldsymbol{Q}^{(n+1)}\right] = \frac{[L]}{[t]}\left[\boldsymbol{Q}^{(n)}\right]$, where $[L]$ is the unity in length. As a direct consequence of the property that $\boldsymbol{Q}^{(2)}$ is the flux of $\boldsymbol{q}$, one has in Eq. (1.8) that $\frac{\beta_1}{\gamma_1} = -1$ or $\beta_1 = -\gamma_1 = -\tau_1\nu_1$. Let us move to the second-order approximation. Defining the length scale as given by the mean free path of the phonons ℓ and the unit time by τ (making no longer a difference between the higher orders of the mean free paths and the relaxation times) leads to $\nu_2 = -\frac{\tau}{\ell^2}\beta_1 = -\frac{\tau^2}{\ell^2}\nu_1$ and further to $\beta_2 = -\gamma_2 = -\tau\nu_2$. At the nth order, it is directly inferred that

$$\begin{aligned}\gamma_n &= \tau\left(\frac{\tau}{\ell}\right)^{2n-2}\nu_1 \\ \beta_n &= -\tau\left(\frac{\tau}{\ell}\right)^{2n-2}\nu_1 \\ \nu_n &= \left(\frac{\tau}{\ell}\right)^{2n-2}\nu_1\end{aligned} \tag{1.16}$$

Substituting (1.16) in Eq. (1.15) and performing some rearrangements, one obtains the final result as follows:

$$\tau\partial_t\boldsymbol{q} + \boldsymbol{q} = -\lambda\nabla T + \sum_{n=1}^{N}\left(\ell^2\tau^{n-1}\frac{\partial^{n-1}}{\partial t^{n-1}}\nabla^2\boldsymbol{q} - \tau^{n+1}\frac{\partial^{n+1}}{\partial t^{n+1}}\boldsymbol{q}\right) \tag{1.17}$$

Within the perspective of practical applications, it is interesting to consider the three first-order approximations of Eq. (1.17) given by

$$\tau\partial_t\boldsymbol{q} + \boldsymbol{q} = -\lambda\nabla T + \ell^2\nabla^2\boldsymbol{q} - \tau^2\partial_t^2\boldsymbol{q} \tag{1.18}$$

$$\tau\partial_t\boldsymbol{q} + \boldsymbol{q} = -\lambda\nabla T + \ell^2\nabla^2\boldsymbol{q} - \tau^2\partial_t^2\boldsymbol{q} + \ell^2\tau\partial_t\nabla^2\boldsymbol{q} - \tau^3\partial_t^3\boldsymbol{q} \tag{1.19}$$

$$\tau\partial_t\boldsymbol{q} + \boldsymbol{q} = -\lambda\nabla T + \ell^2\nabla^2\boldsymbol{q} - \tau^2\partial_t^2\boldsymbol{q} + \ell^2\tau\partial_t\nabla^2\boldsymbol{q} - \tau^3\partial_t^3\boldsymbol{q} + \ell^2\tau^2\partial_t^2\nabla^2\boldsymbol{q} - \tau^4\partial_t^4\boldsymbol{q} \tag{1.20}$$

Let us briefly comment Eq. (1.18). Letting τ and ℓ tend to zero, one finds back Fourier's law $\boldsymbol{q} = -\lambda\nabla T$. With $\ell = 0$ and omitting the second-order derivatives in space and time, one recovers Cattaneo's relation, while setting only $\partial_t^2\boldsymbol{q} = 0$, expression (1.18) reduces to a Guyer-Krumhansl-like equation [9].

It may be convenient to reformulate the above results in dimensionless form. Let us, therefore, rescale the heat flux $\boldsymbol{q}$ with $\frac{\lambda(T_H-T_L)}{L}$, the space coordinate with the characteristic length L, the time t with τ, and define the dimensionless temperature by $T \to (T-T_L)/(T_H-T_L)$. Here, T_H and T_L are the high and low temperatures at the respective hot and cold sides of the system. Introducing the Knudsen number $Kn \equiv \frac{\ell}{L}$ and using the same notation for the non-dimensional quantities, Eq. (1.17) becomes

$$\partial_t \boldsymbol{q} + \boldsymbol{q} = -\nabla T + \sum_{n=1}^{N} \left(Kn^2 \frac{\partial^{n-1}}{\partial t^{n-1}} \nabla^2 \boldsymbol{q} - \frac{\partial^{n+1}}{\partial t^{n+1}} \boldsymbol{q} \right) \tag{1.21}$$

1.4 A Simplified Expression of Eq. (1.17)

Equation (1.21) describes both non-local effects and high-frequency processes. Solving it implies the determination of the high-order time derivatives $\partial^n \boldsymbol{q}/\partial t^n$ at $t=0$. However, if we consider only non-local processes, it is then desirable to replace the terms $\partial^n/\partial t^n$ $(n>2)$ from relation (1.17) by space-dependent variables. This is easily achieved by assuming that $\gamma_n \ll \gamma_1$, which amounts to neglect the higher relaxation times τ_n $(n \geq 2)$ in comparison with τ in Eq. (1.12). Under these assumptions and limiting the developments to $n=3$, the latter is expressed as

$$\nabla T^{-1} - \gamma_1 \partial_t \boldsymbol{q} + \frac{\beta_1}{\mathrm{v}_2} \left(\beta_1 \nabla \cdot \nabla \boldsymbol{q} + \frac{\beta_2}{\mathrm{v}_3} \left(\beta_2 \nabla \cdot \nabla \cdot \nabla \boldsymbol{Q}^{(2)} + \frac{\beta_3}{\mathrm{v}_4} \left(\beta_3 \nabla \cdot \nabla \cdot \nabla \cdot \nabla \boldsymbol{Q}^{(3)} \right) \right) \right) = \mathrm{v}_1 \boldsymbol{q} \tag{1.22}$$

Moreover, relation (1.9), wherein $n=2$ and $\gamma_2=0$, may be written as

$$\boldsymbol{Q}^{(2)} = \frac{\beta_1}{\mathrm{v}_2} \nabla \boldsymbol{q} + \frac{\beta_2}{\mathrm{v}_2} \nabla \cdot \boldsymbol{Q}^{(3)} \tag{1.23}$$

Substituting (1.23) in Eq. (1.22), making use of Eq. (1.16) and omitting high-order flux terms of $O\left(\boldsymbol{Q}^{(3)}\right)$ yields

$$\nabla T^{-1} - \gamma_1 \partial_t \boldsymbol{q} + \frac{\beta_1}{\mathrm{v}_2} \beta_1 \nabla^2 \boldsymbol{q} - \frac{\ell^2}{\tau} \frac{\beta_1}{\mathrm{v}_2} \frac{\beta_2}{\mathrm{v}_3} \beta_2 \nabla^4 \boldsymbol{q} = \mathrm{v}_1 \boldsymbol{q}, \tag{1.24}$$

which, in virtue of the relation (1.16), leads to

$$\tau \partial_t \boldsymbol{q} + \boldsymbol{q} = -\lambda \nabla T + \ell^2 \nabla^2 \boldsymbol{q} + \ell^4 \nabla^4 \boldsymbol{q}. \tag{1.25}$$

It is easy to verify that performing the same exercise for $n = 1, 2, 3, \ldots, N$, using the same rescaling notation for the non-dimensional quantities as for (1.21), results into

$$\partial_t \boldsymbol{q} + \boldsymbol{q} = -\nabla T + \sum_{n=1}^{N} \left(Kn^{2n} \nabla^{2n} \boldsymbol{q} \right) \tag{1.26}$$

1.5 One-Dimensional Numerical Illustration

As a numerical application, we consider the problem of heat conduction in a one-dimensional (1D) rigid body of length L at rest, subject to two different kinds of boundary conditions.

To assess the importance of the high-order fluxes, we calculate the temperature distribution in a 1D rigid body subject to a temperature difference $\nabla T = T_H - T_L$ between the hot and cold faces located at $z = 1$ and $z = 0$, respectively. The governing equations are expression (1.21) for $N = 2$, and the dimensionless energy balance written as

$$\partial_t T = -\tau \frac{\kappa_T}{L^2} \nabla \cdot \boldsymbol{q} = -M \nabla \cdot \boldsymbol{q} \tag{1.27}$$

where $\kappa_T (= \lambda / c_v)$ is the thermal diffusivity, c_v the heat capacity per unit volume and $M \equiv \frac{\tau}{\tau_T}$ a dimensionless number, with $\tau_T = L^2/\kappa_T$ the thermal characteristic time. The quantity M allows to compare ballistic to diffusion heat transfer; for $M \gg 1$, heat propagation is mainly of diffusive rather than ballistic nature, while the opposite holds for $M \ll 1$. Here, we will select $\tau \equiv \frac{1}{2}\tau_T$, to visualize the relative importance of both heat transfer mechanisms. The results provided by Eq. (1.21) will be compared to those obtained from Fourier's law through two different boundary conditions. Figure 1.1 presents a 3D plot of the temperature as a function of the dimensionless space coordinate z and time t. The cold side $z = 0$ is at a fixed temperature T_L, and at the hot side $z = 1$, two different boundary conditions are imposed: first, a fixed temperature $T = T_H$ and second, a periodic one

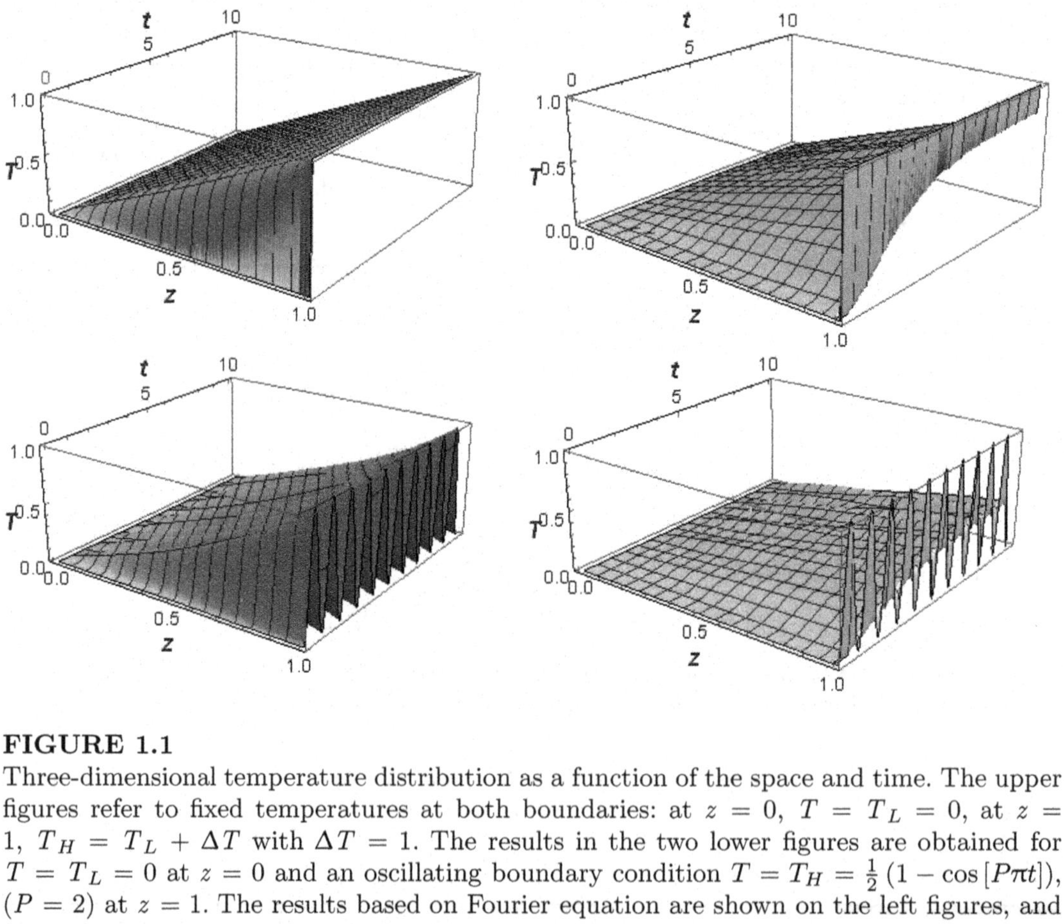

FIGURE 1.1

Three-dimensional temperature distribution as a function of the space and time. The upper figures refer to fixed temperatures at both boundaries: at $z = 0$, $T = T_L = 0$, at $z = 1$, $T_H = T_L + \Delta T$ with $\Delta T = 1$. The results in the two lower figures are obtained for $T = T_L = 0$ at $z = 0$ and an oscillating boundary condition $T = T_H = \frac{1}{2}(1 - \cos[P\pi t])$, $(P = 2)$ at $z = 1$. The results based on Fourier equation are shown on the left figures, and the results from this chapter $(N = 2)$ are on the right ones; the Knudsen number is taken as $Kn = 1$. (From: Machrafi, H., Lebon, G. 2016. General constitutive equations of heat transport at small length scales and high frequencies with extension to mass and electrical charge transport. *Applied Mathematics Letters* 52:30–37.)

$T_H = \frac{1}{2}(1 - \cos[P\pi t])$ with $P = 2/\tau$ Hz. Initially at $t = 0$, while the heat flux and its time derivatives are assumed to be zero.

By imposing an oscillatory temperature at one boundary, one observes that, with Fourier's law, "the heat signal reaches instantaneously the opposite boundary of the system, indicating that heat propagates at infinite speed. In contrast, with the (1.21) model, much of the oscillations are strongly attenuated (should M have been chosen much smaller than 1 or Kn much larger than one, the oscillations would have been dimmed instantaneously) and are not reminiscent of the oscillations imposed at the border, reflecting the property that heat propagates at finite speed. The same behaviour is noticed in the case of fixed temperatures at both sides. With Fourier's law, the temperature is felt everywhere in the bulk after application of the temperature difference at the boundaries, while with (1.21), the temperature is increasing gradually in the course of time and space. Typically, one also observes that (1.21) predicts a small persisting temperature jump, which is a property that is pointed out by several authors, e.g. [1,10].

1.6 Extension to Other Constitutive Laws

The previous sections were devoted to the problem of heat rigid conductors, and general constitutive equations for the heat flux were derived in the presence of strong non-localities and high frequencies. Formally, the same procedure can be extended to any kind of phenomena whose behaviour is governed by constitutive relations having the structure of Fourier's law, such as Fick's or Ohm's laws. Table 1.1 gives some examples of how such classical constitutive equations can be extended to small-length-scale systems and high-frequency processes. Note that thermoelectric effects like the Seebeck effect (expressing that a temperature gradient ∇T gives rise to an electric current density $\boldsymbol{I}$) or the Peltier effect (an electrical potential gradient ∇V produces a heat flux $\boldsymbol{q}$) can be treated in the same way by combining the expressions of the first and last equations of Table 1.1. Of course, similar generalizations can also be formulated from Eq. (1.26).

Generalizing Fick's and Ohm's equations requires specifying the corresponding set of state variables and their evolution equations. In case of matter diffusion in a two-component system, the state variables are $\boldsymbol{V}_F = (C, \boldsymbol{J})$, where C stands for the mass fraction of, say, the first constituent and $\boldsymbol{J}$ for the flux of matter. In case of electrical conduction, one has as state variables $\boldsymbol{V}_0 = (z, \boldsymbol{I})$, with z the electrical charge per unit mass (related to the electrical potential V by Poisson's equation $\rho z = -\varepsilon_0 \nabla^2 V$, where ε_0 is the electrical permittivity) and $\boldsymbol{I}$ the electrical current density. Finally, in Table 1.1, D denotes the coefficient of diffusion and σ the electrical conductivity. The balance equations for C and z are, respectively, $\rho\partial_t C = -\nabla \cdot \boldsymbol{J}$ and $\rho\partial_t z = -\nabla \cdot \boldsymbol{I}$.

TABLE 1.1
Generalized Fourier's, Fick's, and Ohm's Laws.

Fourier	$\boldsymbol{q} = -\lambda\nabla T_q$	$\tau\partial_t \boldsymbol{q} + \boldsymbol{q} = -\lambda\nabla T + \sum_{n=1}^{N}\left(\ell^2\tau^{n-1}\frac{\partial^{n-1}}{\partial t^{n-1}}\nabla^2\boldsymbol{q} - \tau^{n+1}\frac{\partial^{n+1}}{\partial t^{n+1}}\boldsymbol{q}\right)$
Fick	$\boldsymbol{J} = -\rho D\nabla C$	$\tau\partial_t \boldsymbol{J} + \boldsymbol{J} = -\rho D\nabla C + \sum_{n=1}^{N}\left(\ell^2\tau^{n-1}\frac{\partial^{n-1}}{\partial t^{n-1}}\nabla^2\boldsymbol{J} - \tau^{n+1}\frac{\partial^{n+1}}{\partial t^{n+1}}\boldsymbol{J}\right)$
Ohm	$\boldsymbol{I} = -\sigma\nabla V$	$\tau\partial_t \boldsymbol{I} + \boldsymbol{I} = -\sigma\nabla V + \sum_{n=1}^{N}\left(\ell^2\tau^{n-1}\frac{\partial^{n-1}}{\partial t^{n-1}}\nabla^2\boldsymbol{I} - \tau^{n+1}\frac{\partial^{n+1}}{\partial t^{n+1}}\boldsymbol{I}\right)$

1.7 Conclusions

The aim of this chapter was to generalize the usual transport equations of continuum physics by considering the problem of heat conduction in rigid bodies as a case study. This is achieved in the framework of extended non-equilibrium thermodynamics by using an infinite set of flux variables. The approach is original, and the two most important results are embodied in Eqs. (1.17) and (1.25). Including an infinite number of fluxes into the description is not merely a pure academic game but is of importance to describe very high-frequency processes and low scale phenomena as those occurring at micro- and nanoscales. The main advantage of relations (1.17) and (1.25) is that they do not ignore a priori the influence of the high-order fluxes. However, for convenience and within the perspective of practical applications, the relevant constitutive laws are not formulated in terms of the whole set of high-order fluxes but instead in terms of the sole heat flux $\boldsymbol{q}$, which is directly accessible to measurements, contrary to the higher order fluxes. The analysis is also easily transposable to other situations involving electrical current (Ohm's law) and matter diffusion (Fick's law), keeping in mind the corresponding balance equations. As an application, we have calculated heat propagation in a 1D rigid rod submitted to a fixed temperature at one side, the other side being subject to either an oscillatory or a fixed temperature. The numerical results show explicitly that the proposed model predicts a finite velocity of propagation of heat signals, in opposition to Fourier's law, which leads to temperature disturbances propagating with boundless speed within the limit of high frequencies.

References

[1] Machrafi, H., Lebon, G. 2016. General constitutive equations of heat transport at small length scales and high frequencies with extension to mass and electrical charge transport. *Applied Mathematics Letters* 52:30–37.

[2] Hill, T.L. 1994. *Thermodynamics of Small Systems.* New York: Dover.

[3] Koh, Y.K., Cahill, D.G., Sun, B. 2014. Nonlocal theory for heat transport at high frequencies. *Physical Review B* 90:205412.

[4] Tzou, D.Y. 1997. *Macro to Microscale Heat Transfer. The Lagging Behaviour.* New York: Taylor & Francis Group.

[5] Sabliov, C.M., Salvi, D.A., Boldor, D. 2007. High frequency electromagnetism, heat transfer and fluid flow coupling in ANSYS multiphysics. *Journal of Microwave Power and Electromagnetic Energy* 41:5–17.

[6] Jou, D., Casas-Vazquez, J., Lebon, G. 2010. *Extended Irreversible Thermodynamics*, fourth ed. Berlin: Springer-Verlag.

[7] Cattaneo, C. 1948. Sulla conduzione del calore, Atti del Seminario Matematico e Fisico dell' Universita di Modena 3:83–101.

[8] De Groot, S.R., Mazur, P. 1962. *Non-equilibrium Thermodynamics.* Amsterdam: North-Holland Publishing Company.

[9] Guyer, R.A., Krumhansl, J.A. 1966. Solution of the linearized Boltzmann phonon equation. *Physical Review* 148:766–778.

[10] Swartz, E.T., Pohl, R.O. 1989. Thermal boundary resistance. *Reviews of Modern Physics* 61:605–668.

2

Heat Transfer in Nanomaterials

2.1 Transient Heat Transport in Nanofilms

Micro- and nanomaterials are characterized by the property that the ratio of the mean free path l of the heat carriers to the mean dimension L of the system, the Knudsen number $Kn = l/L$, is comparable or larger than unity. In the present approach, we assume the coexistence of two kinds of heat carriers: diffusive phonons which undergo multiple collisions within the core of the system, and ballistic phonons which originate at the boundaries and experience collisions with the walls only. This model is called the ballistic–diffusion one. Accordingly, the internal energy and the heat flux are decomposed into a ballistic and a diffusive component in such a way that

$$u = u_b + u_d, \tag{2.1}$$

$$\boldsymbol{q} = \boldsymbol{q}_b + \boldsymbol{q}_d \tag{2.2}$$

where the indexes b and d refer to ballistic and diffusive contributions, respectively. The construction of the model proceeds in four steps.

2.1.1 Definition of the Space of State Variables

According to the above decomposition of u and $\boldsymbol{q}$, the state variables are selected as follows:

i. the couple $u_d - \boldsymbol{q}_d$ in order to account for the diffusive behaviour of the heat carriers;

ii. the couple $u_b - \boldsymbol{q}_b$ so as to provide a description of the ballistic motion of the carriers.

To summarize, the space of state variables is constituted by the quantities u_d, u_b, $\boldsymbol{q}_d$ and $\boldsymbol{q}_b$. For future use, we introduce also the so-called diffusive and ballistic temperatures T_d and T_b, defined, respectively, by $T_d = u_d/c_d$ and $T_b = u_b/c_b$, where c_d and c_b have the dimension of a heat capacity per unit volume and are positive quantities to guarantee stability of the equilibrium state. Admitting that $c_d = c_b = c$, the total temperature is defined by $T = u/c$ or, equivalently, $T = T_d + T_b$. Although the quantities T_d, T_b and T bear some analogy with the classical definition of the temperature, it should however be realized that, strictly speaking, these quantities do not represent temperatures in the usual sense but must be considered as a measure of the internal energies.

2.1.2 Establishment of the Evolution Equations

After having defined the sate variables, one must specify their behaviour in the course of time and space. The evolutions of the internal energies u_d and u_b are governed by the classical energy balance laws:

$$\frac{\partial u_d}{\partial t} = -\nabla \cdot \boldsymbol{q}_d + r_d \tag{2.3}$$

$$\frac{\partial u_b}{\partial t} = -\nabla \cdot \boldsymbol{q}_b + r_b \tag{2.4}$$

while the total internal energy $u = u_d + u_b$ satisfies the first law of thermodynamics:

$$\frac{\partial u}{\partial t} = -\nabla \cdot \boldsymbol{q} + r \tag{2.5}$$

The quantities r_d, r_b and r designate source terms, which may be either positive or negative. In virtue of Eq. (2.5), one has to satisfy $r_d + r_b = r$. In the absence of energy sources ($r = 0$), one simply has that $r_d = -r_b$. Based on kinetic theory considerations [1], it is shown that

$$r_b = -u_b/\tau_b \tag{2.6}$$

where the minus sign indicates that ballistic carriers can be converted into diffusive ones, but that the inverse is not possible. Moreover, τ_b is the relaxation time of the ballistic energy flux $\boldsymbol{q}_b$. Now, it remains to derive the evolution equation for the fluxes. Concerning the diffusive phonons, it is assumed that they satisfy Cattaneo's equation to cope with their high-frequency properties. As a consequence, we are allowed to write

$$\tau_d \frac{\partial \boldsymbol{q}_d}{\partial t} + \boldsymbol{q}_d = -\lambda_d \nabla T_d \tag{2.7}$$

wherein the relaxation time τ_d and the thermal conductivity coefficient λ_d are positive quantities to meet the requirements of stability of equilibrium and positivity of the entropy production, respectively [2,3]. However, expression (2.7) is not able to describe the ballistic regime, which is mainly influenced by non-local effects as most of the ballistic carriers cross the system without experiencing collisions except with the boundaries. As shown before [2,4,5], this situation is satisfactorily described through Guyer–Krumhansl's equation:

$$\tau_b \frac{\partial \boldsymbol{q}_b}{\partial t} + \boldsymbol{q}_b = -\lambda_b \nabla T_b + l_b^2 \left(\nabla^2 \boldsymbol{q}_b + 2\nabla\nabla \cdot \boldsymbol{q}_b\right) \tag{2.8}$$

where l_b is related to the correlation length of the ballistic phonons. Note that the terms involving the space derivatives of the heat flux vector account for the non-local effects and are important when the spatial scale of variation of the heat flux is comparable to the mean free path of the heat carriers. From the kinetic point of view, Guyer and Krumhansl have shown that τ_b can be identified with the collision time τ_R of the resistive phonons collisions (non-conserving momentum collisions) and that $l_b^2 = \frac{1}{5} v_{ph}^2 \tau_R \tau_N$ with v_{ph} the mean velocity of phonons and τ_N the collision time of normal (momentum conserving) phonons collisions. Let us also mention that the relaxation times, the mean free paths and the heat conductivities are not independent but according to the phonon kinetic theory kinetic, they are related as follows:

$$\lambda_{d,b} = \frac{1}{3} c_{d,b} v_{d,b}^2 \tau_{d,b} \tag{2.9}$$

wherein $v_{d,b} = l_{d,b}/\tau_{d,b}$ is the mean velocity of phonons. Expressions (2.3), (2.4), (2.7) and (2.8) provide the basic set of eight scalar evolution equations for the eight unknowns u_d, u_b, $\boldsymbol{q}_d$ and $\boldsymbol{q}_b$.

2.1.3 Elimination of the Fluxes

The purpose is now to eliminate $\boldsymbol{q}_d$ and $\boldsymbol{q}_b$, assuming that all the transport coefficients are constant. Application of operator $\nabla\cdot$ on Cattaneo's equation (2.7) and use of $T_d = u_d/c_d$ yields

$$\tau_d \nabla \cdot (\partial_t \boldsymbol{q}_d) + \nabla \cdot \boldsymbol{q}_d = -\frac{\lambda_d}{c_d} \nabla^2 u_d \tag{2.10}$$

with ∂_t the time derivative. Moreover, the balance of the total energy can be written in the form:

$$\nabla \cdot \boldsymbol{q}_d = -\partial_t u - \nabla \cdot \boldsymbol{q}_b = -\partial_t u_d - \partial_t u_b - \nabla \cdot \boldsymbol{q}_b \tag{2.11}$$

After differentiating (2.11) with respect to time and substituting in Eq. (2.10), the following expression is obtained:

$$-\tau_d \partial_t^2 (u_d + u_b) - \tau_d \nabla \cdot (\partial_t \boldsymbol{q}_b) = \nabla \cdot \boldsymbol{q}_b + \partial_t u_d + \partial_t u_b - \frac{\lambda_d}{c_d} \nabla^2 u_d \tag{2.12}$$

wherein $\boldsymbol{q}_d$ has been eliminated. We now eliminate $\tau_d \partial_t^2 u_b$ by taking the time derivative of Eq. (2.4) with $r_b = -u_b/\tau_b$, and we multiply this equation by τ_d. The result is

$$\tau_d \partial_t^2 u_b = -\tau_d \nabla \cdot (\partial_t \boldsymbol{q}_b) - \frac{\tau_d}{\tau_b} \partial_t u_b \tag{2.13}$$

Substituting this result in Eq. (2.12) and replacing in the right-hand side of Eq. (2.12) the terms $\nabla \cdot \boldsymbol{q}_b + \partial_t u_b$ by $-u_b/\tau_b$ in virtue of Eq. (2.4), we finally find the equation for u_d:

$$\tau_d \partial_t^2 u_d + \partial_t u_d - \frac{\tau_d}{\tau_b} \partial_t u_b - u_b/\tau_b = \frac{\lambda_d}{c_d} \nabla^2 u_d \tag{2.14}$$

It remains to find the equation for u_b. To derive this, we start from the time derivative of the energy balance (2.4), which is multiplied by τ_b, taking the form:

$$\tau_b \partial_t^2 u_b = -\tau_b \nabla \cdot (\partial_t \boldsymbol{q}_b) - \partial_t u_b \tag{2.15}$$

Now, we use Eq. (2.8), to which we apply the operator $\nabla \cdot$, leading to

$$\tau_b \nabla \cdot (\partial_t \boldsymbol{q}_b) + \nabla \cdot \boldsymbol{q}_b = -\frac{\lambda_b}{c_b} \nabla^2 u_b + l_b^2 \nabla \cdot (\nabla^2 \boldsymbol{q}_b + 2 \nabla \nabla \cdot \boldsymbol{q}_b) \tag{2.16}$$

After substitution of Eq. (2.16) in Eq. (2.15) and use of Eq. (2.4) to eliminate $\nabla \cdot \boldsymbol{q}_b$, one obtains, after some elementary arithmetic, the relation for u_b:

$$\tau_b \partial_t^2 u_b + 2 \partial_t u_b + u_b/\tau_b = \frac{\lambda_b}{c_b} \nabla^2 u_b + 3 l_b^2 \nabla \cdot \left(\nabla (\partial_t u_b) + \frac{1}{\tau_b} \nabla u_b \right) \tag{2.17}$$

Expressions (2.14) and (2.17) are the key relations of the temperature model. Making use of Eq. (2.4) and setting $l_b = 0$, $\tau_d = \tau_b = \tau$, one recovers the basic result from expression (2.14), that is

$$\tau \partial_t^2 u_d + \partial_t u_d + \nabla \cdot \boldsymbol{q}_b = \frac{\lambda_d}{c_d} \nabla^2 u_d \tag{2.18}$$

which differs from the telegraph equation that should be obtained from Cattaneo's law by the presence of the term $\nabla \cdot \boldsymbol{q}_b$. The heat flux vector $\boldsymbol{q}_b$ can be obtained by using the kinetic definition of the heat flux and by solving Boltzmann's equation. Here, we do not refer to a kinetic approach but solve the problem exclusively at the macroscopic level. It should also be underlined that this renders our model more general instead of introducing, without any justification, the simplifying assumptions that $\tau_d = \tau_b$. As illustration, we will apply the above considerations to the study of transient heat transport through a thin film whose two faces are subject to a temperature difference.

2.2 Transient Temperature Distribution in Thin Films

Thin films are defined here as films in which the thickness L may be of the same order of magnitude or even smaller than the mean free path l of the phonons. Heat capacity and thermal conductivity are assumed to be constant, and to take the same values for the diffusive and ballistic phonons, internal energy sources are absent ($r = 0$). Initially, the system is at uniform energy u_0 or, using an equivalent terminology, at the "quasi-temperature" T_0 related to u_0 by $u_0 = cT_0$. Often, for thin films in real applications, the other dimensions are much larger, so that we can neglect their effect and consider only a one-dimensional model, with coordinate z. The lower surface $z = 0$ is suddenly brought at $t = 0$ to the "quasi-temperature" $T_l = T_0 + \Delta T$, whereas the upper surface $z = L$ is kept at "quasi-temperature" T_0. For further purpose, we introduce the Knudsen numbers $Kn_i = l_i/L$ ($i = d$, b), which, in virtue of expression (2.9), taking $\lambda_d = \lambda_b = \lambda$, can be given the more general form:

$$Kn_i^2 = 3\lambda\tau_i/cL^2 \quad (i = d, b) \tag{2.19}$$

Having in mind numerical solutions, it is convenient to use dimensionless quantities with the following scalings (the asterisk denoting the dimensionless parameter):

$$\left.\begin{aligned} t^* &= t/\tau_b \\ z^* &= z/L \\ \theta_d &= \tfrac{u_d - u_d(z=L)}{c\Delta T} \\ \theta_b &= \tfrac{u_b - u_b(z=L)}{c\Delta T} \\ \theta &= \tfrac{u - cT(z=L)}{c\Delta T} \end{aligned}\right\} \tag{2.20}$$

with θ_d, θ_b and $\theta(=\theta_d + \theta_b)$ designating the non-dimensional energy (or temperature) associated with the diffusive, ballistic and total energy, respectively. The corresponding evolution Eqs. (2.14) and (2.17) take now the form:

$$\frac{Kn_d^2}{Kn_b^2}\left(\frac{\partial^2\theta_d}{\partial t^{*2}} - \frac{\partial\theta_b}{\partial t^*}\right) - \frac{Kn_b^2}{3}\frac{\partial^2\theta_d}{\partial z^{*2}} + \frac{\partial\theta_d}{\partial t^*} - \theta_b = 0 \tag{2.21}$$

$$\frac{\partial^2\theta_b}{\partial t^{*2}} + 2\frac{\partial\theta_b}{\partial t^*} - \frac{10Kn_b^2}{3}\frac{\partial^2\theta_b}{\partial z^{*2}} - 3Kn_b^2\frac{\partial^3\theta_b}{\partial t^*\partial z^{*2}} + \theta_b = 0 \tag{2.22}$$

2.2.1 Initial Conditions

At $t = 0$, the sample is at uniform temperature T_0, which implies that the total energy is given by $u(z, 0) = u_d(z, 0) + u_b(z, 0) = cT_0$. However, it is reasonable to suppose that at short times, the ballistic phonons are dominant so that the initial energy will be essentially of ballistic nature leading to $u_b(z, 0) = cT_0$ or in dimensionless notation:

$$\left.\begin{aligned} \theta_b\,(z^*, 0) &= 0 \\ \theta_d\,(z^*, 0) &= 0 \end{aligned}\right\}. \tag{2.23}$$

Throughout the sample, at time $t = 0$, the heat flux $\boldsymbol{q}$ is also zero; as a consequence of the energy balance (2.3)–(2.4), it is checked that initially $\frac{\partial\theta(z^*,t^*)}{\partial t^*} = 0$. This result remains, in particular, satisfied under the assumptions

$$\left.\begin{aligned}\left.\frac{\partial\theta_d(z^*,t^*)}{\partial t^*}\right|_{t^*=0} &= 0\\ \left.\frac{\partial\theta_b(z^*,t^*)}{\partial t^*}\right|_{t^*=0} &= 0\end{aligned}\right\}. \tag{2.24}$$

2.2.2 Boundary Conditions

The formulation of the boundary conditions is a more delicate problem. Their importance has to be underlined because in nanomaterials, their influence is felt throughout the whole system. To satisfy the conditions $\theta(0,\, t^*) = 1$ and $\theta(1,\, t^*) = 0$, the simplest tentative would be to suppose that, at $z^* = 0$, $\theta_b(0,\, t^*) = 1$ together with $\theta_d(0,\, t^*) = 0$, whereas at $z^* = 1$, the temperature of both the ballistic and diffusive constituents would be zero. However, such expressions are too simple and do not, in particular, cope with temperature jumps due to thermal boundary resistance. This is the reason why we have considered the following boundary conditions for the ballistic carriers:

$$\left.\begin{aligned}\theta_b\,(0,t^*) &= a\\ \theta_b\,(1,t^*) &= 0\end{aligned}\right\}. \tag{2.25}$$

The quantity a, which represents the temperature jump of the ballistic phonons at the face $z^* = 0$ at $t^* = 0$, is taken equal to $\frac{1}{2}$. This value may be understood statistically. Since the temperature boundary condition at $z^* = 0$ actually represents an internal energy boundary condition, it can be said that the ballistic phonons which are generated at the heated face are formed by half of the carriers at the initial internal energy $\theta_b = 0$ and the other half at the value $\theta_b = 1$, corresponding to the energy at the face where the temperature is suddenly increased. A discussion on this result can be found in [1] where it was mentioned that, by solving Boltzmann's equation, $\theta_b(0, t^*) = \frac{1}{2}$ at the heated boundary $z^* = 0$. A posteriori, it is shown later on that this value leads to results, which match satisfactory well with other different approaches. Concerning the diffusive carriers, we assume that both of the interfaces are black phonon emitters and absorbers, implying that the boundaries are made of incident diffusive carriers only. Combining Cattaneo's equation [6] and Marshak's boundary condition [7,8] for black body thermal radiation, one obtains for $z^* = 1$, 0:

$$\frac{Kn_d^2}{Kn_b^2}\frac{\partial\theta_d}{\partial t^*} + \theta_d = \pm\frac{2}{3}Kn_d\frac{\partial\theta_d}{\partial z^*} \tag{2.26}$$

the positive and negative signs at the right-hand side correspond to the upper $z^* = 1$ and lower $z^* = 0$ faces, respectively, while the factor $(Kn_d/Kn_b)^2$ stems from the non-equality of relaxation times.

2.2.3 Discussion of the Results

Here, we assume $Kn_d = Kn_b = Kn$, because it is wanted to check the validity of our model by comparing with previous different approaches. In particular, we have compared our results with those of Cattaneo's and Fourier's laws. For this, we have solved the hyperbolic Cattaneo and the parabolic Fourier equations for the identical geometry and boundary conditions.

Figures 2.1 and 2.2 represent the non-dimensional temperature profiles for $Kn = 1$ & 10 versus the distance at different times. To emphasize the specific roles of the two constituents, we have made explicit the contributions of the total, ballistic and diffusive components. The region close to the hot side is mainly dominated by the ballistic component contribution which decreases with space while the diffusive one is increasing up to a maximum, after which one observes a descent towards zero, and the descent is the steepest for a smaller Kn.

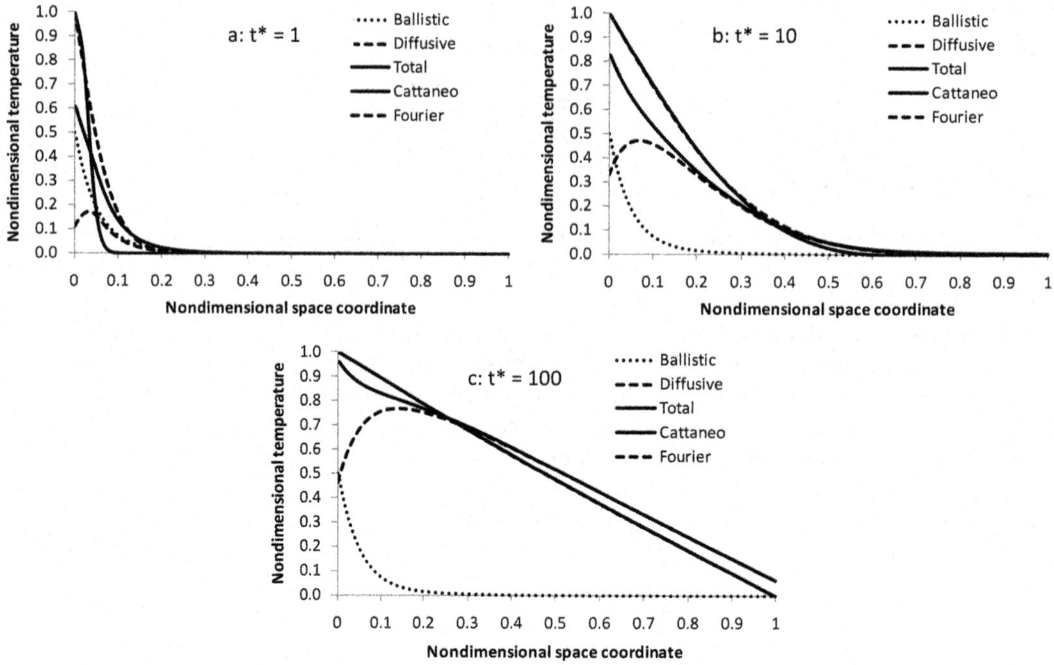

FIGURE 2.1
Non-dimensional temperature profiles $\theta(z^*, t^*)$ as a function of distance $z^* = z/L$ at different times $t^* = t/\tau_b$ ($t^* = 1$, 10 and 100, respectively) for $Kn_d = Kn_b = Kn = 0.1$. The respective contributions of the ballistic, diffusive and total temperatures are shown and compared to the ones obtained from Cattaneo's and Fourier's equations. (Modified from: Lebon, G., Machrafi, H., Grmela, M., Dubois, C. 2011. An extended thermodynamic model of transient heat conduction at sub-continuum scales. *Proceedings of the Royal Society A* 467:3241–3256.)

As expected, the influence of the ballistic constituent becomes more important as Kn is larger while the role of the diffusive one is dominant for small Kn and is growing with time. This observation reflects the conversion of the ballistic internal energy into the diffusive one as time is going on. It is also shown that for $Kn = 10$, the steady state is reached rather soon (after $t^* = 1$) and is decreasing linearly with space (see Figure 2.2). The results are in qualitative accord with the aforementioned formalisms with however small discrepancies at small times ($t^* < 0.1$) especially for $Kn = 10$. It is concluded that our description matches the results derived from various points of view, ranging from macroscopic, microscopic and mixed micro–macro approaches. Note also that for a higher value of Kn (especially $Kn = 10$), a temperature jump is observed at the cold face. This indicates that the ballistic part exhibits a strong wall resistance not only at the hot but also at the cold face. The small bump just before the temperature jump is caused by numerical errors due to the abrupt temperature change. It is clearly seen that both Cattaneo and Fourier descriptions lead to unrealistic results. Neither of these models predicts the temperature jumps at the boundary; moreover, they yield overestimated values for the temperature profiles as they do not integrate the specific properties of heat transport at nanoscales, and this is particularly true at a large value of Kn.

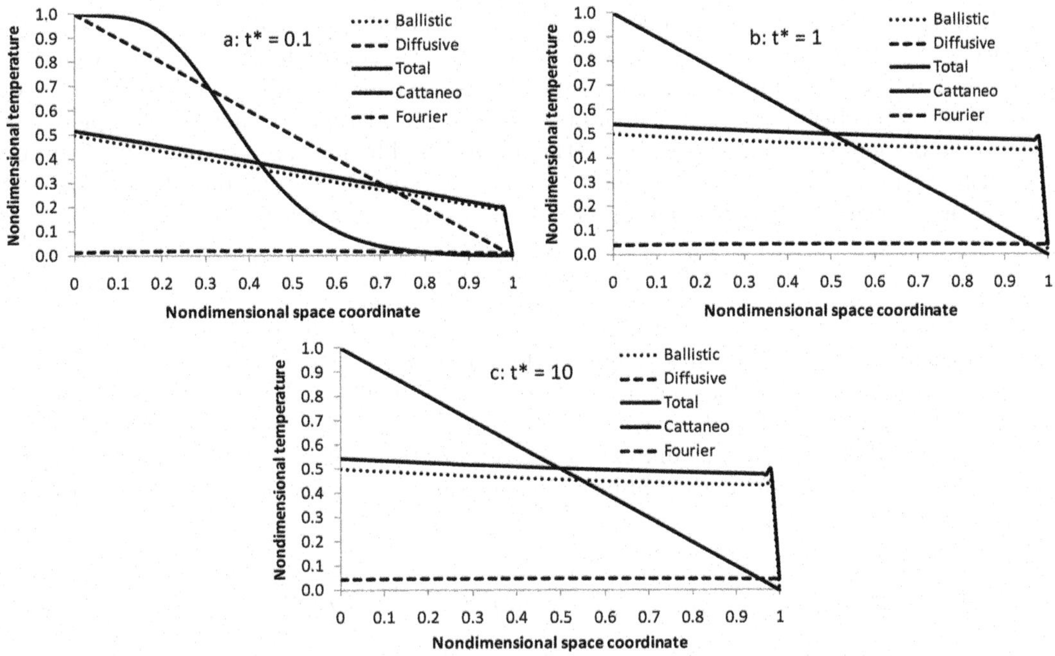

FIGURE 2.2
Non-dimensional temperature profiles $\theta(z^*, t^*)$ as a function of distance $z^* = z/L$ at different times $t^* = t/\tau_b$ ($t^* = 0.1$, 1 and 10, respectively) for $Kn_d = Kn_b = Kn = 10$. The respective contributions of the ballistic, diffusive and total temperatures are shown and compared to the ones obtained from Cattaneo's and Fourier's equations. (Modified from: Lebon, G., Machrafi, H., Grmela, M., Dubois, C. 2011. An extended thermodynamic model of transient heat conduction at sub-continuum scales. *Proceedings of the Royal Society A* 467:3241–3256.)

2.3 Heat Conduction in Nanoparticles Through an Effective Thermal Conductivity

One way of studying heat conduction in a nanoscaled material is to consider extended constitutional equations and solving them with appropriate boundary conditions. This has been done in the previous sections. Another way to investigate the heat conduction is to use Fourier's law, but with an effective thermal conductivity that takes into account the size of the material. We will study two geometries: nanowires (in this section) and nanoparticles (in the next one). To emphasize the role of the size effects on the thermal conductivity λ_p of the nanowires, we write λ_p in the form:

$$\lambda_p = \lambda_p^0 f_p^c \tag{2.27}$$

wherein λ_p^0 contains all the contributions except those linked to the sizes of the nanoparticles which are described by the correcting factor f_p^c. The quantity λ_p^0 is given by a relation similar to (2.9) with sub-index p

$$\lambda_p^0 = \frac{1}{3} c_p \ell_p v_p, \tag{2.28}$$

but wherein l_p refers only to the bulk contribution as the effects of ballistic collisions will be included in f_p^c. To determine f_p^c, we upgrade the heat flux and its higher order fluxes to the rank of independent variables at the same footing as the energy or the temperature. Part of this procedure has been presented in Chapter 1, but is partly repeated here for completeness and also to demonstrate the versatility of the approach. The first step consists in assuming that the entropy $\mathcal{S}(e, \boldsymbol{q})$ is depending not only on the internal energy e but also on the heat flux vector $\boldsymbol{q}$ so that the corresponding Gibbs equation can be written as

$$d_t\mathcal{S}(e, \boldsymbol{q}) = T^{-1}d_te - \gamma_1\boldsymbol{q}\cdot d_t\boldsymbol{q}, \tag{2.29}$$

wherein $\mathcal{S}$ and e are measured per unit volume, and T is the temperature and γ_1 a phenomenological coefficient identified later on. Furthermore, d_t denotes the time derivative. However, expression (2.29) does not account for non-local effects. To introduce such effects, it is appealed to a hierarchy of fluxes $\boldsymbol{Q}^{(1)}$, $\boldsymbol{Q}^{(2)}$, $\boldsymbol{Q}^{(N)}$ with $\boldsymbol{Q}^{(1)}$ identified with the heat flux vector $\boldsymbol{q}$, $\boldsymbol{Q}^{(2)}$ (a tensor of rank two) is the flux of the heat flux, $\boldsymbol{Q}^{(3)}$ the flux of $\boldsymbol{Q}^{(2)}$, etc. Up to the nth-order flux, the Gibbs equation generalizing relation (2.29) becomes

$$\begin{aligned} &d_t\mathcal{S}\left(e, \boldsymbol{q}, \boldsymbol{Q}^{(1)}, \ldots, \boldsymbol{Q}^{(n)}\right) \\ &\quad = T^{-1}d_te - \gamma_1\boldsymbol{q}\cdot d_t\boldsymbol{q} - \gamma_2\boldsymbol{Q}^{(2)}\otimes d_t\boldsymbol{Q}^{(2)} - \cdots - \gamma_n\boldsymbol{Q}^{(n)}\otimes d_t\boldsymbol{Q}^{(n)}, \end{aligned} \tag{2.30}$$

where the symbol $\otimes$ denotes the inner product of the corresponding tensors. The second step is the formulation of the entropy flux J^s. It is natural to expect that it is not simply given by the classical expression $T^{-1}\boldsymbol{q}$, but that it will depend on higher order fluxes, as

$$J^s = T^{-1}\boldsymbol{q} + \beta_1\boldsymbol{Q}^{(2)}\cdot\boldsymbol{q} + \cdots + \beta_{n-1}\boldsymbol{Q}^{(n)}\otimes\boldsymbol{Q}^{(n-1)}, \tag{2.31}$$

The next step is the derivation of the rate of entropy production per unit volume η^s which is defined by

$$\eta^s = d_t\mathcal{S} + \nabla\cdot J^s \geq 0 \tag{2.32}$$

and is a positive definite quantity according to the second law of thermodynamics. It is checked that, after substituting the expressions of $d_t\mathcal{S}$ and J^s from Eqs. (2.30) and (2.31) in Eq. (2.32), respectively, and eliminating d_te via the energy conservation law for rigid heat conductors ($d_te = -\nabla\cdot\boldsymbol{q}$), one obtains

$$\begin{aligned} \sigma^s = &-\left(-\nabla T^{-1} + \gamma_1 d_t\boldsymbol{q} - \beta_1\nabla\cdot\boldsymbol{Q}^{(2)}\right)\cdot\boldsymbol{q}\cdots \\ &-\sum_{n=2}^{N}\boldsymbol{Q}^{(n)}\otimes\left(\gamma_n d_t\boldsymbol{Q}^{(n)} - \beta_n\nabla\cdot\boldsymbol{Q}^{(n+1)} - \beta_{n-1}\nabla\boldsymbol{Q}^{(n-1)}\right) \geq 0 \end{aligned} \tag{2.33}$$

The above bilinear expression in fluxes and forces (the quantities between parentheses) suggests the following hierarchy of linear flux–force relations:

$$\nabla T^{-1} - \gamma_1 d_t\boldsymbol{q} + \beta_1\nabla\cdot\boldsymbol{Q}^{(2)} = \nu_1\boldsymbol{q} \tag{2.34}$$

$$\beta_{n-1}\nabla\boldsymbol{Q}^{(n-1)} - \gamma_n d_t\boldsymbol{Q}^{(n)} + \beta_n\nabla\cdot\boldsymbol{Q}^{(n+1)} = \nu_n\boldsymbol{Q}^{(n)} \quad (n = 2, 3, \ldots, N) \tag{2.35}$$

With ν_1, $\nu_2, \ldots$, ν_n being positive phenomenological coefficients to meet the property that σ^s is positive definite. Equations (2.34) and (2.35) can also be seen as time evolution equations for the fluxes $\boldsymbol{q}$, $\boldsymbol{Q}^{(2)}, \ldots$, $\boldsymbol{Q}^{(n)}$. In order to gain insight about the physical meaning of the various phenomenological coefficients, let us assume the absence of non-locality so that the term in $\nabla\cdot\boldsymbol{Q}^{(2)}$ will not appear in Eq. (2.34) which reduces to Cattaneo's relation.

If in addition, one considers steady situations, the term in $d_t\boldsymbol{q}$ vanishes and one recovers Fourier's law. These observations lead to the following identities:

$$\nu_1 = \frac{1}{\lambda T^2}, \ \gamma_1 = \frac{\tau}{\lambda T^2}, \tag{2.36}$$

indicating that ν_1 is related to the thermal conductivity λ and γ_1 to the relaxation time τ of the heat flux $\boldsymbol{q}$. The identification of the higher order coefficients demands to compare with higher order evolution equations, but it is expected by analogy that the parameters ν_n and γ_n are related to coefficients of thermal conductivity λ_n and relaxation times τ_n of order n, respectively. Moreover, since $\boldsymbol{Q}^{(n+1)}$ is the flux of $\boldsymbol{Q}^{(n)}$, this implies, by the very definition of a flux, that $d_t\boldsymbol{Q}^{(n)} = -\nabla \cdot \boldsymbol{Q}^{(n+1)}$. Now, when dividing (2.34) by γ_1 and (2.35) by γ_n ($n = 2, 3, \ldots$), it follows that $\frac{\beta_1}{\gamma_1} = -1$ and $\frac{\beta_n}{\gamma_n} = -1$ or, more generally, $\gamma_n = -\beta_n$, which reduces considerably the number of undetermined coefficients.

We assume now that the system is described by an infinite number of flux variables. Applying the spatial Fourier transform $\boldsymbol{q}(\boldsymbol{k}, t) = \int_{-\infty}^{+\infty} \boldsymbol{q}(\boldsymbol{r}, t)e^{-i\boldsymbol{k}\cdot\boldsymbol{r}}d\boldsymbol{r}$ to relations (2.34) and (2.35), with $\boldsymbol{k}$ the wave number vector and $\boldsymbol{r}$ the position vector, the following Cattaneo-type evolution equation of the Fourier transformed heat flux is obtained:

$$\tau d_t\hat{\boldsymbol{q}}(\boldsymbol{k}) + \hat{\boldsymbol{q}}(\boldsymbol{k}) = -i\boldsymbol{k}\lambda(\boldsymbol{k})\hat{T}(\boldsymbol{k}), \tag{2.37}$$

where $\lambda(\boldsymbol{k})$ is a wavelength-dependent thermal conductivity taking the form of a continued fraction expansion [3,9,10], so that the correction factor (this continued fraction divided by λ_p^0) becomes finally

$$f_p^c(\boldsymbol{k}) = \cfrac{1}{1 + \cfrac{\boldsymbol{k}^2 l_1^2}{1 + \cfrac{\boldsymbol{k}^2 l_2^2}{1 + \cfrac{\boldsymbol{k}^2 l_3^2}{1 + \cdots}}}}, \tag{2.38}$$

where l_n is the correlation length of order n defined by $l_n^2 = \beta_n^2/(\mu_n\mu_{n+1})$. Here, it is assumed that the relaxation times τ_n ($n > 1$) corresponding to higher order fluxes are negligible with respect to τ_1, which is a hypothesis generally admitted in kinetic theories. In the present problem, there is only one dimension, namely, the radius $a_{p,s}$ of the spheres, so that it is natural to define $\boldsymbol{k} = k \equiv 2\pi/a_{p,s}$. The correlation lengths selected as $l_n^2 = a_{n+1}l^2$, with $a_n = n^2/(4n^2 - 1)$ and l identified as the mean free path independently of the order of approximation. This is a rather natural choice for phonons [1]. With these results in mind, the continued fraction (15) reduces to an asymptotic limit [11], leading finally to the following expression for f_p^c:

$$f_p^c = \frac{3}{4\pi^2 Kn^2}\left[\frac{2\pi Kn}{\arctan(2\pi Kn)} - 1\right] \tag{2.39}$$

where the Knudsen number is defined as $Kn = \ell_p/a_{p,s}$. After combining expressions (2.29) and (2.39), one obtains the final expression of the thermal conductivity of the nanoparticles, that is

$$\lambda_p = \frac{3\lambda_p^0}{4\pi^2 Kn^2}\left[\frac{2\pi Kn}{\arctan(2\pi Kn)} - 1\right] \tag{2.40}$$

with λ_p^0 from Eq. (2.28).

2.4 Heat Conduction in Nanowires Through an Effective Thermal Conductivity

The heat conduction in a nanowire can be obtained in the same way as for the nanoparticle [12–14]. We take it from Eq. (2.38). The difference is now that for particles that have two distinct dimensions, which is the case for nanowires. So, l_n should be taken as coefficients analogous to the mean free paths associated with the heat flux of order n. Suppose for simplicity [11] that all the l_n's are equal ($l_n = \ell_p/2$) with l_p the mean free path of the phonons and that the wave number is identified as $\boldsymbol{k} = k = 2\pi/\sqrt{r^{-2} + L^{-2}}$, with r the nanowire radius and L the nanowire's length. Defining the Knudsen number as $Kn = \ell_p/\sqrt{r^{-2} + L^{-2}}$, expression (2.38) has the asymptotic limit [11]

$$f_p^c = \frac{1}{2\pi^2 Kn^2}\left(\sqrt{1 + 4\pi^2 Kn^2} - 1\right) \tag{2.41}$$

Note that for sizes of the order of the "de Broglie" wavelength [about 1 nm for Si, which is the limit size for the validity of Eq. (2.39)], quantum confinement effects should be taken into account. For small values of Kn, heat transport is governed by the diffusive regime, i.e. by Fourier's law, and f_p^c tends to one, which confirms that λ_p can be identified with its bulk value. For $Kn \geq 1$, which is typical of nano-configurations with heat transport of ballistic nature, within the limit $Kn \to \infty$, f_p^c increases linearly with the radius of the sample in agreement with experimental observations. After combining expressions (2.29) and (2.41), one obtains the final expression of the thermal conductivity of the nanotubes, that is

$$\lambda_p = \lambda_p^0 \frac{\sqrt{1 + 4\pi^2 Kn^2} - 1}{2\pi^2 Kn^2} \tag{2.42}$$

with λ_p^0 from Eq. (2.28).

References

[1] Lebon, G., Machrafi, H., Grmela, M. 2011. An extended irreversible thermodynamic modelling of size-dependent thermal conductivity of spherical nanoparticles dispersed in homogeneous media. *Proceedings of the Royal Society A* 471:20150144.

[2] Lebon, G., Jou, D., Casas-Vazquez, J. 2008. *Understanding Non-equilibrium Thermodynamics.* Berlin: Springer.

[3] Jou, D., Casas-Vazquez, J., Lebon, G. 2010. *Extended Irreversible Thermodynamics*, fourth ed. Berlin: Springer.

[4] Guyer, R.A., Krumhansl, J.A. 1966. Solution of the linearized phonon Boltzmann equation. *Physical Review* 148:766–777.

[5] Guyer, R.A., Krumhansl, J.A. 1966. Thermal conductivity, second sound, and phonon hydrodynamic phenomena in nonmetallic crystals. *Physical Review* 148:778–788.

[6] Cattaneo, C. 1948. Sulla conduzione del calore. Atti del Seminario Matematico e Fisico delle Università di Modena 3:83–101.

[7] Modest, M.F. 1993. *Radiative Heat Transfer.* New York: McGraw-Hill.

[8] Alvarez, F.X., Jou, D. 2010. Boundary conditions and evolution of ballistic heat transport. *ASME Journal of Heat Transfer* 132:0124404.

[9] Lebon, G., Jou, D., Casas-Vazquez, J. 2008. *Understanding Non-equilibrium Thermodynamics*. Berlin: Springer.

[10] Lebon, G. 2014. Heat conduction at micro and nanoscales: A review through the prism of extended irreversible thermodynamics. *Journal of Non-Equilibrium Thermodynamics* 39:35–59.

[11] Hess, S. 1977. On nonlocal constitutive relations, continued fraction expansion for the wave vector dependent diffusion coefficient. *Zeitschrift für Naturforschung* 32a:678–684.

[12] Lebon, G., Machrafi, H., Grmela, M., Dubois, C. 2011. An extended thermodynamic model of transient heat conduction at sub-continuum scales. *Proceedings of the Royal Society A* 467:3241–3256.

[13] Machrafi, H. 2016. An extended thermodynamic model for size-dependent thermoelectric properties at nanometric scales: Application to nanofilms, nanocomposites and thin nanocomposite films. *Applied Mathematical Modelling* 40:2143–2160.

[14] Lebon, G., Machrafi, H. 2015. Thermal conductivity of tubular nanowire composites based on a thermodynamical model. *Physica E* 71:117–122.

3

Heat Conduction in Nanocomposites

3.1 Theoretical Models

After giving a general description of the effective medium approach, we present a modified version taking into account the possibility of formation of agglomeration of nanoparticles Afterwards, we briefly recall the derivation of the effective thermal conductivity of the host matrix and the individual nanoparticles. The ensemble are efficient tools for the prediction of the effective thermal conductivity of nanocomposites.

3.1.1 Effective Medium Approach

Our main purpose is to model heat transport associated with the dispersion of nanoparticles in a bulk material, called the matrix. The description of such a heterogeneous two-component medium can be simplified by appropriately homogenizing it, as described within the effective medium approach, first introduced by Maxwell [1] in the framework of electrical conductivity. Following the lines of thought of Hasselman [2] and later on by Nan et al. [3], the effective thermal conductivity λ^{eff} of the homogenized nanocomposite is given by the following:

$$\lambda^{eff} = \lambda_m \frac{2\lambda_m + (1 + 2\alpha_i)\,\lambda_p + 2\varphi\,[(1 - \alpha_i)\,\lambda_p - \lambda_m]}{2\lambda_m + (1 + 2\alpha_i)\,\lambda_p - \varphi\,[(1 - \alpha_i)\,\lambda_p - \lambda_m]} \tag{3.1}$$

In this equation, λ is the thermal conductivity coefficient, subscripts m and p refer to the matrix, and suspended particle, respectively, φ is the volume fraction of the particles, and α_i is a dimensionless parameter related to the particle–matrix interface which is given by

$$\alpha_i = R_i \lambda_m / a_p, \tag{3.2}$$

where a_p is the radius of the nanoparticle, R_i is the thermal boundary resistance coefficient, and $R_i\lambda_m$ is the so-called Kapitza radius. Throughout the present analysis, it is assumed that the nanoparticles are characterized by a *diffusive* surface, meaning that the direction of phonons after impact is independent of the direction of the impacting phonons, and this is justified as the interface between matrix and agglomerates is generally rougher than that between the individual particles and the matrix. The roughness of the surface can be macroscopically simulated by introducing a surface parameter, called the specularity, s. We would use instead of the nanoparticle radius a so-defined specular nanoparticle radius: $a_{p,s} = \frac{1+s}{1-s} a_p$. Purely diffusive surfaces, which is the case in our work, are characterized by $s = 0$. If the surface is perfectly smooth, one would have $s \to 1$. In the latter case, the thermal boundary resistance would be completely negligible. In the case of diffusive interfaces ($s = 0$), R_i can be written as follows: [3,4]

$$R_i = 4/c_m v_m + 4/c_p v_p \tag{3.3}$$

with c standing for the volumetric heat capacity and v for the group velocity.

3.1.2 Effect of Agglomeration

To account for the formation of nanoparticles in nanofluids or nanocomposites in the form of aggregates, a modification is to be introduced in the conventional Hamilton–Crosser model [5], by first substituting φ by the agglomerate volume fraction φ_a, defined as

$$\varphi_a = \varphi \left(\frac{a_{p,a}}{a_p} \right)^{3-D} \tag{3.4}$$

where $a_{p,a}$ is the agglomerate radius and D a fractal index. Note, by the way, that the specular agglomerate radius would be $a_{p,as} \equiv \frac{1+s}{1-s} a_{p,a}$. D has typical values of 1.6~2.5 for the aggregates of spherical nanoparticles and 1.5~2.45 for those of rod-like nanoparticles depending on the type of aggregation, chemistry environment, particle size and shape, and shear flow conditions [6]. The value for D is often taken equal to 1.8, and since the thermal conductivity appears to depend only weakly on its value [6–9], we will here work with this value.

In the presence of agglomerates, expression (3.1) of the effective thermal conductivity is modified as follows:

$$\lambda^{eff} = \lambda_m \frac{2\lambda_m + (1 + 2\alpha_i)\, \lambda_a + 2\varphi_a \left[(1 - \alpha_i)\, \lambda_a - \lambda_m \right]}{2\lambda_m + (1 + 2\alpha_i)\, \lambda_a - \varphi_a \left[(1 - \alpha_i)\, \lambda_a - \lambda_m \right]}, \tag{3.5}$$

wherein φ has been replaced by φ_a and λ_p by λ_a, the agglomerate thermal conductivity. For a binary mixture of homogeneous spherical inclusions (recall that we approximate the nanoparticles in this study as spheres), the mean field approach [10] leads to the following result:

$$\lambda_a = \frac{1}{4} \left[3\varphi_s\, (\lambda_p - \lambda_m) + (2\lambda_m - \lambda_p) + \sqrt{8\lambda_m \lambda_p + \left(3\varphi_s\, (\lambda_m - \lambda_p) + (\lambda_p - 2\lambda_m) \right)^2} \right]. \tag{3.6}$$

where $\varphi_s = \frac{\varphi}{\varphi_a}$ is the volume fraction of particles in the aggregates. In the absence of agglomeration, for which there is only one particle per aggregate, one has $\varphi_s = 1$ and λ_a reduces to λ_a, as it should.

3.1.3 Effective Thermal Conductivity of the Matrix and the Nanoparticles

To close the problem, it remains to determine the expressions of λ_m and λ_p. According to the classical Boltzmann–Peierls kinetic theory, the thermal conductivity of the matrix is given, at fixed reference temperature (T_{ref}), by

$$\lambda_m = \frac{1}{3} \left(c_m v_m \ell_m \right)|_{T_{ref}}. \tag{3.7}$$

where ℓ_m is the mean free path of the phonons in the matrix. Following Matthiesen's rule, it is of the form:

$$\frac{1}{\ell_m} = \frac{1}{\ell_{m,b}} + \frac{1}{\ell_{m,coll}} \tag{3.8}$$

with $\ell_{m,b}$ designating the contribution from the bulk and $\ell_{m,coll}$ the supplementary contribution arising from collisions at the agglomerate–matrix interface, the latter being given by,

$$\ell_{m,coll} = 4a_{p,a}/3\varphi. \tag{3.9}$$

We are now left with the determination of λ_p. Instead of using an expression similar to Eq. (3.9) for k_p, we propose a new closed-form formula:

$$\lambda_p = \lambda_p^0 f(Kn), \tag{3.10}$$

consisting in a constant value λ_p^0 multiplied by a correction factor $f(Kn)$, which takes into account the nanoscale of the particles through the Knudsen number Kn defined next. The quantity λ_p^0 is the thermal conductivity, at a given reference temperature, of the bulk material of which the nanoparticle is composed of:

$$\lambda_p^0 = \lambda_{p,b}\big|_{T_{ref}}, \tag{3.11}$$

its expression being analogous to (3.9) for the matrix, and equivalent to Eq. (2.28), i.e.

$$\lambda_p^0 = \frac{1}{3}\left(c_p \ell_p v_{p,b}\right)\big|_{T_{ref}}. \tag{3.12}$$

With the difference that now the mean free path is the bulk one (so that $\ell_p = \ell_{p,b}$), the contribution of the collisions is hidden in the correction factor $f(Kn)$. The latter has been determined in Chapter 2 and depends on the radius of the particle and on the mean free path of the phonons inside the particle, $\ell_{p,b}$, so that it is rather natural to define the Knudsen number as follows:

$$Kn \equiv \ell_{p,b}/a_{p,a}. \tag{3.13}$$

We use the expression developed in Chapter 2 [Eq. (2.40)]:

$$\lambda_p = \frac{3\lambda_p^0}{4\pi^2 Kn^2}\left[\frac{2\pi Kn}{\arctan(2\pi Kn)} - 1\right], \tag{3.14}$$

3.1.4 Nanocomposites with Embedded Nanowires

The basic relation is Maxwell's one [1] improved by Hasselman and Johnson [2] to include thermal boundary resistance and by Nan et al. [3] who considered various particle shapes. Accordingly, the overall thermal conductivity of the composite in the direction perpendicular and normal to the heat flux is, respectively, given by

$$\lambda_\perp = \lambda_m \frac{\lambda_m + (1+\alpha_i)\lambda_p + \varphi\left[(1-\alpha_i)\lambda_p - \lambda_m\right]}{\lambda_m + (1+\alpha_i)\lambda_p - \varphi\left[(1-\alpha_i)\lambda_p - \lambda_m\right]}, \tag{3.15}$$

$$\lambda_\parallel = (1-\varphi)\lambda_m + \varphi\lambda_p \tag{3.16}$$

where λ_m is the thermal conductivity of the matrix and λ_p the thermal conductivity of the wires. Unlike λ_m, the quantity λ_p will incorporate explicitly size effects as made explicit later on. Furthermore, φ denotes the volume fraction of the fibres, and α_i, in analogy with (3.2), is given by $\alpha_i = R_i\lambda_m/r_p$, taking into account the interface wire matrix, with r_p the radius of the wire of length L. Moreover, R_i is given by Eq. (3.3). Specularity is taken explicitly into account by introducing a "modified" radius $r_{p,s} = \frac{1+s}{1-s} r_p$ and a "modified" length $L_s = \frac{1+s}{1-s} L$, where the parameter s denotes the probability of specular diffusion of the phonons on the particle–matrix interface in the same manner as for nanoparticles (see Section 3.1.1).

The bulk thermal conductivity of the host medium is given by the classical expression (3.7), where the mean free path is also given by Eq. (3.8). Again, ℓ_m gives the mean free path of the phonons in the bulk matrix material, while $\ell_{m,coll}$, still denoting the supplementary contribution due to the interactions at the particle–matrix interface, is now different from

Eq. (3.9). Note that although the relation between the mean free path in the matrix and the average separation between neighbouring nanowires does not appear explicitly in this formulation, it is implicitly included in the expression of $\ell_{m,coll}$. In the case of wires with their axis oriented normally and parallel to the heat flux, this supplementary contribution is given , respectively, as follows[4]:

$$\frac{1}{\ell_{m,coll}} \equiv \frac{1}{\ell_{m,coll,\perp}} = \frac{2\varphi}{r_{p,s}\pi} \tag{3.17}$$

$$\frac{1}{\ell_{m,coll}} \equiv \frac{1}{\ell_{m,coll,\parallel}} = \varphi\left(\frac{1}{L_s} + \frac{2\zeta}{r_{p,s}\pi}\right) \tag{3.18}$$

where ζ stands for $\zeta = \frac{\sqrt{\varphi}}{\sqrt{\varphi+1}}$.

To emphasize the role of the size effects on the thermal conductivity λ_p of the nanowires, we write λ_p in the form (3.10), which is derived in Chapter 2. We obtain the final expression of the thermal conductivity of nanotubes or nanowires from Eq. (2.42):

$$\lambda_p = \lambda_p^0 \frac{\sqrt{1+4\pi^2 Kn^2} - 1}{2\pi^2 Kn^2} \tag{3.19}$$

with λ_p^0 from Eq. (3.12) and $Kn = \ell_p / \sqrt{r_{p,s}^{-2} + L_s^{-2}}$.

3.1.5 Temperature Dependence

The temperature dependence of the thermal conductivity will appear implicitly through the frequency ω dependence of the various quantities appearing in the general expressions (3.7) and (3.12) for nanoparticles [11]. Therefore, the bulk thermal conductivities can be written as follows:

$$\lambda_i(T,\omega) = \frac{1}{3}\int_0^{\omega_D} C_i(T,\omega) v_i(T,\omega) \ell_i(T,\omega) d\omega \tag{3.20}$$

The determination of the thermal conductivity requires, therefore, the knowledge of $c_i(T,\ \omega)$, $v_i(T,\ \omega)$ and $\ell_i(T,\omega)$ for $i = m,\ p$. Note that $\ell_i^{-1}(T,\omega) = \ell_{i,b}^{-1}(T,\omega) + \ell_{i,coll}^{-1}$ for the matrix and that $\ell_i(T,\omega) = \ell_{i,b}(T,\omega)$ for the nanoparticle. The limit of integration, ω_D, is the Debye frequency cut-off. In agreement with [11,12], we assume that the group velocity v is independent on T and ω, while the heat capacity and the bulk mean free path are given, respectively, as follows:

$$c_i = \frac{3h^2}{2\pi^2 v_i^2 k_B T^2} \frac{\omega^4 e^{\frac{h\omega}{k_B T}}}{\left(e^{\frac{h\omega}{k_B T}} - 1\right)^2}, \tag{3.21}$$

$$\frac{1}{\ell_{i,b}} = B_i T \omega^2 e^{-\frac{\theta_i}{T}}, \tag{3.22}$$

where B_i and θ_i are constant quantities obtained by fitting experimental data measured by Glassbrenner & Slack [13]. We are now in possession of all the elements needed to evaluate the effective thermal conductivity of the nanocomposite as expressed by relations (3.1), (3.15) and (3.16). To be explicit, the thermal conductivity $\lambda_m(T,\ \omega)$ of the matrix element is directly derived from Eq. (3.7) [the temperature dependence coming from Eq. (3.20)] with the mean free path in the matrix $\ell_m(T,\omega)$, while, accordingly to (3.12) and (3.14), λ_p for the nanoparticles can be written as follows:

$$\lambda_p(T,\omega) = \frac{1}{4\pi^2}\int_0^{\omega_D} c_p(T,\omega) v_p \ell_{p,b}(T,\omega) \frac{1}{Kn(T,\omega)^2}\left[\frac{2\pi Kn(T,\omega)}{\arctan(2\pi Kn(T,\omega))} - 1\right] d\omega \tag{3.23}$$

with $n(T,\omega) = \ell_{p,b}(T,\omega)/\alpha_{p,s}$. For the nanowires, λ_p is given by

$$\lambda_p(T,\omega) = \frac{1}{6\pi^2}\int_0^{\omega_D} c_p(T,\omega)v_p\ell_{p,b}(T,\omega)\frac{1}{Kn(T,\omega)^2}\left[\sqrt{1+4\pi^2 Kn(T,\omega)^2}-1\right]d\omega \quad (3.24)$$

with $Kn(T,\omega) = \ell_{p,b}(T,\omega)/\sqrt{r_{p,s}^{-2}+L_s^{-2}}$.

3.2 Polymeric Nanocomposites

Polymeric nanocomposites are used in a broad variety of applications and industrial domains such as microelectronic packaging, coatings, adhesives and fire retardant. In thermal applications, the often low thermal conductivity of the polymeric matrix is typically increased by dispersing in the host matrix inorganic fillers, such as aluminium nitride (AlN) [14,15], boron nitride [16] and carbon nanotubes [17], or more specifically, ceramic fillers, such as aluminium oxide (Al_2O_3) [18]. Another way is to design a new material where the material orientation is controlled [19,20]. When fillers are used, to determine their influence on the thermal conductivity of nanocomposites, it is required to set up models that predict the behaviour of the thermal conductivity as a function of several parameters [21].

In the case of micro-particles, Fourier's law of heat conduction provides a valuable approach. However, in the presence of nanoparticles, Fourier's law is no longer applicable and new models should be developed to include small space scales. Several formalisms have been proposed which describe how bulk thermal properties are influenced by the addition of nanoparticles. Principally, molecular dynamics approaches based on phonon's Boltzmann transport equation [22–24] or ad-hoc semi-analytical formulations [25,26] have been developed.

In the foregoing, we examine the significance of various effects on the effective thermal conductivity of the system, namely, the particle's shape and size, the volume fraction of particles, and the boundary matrix–particle interface resistance. We focus also on the influence of nanoparticles' clusters and their progressive agglomeration. For the sake of simplicity, the particles are supposed to be spherical and monodisperse. Many experiments have been performed on investigating the role of agglomeration [27–31], which will therefore be given a special attention. The majority of models taking into account nanoparticles' agglomeration introduce an agglomerate radius that is kept fixed or consider a change in agglomerate's size due to aging [30], without examining the influence of volume fraction on the degree of agglomeration. This point will receive a particular attention in this work, and a relation between the degree of agglomeration and the particle volume fraction will be established experimentally. This relation will be used as one of the inputs in our model in order to predict the effective thermal conductivity.

3.2.1 Volume-Fraction-Dependent Agglomeration

When the density of nanoparticles distributed in a matrix becomes important, the particles have a tendency to coagulate and, as in colloid suspensions, lead to the formation of clusters. Assuming that such clusters fit into a virtual sphere, we can define the agglomeration gyration radius $a_{p,a}$ as the radius of such a virtual sphere. Using Eq. (4), we are able to find the agglomerate radius as a function of the particle radius.

In the case of Al_2O_3 particles dispersed in water, Chen et al. [28] propose $a_{p,a} \approx 3a_p$, whereas Anoop et al. [31] take $a_{p,a} \approx 5a_p$. Other authors [30] reported for $a_p < a_{p,a} < 4a_p$.

Rheology experiments [27] predict $a_{p,a} = 3.34a_p$ for TiO_2 dispersed in ethylene glycol with similar values obtained for Al_2O_3 dispersed in the same alcohol [29]. In [32], AlN and MgO nanoparticles were considered for which experimental gyration radii have been measured. The results suggested that there exists a strong correlation between the gyration radius $a_{p,a}$ and the volume fraction φ. Note that φ is determined at the stage before the polymerization step occurs [29] by dispersing the nanoparticles into the fluid matrix, which is often water based. To determine the validity of the relation $a_{p,a}(\varphi)$ expressing the volume-fraction dependence of the gyration radius, they followed a simplified protocol, which consists in dispersing the nanoparticles in a solvent, say ethanol or water without any dispersion agent, and measuring the agglomerate radius distribution versus the initial particle volume fraction (the volume fraction's values shows an error less than 1%). Their procedure refers to [29] wherein the AlN and MgO nanoparticles are first dissolved in ethanol and then sonicated to break up in large agglomerates. The nanoparticles are then dispersed in the epoxy resin by shear force mixing. The solvent is afterwards evaporated, and the composite is mixed with a hardener via mechanical stirring prior to degassing. Finally, the mixture is cured, obtaining the polymer nanocomposite. What is of importance is that the nanoparticles are first dissolved in a solvent with a given volume fraction. In [29], the solvent is ethanol, but water is also often used, presenting similar characteristics [33] (see also [34] for polymer/clay nanocomposites). It appears that after dispersing the nanoparticle/solvent mixture within the epoxy resin, the size of the particles does not change significantly during curing [34]. It has been shown that the particle size distribution can be safely approximated as still being mainly established in the solvent before curing. Therefore, it is sufficient to analyse the agglomeration behaviour of the nanoparticles in the solvent at different volume fractions. By means of particle size analysis, it had been established [32] that the agglomerate radius increases with the volume fractions. Knowing the initial particle size (a_p), we can define the degree of agglomeration $m \equiv a_{p,a}/a_p$ dependent on the volume fraction ($m = 1$ means no agglomeration). Such an influence of the suspension's concentration on the particle size distribution, and consequently the degree of agglomeration, is also observed in [35]. The question that arises is how does this degree of agglomeration depend on the volume fraction. Let us consider the next qualitative demonstration of the most suitable mathematical form for the volume-fraction dependence of the degree of agglomeration. We start by considering a simple kinetic mechanism called "aggregative growth" [36]. Accordingly, the kinetics of agglomeration are governed by the following equations:

$$P + P \to A, \tag{3.25}$$

$$P + A \to 1.5A, \tag{3.26}$$

where P stands for the nanoparticles and A for the agglomerate. The "nucleation rate," defined as the rate of the first agglomerate formed from the initial nanoparticles, is given by the following:

$$-\frac{1}{2}\frac{dp}{dt} = k_n[P]^2, \tag{3.27}$$

where k_n denotes the rate constant of nucleation, and the square brackets denote the molar concentration. Note that the term $\frac{dP}{dt}$ in Eq. (3.27) stands for the consumption of P due to nucleation only. The "agglomerative growth rate," defined as the rate of the agglomerates growing due to more adhering nanoparticles, is given by the following:

$$-\frac{dP}{dt} = k_g[P][A], \tag{3.28}$$

where k_g is the rate constant of agglomerate growth. Note that $\frac{dP}{dt}$ in Eq. (3.28) stands for the consumption of P due to agglomerates' growth only. Similar relations can be written for the formation rate of the agglomerate. From Eqs. (3.25) and (3.26), we deduct the mass balance:

$$[P] = [P]_0 - 2[A], \tag{3.29}$$

where the subscript "0" stands for the initial value, noting that at the beginning, no agglomerate is present, i.e. $[A]_0 = 0$. The factor "2" stems from the ratio P: $A \equiv 2{:}\,1$, so that conservation of mass imposes to double the concentration of A. It follows from Eq. (3.29) that $\frac{dA}{dt} = -\frac{1}{2}\frac{dP}{dt}$, so that the total rate law for the agglomeration mechanism is given by

$$\frac{dA}{dt} = k_n[P]^2 + \frac{1}{2}k_g[P][A]. \tag{3.30}$$

Making use of the initial conditions $[A]_{t=0} = [A]_0 = 0$ and $[P]_{t=0} = [P]_0$, and relation (3.29), the solution of Eq. (3.30) is given by

$$[A] = \frac{2k_n[P]_0\left(1 - \mathrm{e}^{\frac{k_g[P]_0 t}{2}}\right)}{4k_n\left(1 - \mathrm{e}^{\frac{k_g[P]_0}{2}t}\right) - k_g}. \tag{3.31}$$

It is easy to verify from Eq. (3.31) that for $t \to \infty$, $[A]_\infty = \frac{1}{2}[P]_0$. This shows that the concentration of the agglomerates presents a maximum and this is also true for its size. More interesting, it is seen that the final agglomerate concentration depends on the initial nanoparticle concentration. This indicates that the agglomerate size (and therefore also the degree of agglomeration) is a function of the initial volume fraction of the nanoparticles. We can define the number of nanoparticles in an agglomerate at a given time, N_t, by the proportionality law:

$$N_t = \frac{2\,[A]}{[P]_0}N_\infty, \tag{3.32}$$

where N_∞ is the final number of nanoparticles in the agglomerate. Of course, for $t \to 0$, $N_t = 0$ and for $t \to \infty$, $N_t = N_\infty$. The factor 2 stems from the mass balance (3.29). However, the limit $t \to \infty$ is not realistic and (3.32) is not valuable for this limit. Especially, when sonicated, N_t will not reach N_∞ but rather an intermediate value N_{int} so that Eq. (3.32) should rather be written as follows:

$$N_t = \frac{2\,[A]^*}{[P]_0}N_{int}, \tag{3.33}$$

where $[A]^* < [P]_0/2$. A relation between the agglomerate's size D and the number of nanoparticles in the agglomerate can be postulated as

$$D \propto CN^b, \tag{3.34}$$

with $b \neq 1/3$ ($b = 1/3$ corresponds to spherical particles and the agglomerate of spherical form as well), meaning that the agglomeration is fractal [37]. Moreover, C is a constant composed out of material constants and geometrical data, being of no importance for the present development. Coupling Eqs. (3.33) and (3.34) results in

$$\frac{D_{int}}{D_t} \propto C\left(\frac{N_{int}}{N_t}\right)^b \propto C\left(\frac{[P]_0}{2\,[A]^*}\right)^b. \tag{3.35}$$

We note also that $[P]_0 \propto \varphi$ (which is indirectly also observed by [38] as $N \propto \varphi^d$) and that $m \propto \frac{D_{int}}{D_t}$. This brings us to the conclusions that the most suitable fitting for the experimental results in Equation (3.35) is a power law where $m \propto C\varphi^b$. It is then shown that a suitable expression for the volume-fraction dependence of the degree of agglomeration is provided by a power law in the form:

$$m = \beta\varphi^{\gamma}. \tag{3.36}$$

It is observed that the thermal conductivity first increases linearly with the degree of agglomeration, reaches a maximum at the percolation threshold and finally decreases. The raising of conductivity with size (at nanoscale) was confirmed by many authors [39–44] in the case of epoxy resin with various fillers and by Prasher et al. [9,45] in the case of nanofluids. The increase in λ^{eff} at small m-values may be explained because of the weak agglomeration of the particles so that the regime is that of dispersed primary particles. In this case, the effective thermal conductivity is increasing with the size of the nanoparticles [46] because of the smaller thermal interfacial resistance between particles and matrix [see relation (3.2)]. However, after reaching a peak, at larger m-values (say cluster radii larger than the diameter of the original particles), the influence of agglomeration becomes dominating; the contact area of the agglomerates themselves becomes smaller, which leads to less heat conduction between the agglomerates and the matrix. Moreover, the contact area of the particles within the agglomerates is raising, which causes a larger boundary resistance due to more phonon collisions [see relations (3.2) with (3.7)–(3.9)], which reduces the heat conduction within the agglomerates as well. Hence, the effective thermal conductivity decreases altogether with increasing size at larger m-values. These results are in agreement with those reported in Figure 3.1 as well as an experimental study by Moreira et al. [47]. Note that there is a maximum value of m which corresponds to $\varphi_a = 1$, i.e. $\varphi = \varphi_s$. Making use of the result (3.4), it is easily checked that the maximum of m is

$$m_{\max} = \varphi^{-\frac{1}{3-D}} = \varphi^{-\frac{1}{1.2}} \tag{3.37}$$

after that D has been selected as given by $D = 1.8$.

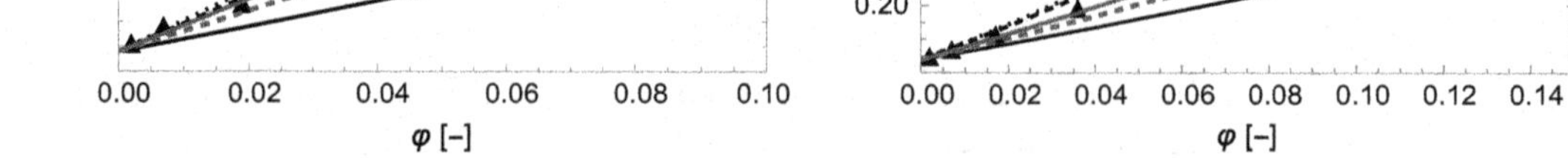

FIGURE 3.1
Effective thermal conductivity versus the original particle volume fraction for AlN/epoxy (left) and MgO/epoxy (right) at several agglomerate radii $a_{p,a}$, with $a_p = 30$ nm for AlN and $a_p = 11$ nm for MgO. The curves represent our model and the symbol ▲ denotes experimental values [29]. (Modified from: Machrafi, H., Lebon, G., Iorio, C.S. 2016. Effect of volume-fraction dependent agglomeration of nanoparticles on the thermal conductivity of nanocomposites: Applications to epoxy resins, filled by SiO_2, AlN and MgO nanoparticles. *Composites Science and Technology* 130:78–87.)

3.2.2 Dependence of the Effective Thermal Conductivity Versus the Volume-Fraction-Dependent Agglomeration

The systems examined in this case study are, respectively, SiO_2, AlN and MgO nanoparticles embedded in an epoxy resin. The material properties of these components are given in Table 3.1.

Note that the values in Table 3.1 have been obtained in the framework of the so-called "dispersion model" [53] where it is admitted that the phonons have different energies and velocities due to their dispersion. In a previous work [54], the SiO_2–epoxy mixture in the absence of agglomeration has been studied, i.e. by using the present model (Section 3.1) with $\varphi_a = \varphi$ and $\lambda_a = \lambda_p$. Satisfactory agreement with experiments [29] was observed. It is clear that for this system, the dependence with respect to agglomeration is negligible and will therefore no longer be discussed in the following. In contrast, in the case of the mixture AlN–epoxy [20], the theoretical model fails to fit the experimental data if $\lambda_a = \lambda_p$. So, we should take $\lambda_a \neq \lambda_p$. The results for the effective thermal conductivity as a function of the particle volume fraction of the original particles for several values of the agglomerate radii are presented in Figure 3.1 for AlN and MgO nanoparticles embedded in epoxy resin, respectively. To assess the role of the agglomeration, we have also drawn the curves corresponding to the absence of clustering for which $a_{p,a} = a_p$. It appears from the selected examples that the thermal conductivity increases with the volume fraction of the nanoparticles as a consequence of a larger specific interface area between particles and matrix. The formation of clusters has a different effect according to the size of the cluster. For a fixed value of the volume fraction, it is a factor of enhancement of the thermal conductivity up to a gyration radius a little bit larger than the diameter of the particles, but beyond this limit, this doping effect is inversed because it corresponds to a reduction in the interface area. A more detailed discussion will follow later on.

By comparison with experimental data, our model is shown to predict satisfactory agreement at low volume fractions <10%. In the case of AlN particles, the best agreement is reached for $a_{p,a} \gtrsim 2a_p$ at low volume fractions ($\varphi < 0.01$) and for $a_{p,a} \approx 3.5a_p$ at higher volume fractions. For MgO particles, a good accord is found for $a_{p,a} \approx 2a_p$ at low volume fractions ($\varphi < 0.02$) and for $a_{p,a} \approx 5a_p$ at higher density. When the volume fraction of the nanoparticles is small, the particles will have less chance to collide with each other and the formation of agglomerates is less important. As the volume fraction is raising, the

TABLE 3.1
Material Properties for Bulk Materials at Room Temperature (T_{ref}).

Material	Heat Capacity [MJ (m^{-3} K^{-1})]	Group Velocity [m s^{-1}]	Mean Free Path [nm]
Epoxy resin	1.91[a]	2400[a]	0.11[b]
SiO_2	1.687[c]	4400[c]	0.558[c]
AlN	2.7[d]	6980[d]	51[e]
MgO	3.32[f]	7028[g]	4.76[h]

[a]Ref. [48].
[b]Calculated from $\ell_b = 3\lambda/cv$ by considering $\lambda = 0.168$ W m^{-1} K^{-1} from Ref. [29].
[c]Ref. [49].
[d]Ref. [50].
[e]Calculated from $\ell_b = 3\lambda/cv$ by considering $\lambda = 319$ W m^{-1} K^{-1} from Ref. [51].
[f]Calculated from experimental correlation given in Ref. [51].
[g]Calculated from $v = 3\lambda/C\ell_b$ by considering $\lambda = 37$ W m^{-1} K^{-1} from Ref. [52].
[h]Interpolation from data at pages 625 and 626 from Ref. [52].

interaction between the nanoparticles is increased and so do the agglomerate radii. A further validation of the model is given next wherein it is compared which experimental data obtained in our laboratory. It is interesting to discuss the morphology of the nanoparticles. In that respect, scanning electron microscopy has been performed, obtaining images of the nanoparticles at different zooms. The nanoparticle powders are first sputter-coated in gold and then placed on a graphite support into a specimen chamber. The observations indicate that AlN exhibits a very compact structure, and many large agglomerates are present with a narrow distribution. The structure of MgO is, on the other hand, less compact. Also the clusters of MgO appear to have a larger distribution with a weaker agglomeration.

The AlN nanoparticles have quasi-regular shapes of cubic-spheroidal type, while the MgO nanoparticles are closer to regular spheroidal shapes. These observations justify that at least within a first approximation, we identify in our theoretical model that the particles are taken as rigid spheres, with a given size distribution. As for the particle sizes, it is found that the mean particle size values (taken as the equivalent diameter $2a_p$) is 54 ± 14 nm for AlN and 53 ± 22 nm for MgO. This confirms the larger size distribution of MgO with respect to AlN.

3.2.3 Final Validation of Dependence of the Effective Thermal Conductivity Versus the Volume-Fraction-Dependent Agglomeration

Mean particle radii were calculated from particle size distributions of aqueous AlN and MgO suspensions. Comparing this with the results of the mean particle size of the dry powders (and thus the initial state), we can calculate the degree of agglomeration: we can trace the degree of agglomeration $m \equiv a_{p,a}/a_p$ with a certain standard deviation (indicated by error bars), against the volume fraction ($m = 1$ means no agglomeration). The results are presented in Figure 3.2.

Fitting the trend line (3.36) with the experimental findings from Figure 3.2 leads to the values of the parameters β and γ, given in Table 3.2. It is worth to stress that the values of the parameters are of the same order of magnitude for both systems.

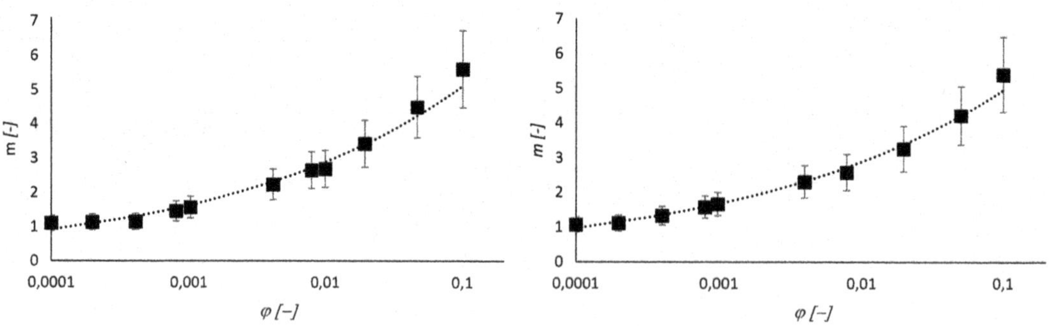

FIGURE 3.2
Degree of agglomeration m against the nanoparticle volume fraction φ for AlN (left) and MgO (right). The symbol ■ denotes the experimental findings, and the dashed line is a trend line. (From: Machrafi, H., Lebon, G., Iorio, C.S. 2016. Effect of volume-fraction dependent agglomeration of nanoparticles on the thermal conductivity of nanocomposites: Applications to epoxy resins, filled by SiO_2, AlN and MgO nanoparticles. *Composites Science and Technology* 130:78–87.)

TABLE 3.2
Fitting Parameters for the Degree of Agglomeration.

	AlN	MgO
β	9.0	8.5
γ	0.25	0.23

The main interest of Eq. (3.36) is that it allows to represent the effective thermal conductivity exclusively in terms of the initial particle volume fraction, without making loose assumptions about the degree of agglomeration. Now, Figure 3.3 shows the effective thermal conductivity, adapted for the volume-fraction-dependent agglomeration gyration radius, as a function of the volume fraction. For comparison, the effective thermal conductivity, not taking into account the agglomeration effect, is given as well.

The results reported on Figure 3.3 indicate a good agreement between our model and experience. Note especially that the sharp increase at $0 < \varphi < 0.01$ for AlN is well represented by our model, using expression (3.36) with the values from Table 3.2. One important conclusion is that it is imperative to take into account the volume-fraction dependence of the agglomeration radius in the study of agglomeration effects on the thermal conductivity of nanocomposites.

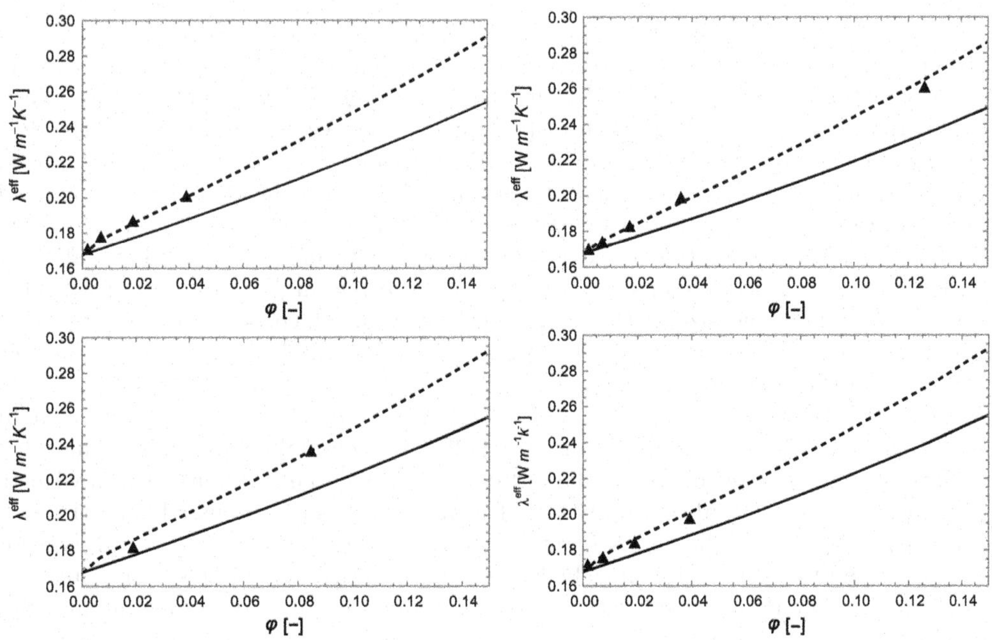

FIGURE 3.3
Effective thermal conductivity as a function of volume fraction (dashed line), using the fitted relation (3.23) between the agglomerate radius and the volume fraction for AlN–epoxy (with a_p = 30 at the upper left [29] and with a_p = 50 nm at the lower left [55] and lower right [56]) and MgO–epoxy (with a_p = 11 nm at the upper right [20]). The solid line refers to the absence of agglomeration in the model, and the symbol ▲ represents the experimental values from [29,55,56]. (Modified from: Machrafi, H., Lebon, G., Iorio, C.S. 2016. Effect of volume-fraction dependent agglomeration of nanoparticles on the thermal conductivity of nanocomposites: Applications to epoxy resins, filled by SiO_2, AlN and MgO nanoparticles. *Composites Science and Technology* 130:78–87.)

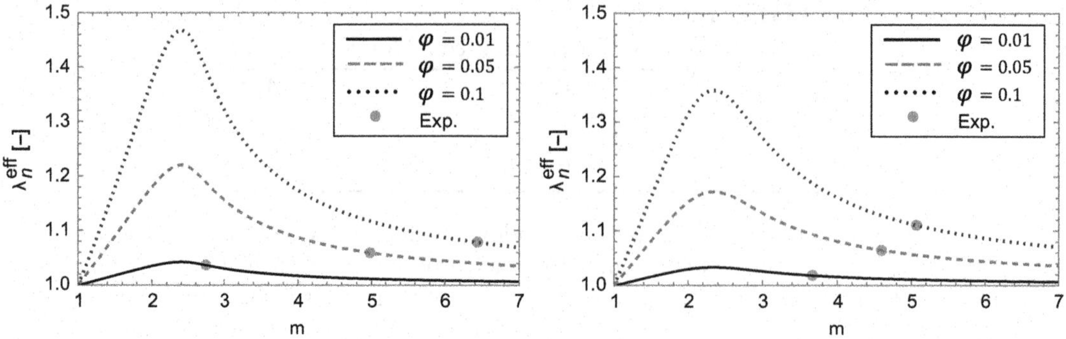

FIGURE 3.4
Effective normalized thermal conductivity, λ_n^{eff}, of the AlN–epoxy (left) and MgO (right) systems as a function of the degree of agglomeration m, for three volume fractions $\varphi = 0.01$, 0.05 and 0.1. Dots represent normalized experimental values (with respect to the value at $m = 1$) at the corresponding volume fractions. (Modified from: Machrafi, H., Lebon, G., Iorio, C.S. 2016. Effect of volume-fraction dependent agglomeration of nanoparticles on the thermal conductivity of nanocomposites: Applications to epoxy resins, filled by SiO_2, AlN and MgO nanoparticles. *Composites Science and Technology* 130:78–87.)

To illustrate the property that agglomeration may either contribute to an increase or a decrease in the thermal conductivity, we have represented in Figure 3.4 the effective thermal conductivity (normalized with respect to $m = 1$), λ_n^{eff}, as a function of the degree of agglomeration m for three values of the volume fraction; $\varphi = 0.01$, 0.05 and 0.1. At fixed φ-values, the effective thermal conductivity first increases with the degree of agglomeration m, and, after reaching a maximum value, at $m \approx 2.5$, it starts decreasing towards a constant value. This may be interpreted by the fact that for weak degrees of agglomeration, the dimension of the particles remains small whence a large Kn and large values of λ_p [see Eq. (3.14)]. By increasing the size of the agglomerate, the interface between the agglomerate and the host matrix is increased, and subsequently, the thermal boundary resistance leads to a decrease in the thermal conductivity.

As expected, the sensibility of k_n^{eff} versus agglomeration is the most important at high particles densities. The dots are the normalized effective thermal conductivity values determined from our theoretical model at the corresponding volume fractions. If it is wished to increase the effective thermal conductivity of the nanocomposites discussed here, the results of Figure 3.4 indicate that the degree of agglomeration (or the agglomeration radius) should be decreased, and this can for instance be achieved in practice by adding surfactants [57,58]. One should however remain cautious to avoid to reduce m beyond the maximum critical value $m \approx 2.5$.

3.3 Semiconductor Nanocomposites

3.3.1 Application to Si/Ge Nanocomposites with Nanoparticle Inclusions

Nanofluids and nanocomposites have been used considerably and in a wide variety of applications that are related to their potential for developing significant modifications of thermal heat transfer properties [59–62]. In what concerns thermoelectric materials [63], the change in thermal conductivity has also been used to obtain enhancements in the figure of merit Z, which behaves as the inverse of the thermal conductivity. A large number of nanostructured materials have been found to overcome the $ZT = 1$ limit. Examples are values $ZT = 2.4$ in Bi_2Te_3 or Sb_2Te_3 [64] and $ZT = 2.6$ in layered SnSe crystals [65].

An increase or decrease in the effective thermal conductivity is especially related to the type of the host matrix and the nanoparticles, their properties. More particularly, the focus is on the volume fraction of nanoparticles, their dimension, the nature of particle–matrix interface, the temperature. In this section, we lead the discussion on how the addition of nanoparticles can modify the effective thermal conductivity of nanocomposites. The temperature dependence is also studied. The model is drawn from Sections 3.1.1, 3.1.3 and 3.1.5. The model shall be illustrated by one case, namely, silicium (Si) particles dispersed in germanium (Ge). The interest in such a case study lies in the observation that Si–Ge systems have been recently the subject of great attention as they have been attested by the works of Wang & Mingo [66] and Kim et al. [67].

The calculations have been performed for the Si–Ge system using three different values for the radius of the Si nanoparticles (a_p = 5, 25, 100 nm) and also for the specularities s = 0, 0.2 and 1. Table 3.3 gives the values for these.

We apply our model to a Si–Ge semiconductor nanocomposite with the values from Table 3.3. The corresponding results are shown in Figure 3.5. Note, that in this figure and thereafter, the volume fraction is limited to $\varphi = \pi/\sqrt{18}$, which corresponds to the maximum packing of hard spherical spheres.

It may seem strange that the thermal conductivity of the composite Si–Ge is smaller than that of the pure matrix Ge when the volume fraction of particles in increased. Indeed, since bulk Si has a larger thermal conductivity than bulk Ge, one should expect that composite conductivity will be higher. The reduction in the conductivity finds its interpretation in the small dimensions of the particles. Indeed, relation (3.14) tells us that the thermal conductivity of nanospheres of radii comparable or smaller than the mean free path of heat carriers may be considerably less than that of the bulk, hence a decrease in the effective thermal conductivity of the composite. Moreover, the smaller is the radius, the smaller the thermal conductivity of the nanoparticles. Similar results are also observed in Si_xGe_{1-x} alloys [66].

As for the temperature dependence of the thermal conductivity, it is shown that λ^{eff} decreases with the temperature irrespective of the radius of the nanoparticles and the s coefficient; at high temperature (T = 500 K) and large a_p-values (around 100 nm), the thermal conductivity remains practically constant. The reduction in λ^{eff} becomes more important as the size of the particles becomes small and the volume fraction large. The decrease in λ^{eff} with temperature may be explained as follows: by increasing the temperature, one causes an increase in the thermal resistance whence a diminution of thermal conductivity. This effect is less pronounced for smaller radii of the particles because of the increase in the particle

TABLE 3.3
Material Parameters (at Room Temperature).

Material	Heat Capacity [MJ (m^{-3} K^{-1})]	Group Velocity [m s^{-1}]	Mean Free Path [nm]
Si[a]	1.66	6,400	40.9
Ge[a]	1.67	3,900	27.5

[a]Ref. [54].

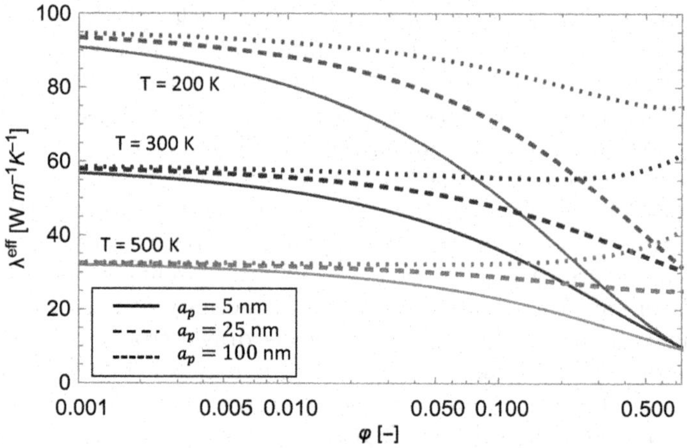

FIGURE 3.5
Effective thermal conductivity of Si–Ge nanocomposite versus the volume fraction at three different temperature: T = 200 K (upper curves), T = 300 K (middle curves) and T = 500 K (bottom curves), different radii $a_p = 5, 2, 100$ nm and $s = 0.5$ (Modified from: Lebon, G., Machrafi, H., Grmela, M. 2015. An extended irreversible thermodynamic modelling of size-dependent thermal conductivity of spherical nanoparticles dispersed in homogeneous media. *Proceedings of the Royal Society A* 471:20150144).

matrix interface. This can be interpreted by saying that phonons will meet more obstacles with, as a consequence, a reduction in heat transport.

The results plotted in Figure 3.5 indicate that the effective thermal conductivity λ^{eff} of Si–Ge composite is decreasing with the nanoparticles density and that at a fixed volume fraction, λ^{eff} is decreasing with decreasing radii. Such an effect may be of interest within the perspective of an optimal conversion of heat transport into electric current. Indeed, the efficiency of this conversion is measured by means of the so-called figure of merit defined by $Z = \sigma_e S^2/\lambda$, with σ_e the electrical conductivity and S the Seebeck coefficient so that a lowering of the thermal conductivity λ will clearly contribute to a better efficiency.

The model presented in section 3.1 is furthermore compared to Monte Carlo simulations, where amorphous Si and Ge are used, instead of crystalline Si and Ge in the earlier case. Table 3.4 gives the material properties for amorphous Si and Ge.

In this case study, two values of the particle radius are considered, i.e. $a_p = 5$ and 100 nm, and four values for the specularity parameter are used, i.e. $s = 0, 0.2, 0.5$ and 1. It

TABLE 3.4
Phonon Properties for Bulk Materials at Room Temperature (T_{ref}).

Material	Heat Capacity [MJ (m^{-3} K^{-1})]	Group Velocity [m s^{-1}]	Mean Free Path [nm]
Si	0.93[a]	1804[a]	268[a]
Ge	0.87[a]	1042[a]	171[a]
Cu	3.47[b]	7723[c]	45[d]

[a]Refs. [54,68,69].
[b]Ref. [54].
[c]Calculated from $v = 3\lambda/c\ell_b$ by considering $\lambda = 402$ W m^{-1} K^{-1} from Ref. [54].
[d]Ref. [54].

is worth noticing that the mean free path of phonons in Si is larger than all the values of the particles' radii. Figure 3.6 reports the results for the effective thermal conductivity versus the volume fraction of the nanoparticle. The results indicating Monte Carlo simulations (from [54]) are added as comparison, using particle radii a_p of 5 and 100 nm. Even though the Si nanoparticles possess a larger bulk thermal conductivity than the one of the Ge matrix, the effective thermal conductivity of the whole system still appears to decrease against initial expectations. The explanation resides in the fact that, according to (3.14), the main cause of variation of the thermal conductivity is, besides the thermal boundary resistance, the dimension of the nanoparticles. Indeed, the smaller the radius, the higher the phonon scattering and the lower the effective thermal conductivity.

The effective thermal conductivity appears to compare well with values from the literature [68,70] for several surface specularities and particle radii. This is, however, less the case for large particle radii and high values of the specularity parameter, for which the effective thermal conductivity is somewhat overpredicted. Depending on the particle radius, the present model agreed well with Monte Carlo simulations for a specularity parameters s ranging from 0 to 0.2. This indicates that for Si–Ge systems, it is more likely that the specularity parameter is between $0 < s < 0.2$. As such, a rather diffuse surface is expected.

Finally, as another case study, the model is applied for the prediction of the effective thermal conductivity of a system consisting of Cu particles dispersed in a Si matrix. The results are shown in Figure 3.7, where the present approach is compared to two other theories, i.e. those of Nan et al. [3] and Ordonez and Alvarado [21]. The material values (specific heat capacity, phonon group velocity and bulk mean free path) are shown in Table 3.4. For the calculations, different values of the particle radius (a_p = 50, 500, 900 and 3000 nm), ranging from nano- to micro-dimensions. The specularity parameter is taken $s = 0$.

Figure 3.7 shows that the model presented in this work has a satisfactory agreement with that of Ordonez and Alvarado 21], whereas less agreement is found with respect to that of Nan et al. [3], especially for small particles, i.e. ($a_p < 500$ nm). An explanation for the latter is to be found in the expression for λ^{eff} obtained by Nan et al., which is based on Fourier's law. The validity of the latter is known to be limited to large length scales and not valid for smaller length scales (see Chapters 1 and 2). It is interesting to note that the model in this work leads to good results, also within the Fourier limit.

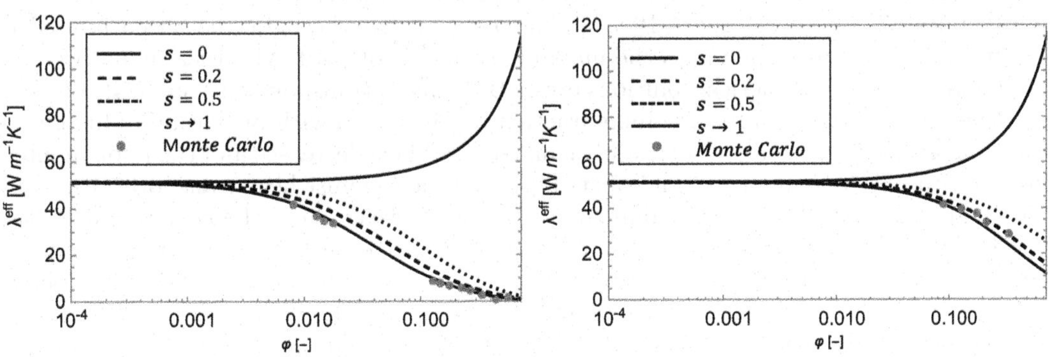

FIGURE 3.6
Effective thermal conductivity of the nanocomposite Si–Ge as a function of the volume fraction (φ) of Si particles for different s-values with radii: a_p = 5 and 100 nm. Monte Carlo simulations (see Ref. [54]) are shown in filled circles. (Modified from: Machrafi, H., Lebon, G. 2014. Effective thermal conductivity of spherical particulate nanocomposites: Comparison with theoretical models, Monte Carlo simulations and experiments. *International Journal of Nanoscience* 13:1450022.)

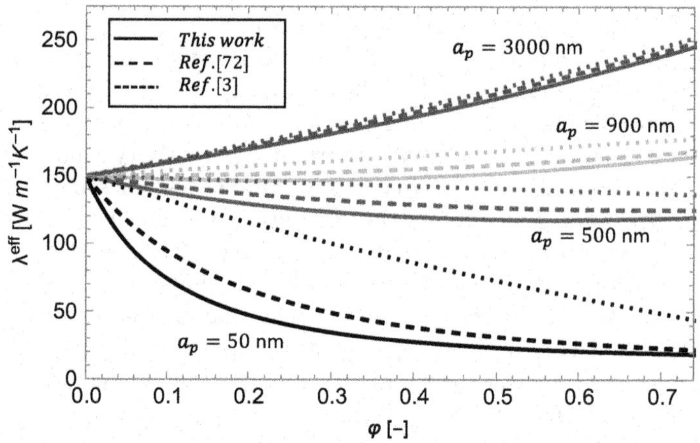

FIGURE 3.7
Effective thermal conductivity of the nanocomposite Cu–Si as a function of the volume fraction (φ) of Cu particles for different particle radii: $a_p = 50$, 500, 900 and 3000 nm ($s = 0$). Comparison with theoretical models (see Refs. [3] and [71]). (Modified from: Machrafi, H., Lebon, G. 2014. Effective thermal conductivity of spherical particulate nanocomposites: Comparison with theoretical models, Monte Carlo simulations and experiments. *International Journal of Nanoscience* 13:1450022.)

3.3.2 Application to Si/Ge Nanocomposites with Nanowire Inclusions

Let us first consider a Ge host matrix with Si cylindrical nanowire inclusions whose axes are aligned normally to the heat flux. In this section, the temperature is first kept fixed, equal to the room temperature, and then the temperature dependence is evaluated later. The values of the bulk parameters used in the calculations are given in Tables 3.3 and 3.4. Figure 3.8 shows the dependence of the effective transversal $\lambda_\perp$ thermal conductivity of wires as a function of the volume fraction for two values of the radius ($r = 5$ and $r = 25$ nm) and $s = 0$ (pure diffusive scattering). It is worth to stress that the figure has been limited to $\varphi = \pi/\sqrt{12}$, which corresponds to the maximum packing of rigid cylindrical nanowires.

Comparison with numerical solutions of the Boltzmann transport equation [24] shows an excellent agreement. For a fixed volume fraction, $\lambda_\perp$ increases with increasing radii. This is easily understood, as in this case, the wire–matrix interface decreases and the phonon interface scattering is less important and offers less resistance against heat transport. Otherwise stated, at larger dimensions, the boundary resistance is weaker and the heat conductivities of both the wires and the matrix become close to their respective bulk values characterized by the absence of boundary effects. When the radius of the wire is fixed, one observes generally a lowering of the thermal conductivity with increasing volume fraction due to the increase in the wire–matrix interface. The predictions from our model are also compared to Fourier's law. In the latter case, the results are not sensitive to size, as it should, and the values of the thermal conductivity are systematically overestimated. The effect of the thermal boundary resistance, measured through the parameter α_i, is assessed, and it is found that the thermal conductivity is larger for $\alpha_i = 0$, which is understandable as it corresponds to the weakest boundary resistance. Moreover, the larger the size of the wire, the greater the difference between a zero and a non-zero α-value, because for larger sizes, a change in the value of α will have larger consequences, due to the larger wire–matrix interface. Since in most nanocomposites, the volume fractions of nanoparticles are relatively small, taking

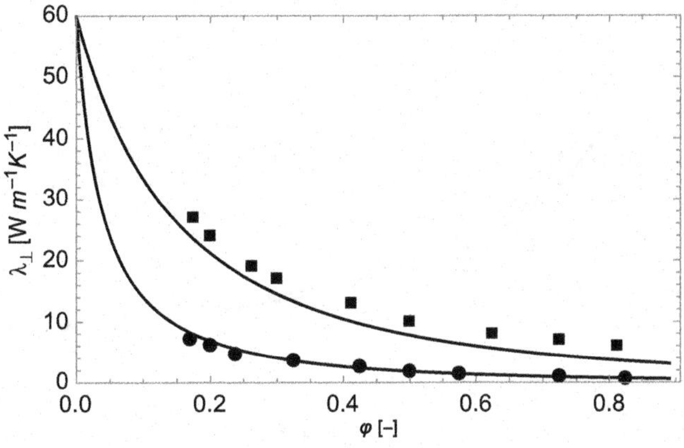

FIGURE 3.8
Transversal effective thermal conductivity of Si/Ge nanocomposite as a function of the volume fraction of the nanowires for two values of the radius (r = 5 and 25 nm) and $s = 0$. The results are compared with Boltzmann's solutions represented by circles for $r = 5$ nm and squares for $r = 25$ nm. (Modified from: Lebon, G., Machrafi, H. 2015. Thermal conductivity of tubular nanowire composites based on a thermodynamical model. *Physica E* 71:117–122.)

$\alpha_i \neq 0$ has little impact in actual applications. In the study of transverse heat conduction, it was admitted that the length of the wires was infinite, in other words much larger than the radii of the wires. To check the validity of this approximation, we have calculated $\lambda_\perp$ for several lengths of the nanowire at various values of the volume fractions and different sizes. It has been found that $\lambda_\perp$ remains practically constant irrespective of the length of the wire and the volume fraction. It is true that for large radii, the length has little influences on the transversal thermal conductivity. Nevertheless, it should be noted that for a radius comparable or larger than the length, there is a more significant influence, albeit this effect is much smaller than that of the volume fraction or the wire radius. In the previous analysis, it was assumed that the interface was perfect, i.e. characterized by $s = 0$. As far as the effect of the s-value on the transversal thermal conductivity is concerned, the only visible differences occur between $s = 0$ and $s = 0.1$. No significant differences are observed between $s = 0.1$ and $s = 1$. The main reason is that the flow of phonons normal to the surface is not greatly affected by the nature of the surface; this is no longer true with the longitudinal phonons moving along the interface. The influence of the s-value has different outcomes. For $s = 0$, thermal conductivity is decreasing with the volume fraction and with decreasing sizes. However, by increasing sufficiently either the radius or the s-value (each independently), the thermal conductivity increases with the volume fraction. Increasing the values of s means a reduction in the roughness of the particles–matrix interface whence less obstacles are experienced by the phonons and therefore a higher thermal conductivity is predicted. Larger wire radii also mean smaller interfaces between the wires and the matrix, with, as a consequence, less obstacles for the phonons and an increase in the thermal conductivity. From a mathematical point of view, a larger wire radius and a higher s-value lead to smaller Knudsen numbers, characteristic of the Fourier regime. The thermal conductivity is then simply a combination of the bulk heat conductivities. Since that of the wires (Si) is larger than that of the matrix (Ge), the thermal conductivity will increase with larger volume fractions. So, whether the longitudinal thermal conductivity increases or decreases as a function of the volume fraction depends on the three factors: wire radius, surface specularity, and bulk heat conductivities.

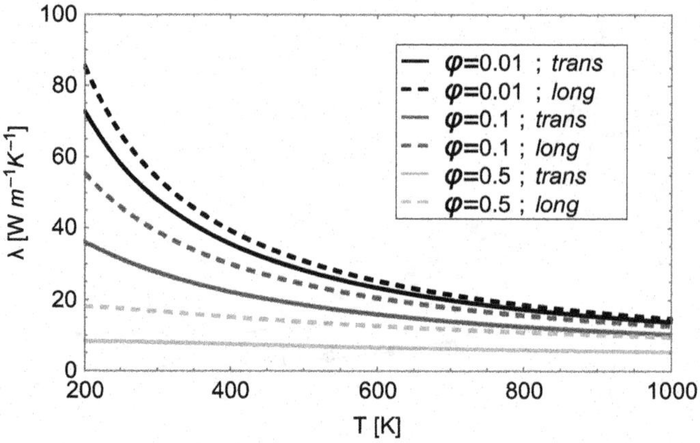

FIGURE 3.9
Comparison between thermal conductivities in the transversal and longitudinal direction, respectively as a function of the temperature for different volumes fractions and for $r = 5$ nm. (Modified from: Lebon, G., Machrafi, H. 2015. Thermal conductivity of tubular nanowire composites based on a thermodynamical model. *Physica E* 71:117–122.)

We have found that the effective thermal conductivity λ^{eff} is seen to decrease significantly with the temperature at fixed radius and volume fraction. This behaviour can be explained by the fact that the thermal boundary resistance is more important at a lower temperature (lower heat capacities) and therefore contributes to a larger reduction in the thermal conductivity. At large volume fraction and small particle size, the thermal conductivity is shown to remain almost constant. This is a consequence of the strong particle–matrix interaction that prevails on such conditions. In Figure 3.9, the overall thermal conductivities in the longitudinal direction $\lambda_{\parallel}$ are compared with those in the transversal direction $\lambda_{\perp}$ as a function of the temperature for several values of φ and a radius of 5 nm. The general behaviour is rather similar in the two configurations with the difference that higher values for the thermal conductivity are observed in the longitudinal direction. This is not surprising as phonons experience less boundary scattering while moving in the longitudinal direction.

3.4 Nanoporous Composites

3.4.1 Nanoporous Materials

The effect of porosity is to decrease considerably the thermal conductivity of crystalline materials, as is observed, for instance, in nanoporous silicon [71–75]. This lowering property has been widely exploited with the objective to reinforce thermal insulation in several systems as microsensors, integrated circuits and semiconductor devices. It has therefore fostered an increasing interest during the last decade. Thermal properties of nanoporous media are generally investigated by referring to, for instance, kinetic theory of phonons [73], molecular simulations [75], Monte Carlo simulations [76], two-component materials [77] or hydrodynamic models [78].

Nanoporous materials are generally made out of a homogeneous matrix wherein nanopores are dispersed. In that respect, such systems are similar to nanocomposites, wherein nanoparticles are embedded in a matrix, the difference being that nanoparticles are replaced by (spherical) nanopores, such as foam. A great deal of papers have been devoted to the problem of heat transport in nanofluids and nanocomposites, see for instance the review paper by Michaelides [70]. In particular, the problem of the overall thermal conductivity of the system as a function of the volume fraction of the nanoparticles and their size has drawn much attention. Our purpose is twofold. Firstly, it is the purpose to study the role of porosity and the size of the pores on the thermal conductivity of porous systems, such as nanoporous silicon, via our thermodynamic model in Section 3.1. Secondly, we examine, using the same approach, the influence of the presence of porous nanoparticles on the thermal conductivity of nanocomposites; as illustration, nanoporous silicon particles dispersed in a germanium matrix are considered. The effect on the thermal conductivity of the size and volume fraction of nanoporous silicon as well as their porosity and pore sizes is investigated. Using (3.1), adapting it to the present case gives

$$\lambda_{pSi}^{eff} = \lambda_m \frac{2\lambda_m + (1+2\alpha_i)\lambda_p + 2\varepsilon\left[(1-\alpha_i)\lambda_p - \lambda_m\right]}{2\lambda_m + (1+2\alpha_i)\lambda_p - \varepsilon\left[(1-\alpha_i)\lambda_p - \lambda_m\right]}. \tag{3.38}$$

where symbol ε stands for the porosity, λ_p for the thermal conductivity of the nanopores and λ_m for the bulk thermal conductivity of the host medium. Furthermore, α_i and its components are given, in analogy with λ_p and λ_m, by Eqs. (3.2), (3.3), (3.7) and (3.8). Instead of Eq. (3.9), we should rather use

$$\ell_{m,coll} = 4a_p\left(1-\varepsilon\right)/3\varepsilon. \tag{3.39}$$

This expression differs from Eq. (3.9) by the presence of the factor $(1-\varepsilon)$ in the numerator. This is justified by the fact that in the limit of a high porosity $[\varepsilon = O(1)]$, the mean free path of phonons colliding with the pores is expected to tend to zero, while the bulk mean free path $\ell_{m,b}$, being a material property, remains of course constant. For λ_p, we use the equivalence of Eq. (3.14):

$$\lambda_p = \frac{3k_p^0}{4\pi^2 Kn^2}\left[\frac{2\pi Kn}{\arctan\left(2\pi Kn\right)} - 1\right]. \tag{3.40}$$

The Knudsen number is here defined by $Kn \equiv \ell_{p,b}/a_{po}$, where a_{po} is the pore radius. Note that the result (3.40) is only valid for a spherical geometry. In the case of cylindrical pores, for instance, a modified expression has to be used, based on section 3.1.4. The previous general considerations will be applied to spherical nano-inclusions in a Si matrix, and the void phase is supposed to be occupied by air. Assuming that the pores are filled with air is by no means limitative of the present approach. Indeed, replacing air by any other gas or liquid will not modify the structure of the model. The material phonon properties are listed in Tables 3.3 and 3.4. Since air is added, and for the sake of completeness, the values for Si and Ge are restated in Table 3.5, next to those of air.

The values for Si and Ge are derived from the dispersion model presented in [78]. The values for Si listed on the second row correspond to the Debye model [78], used by one of the models to which our results will be compared. Our model is applied in this case in comparison with experimental data and five different models [76,78,81,82] briefly described next. The comparison with the aforementioned models will also allow better understanding the underlying difficulties in modelling thermal conductivity of nanoporous materials. The first model, referred to as Eucken's formula (e.g. [76]), is a simplification of expression (3.38) wherein λ_m is replaced by the bulk value λ_m^0, whereas λ_p is neglected:

$$\lambda_{pSi}^{eff,Eucken} = \lambda_m^0 \frac{2-2\varepsilon}{2+\varepsilon}. \tag{3.41}$$

TABLE 3.5
Phonon Properties of Bulk Materials at Room Temperature (T_{ref}).

Material	Heat Capacity [MJ (m^{-3} K^{-1})]	Group Velocity [m s^{-1}]	Mean Free Path [nm]
Si (dispersion)	0.93[a]	1804[a]	268[a]
Si (Debye)	1.66[a]	6400[a]	40.9[a]
Ge	0.87[a]	1042[a]	171[a]
Air	0.00121[b]	347[b]	185.8[c]

[a]Ref. [78].
[b]Ref. [79].
[c]Calculated via $\ell_{p,b} = 3\lambda/(c_p v_p)$, with $\lambda = 0.026$ W m^{-1} K^{-1} [80].

The drawback of such a formulation is that it does not take into account the effects of particle size and phonons interactions.

A more detailed formalism was developed by Alvarez et al. [78] and also used by Criado-Sancho et al. [73,74]. It is based on a phonon hydrodynamics version of Stokes resistance force exerted on a sphere in an infinite medium. It was found that

$$\lambda_{pSi}^{eff,hydro} = \lambda_m^0 \frac{1}{\frac{1}{(1-\varepsilon)^3} + \frac{9}{2}\varepsilon \frac{Kn^2}{1+Kn^*\left(0.864+0.290\mathrm{e}^{-1.25/Kn}\right)}\left(1+\frac{3}{\sqrt{2}}\sqrt{\varepsilon}\right)} \tag{3.42}$$

In this model, the particle size and mean free path of phonons are not considered in the bulk material, but only in the correcting factor through the Knudsen number.

The third model used for comparison consists in letting Kn in Eq. (3.42) tend to zero so that the expression reduces to the so-called percolation model:

$$\lambda_{pSi}^{eff,perco} = \lambda_m^0 (1-\varepsilon)^3 . \tag{3.43}$$

The fourth model, proposed by Lysenko et al. [81], consists in modifying the percolation model formula by a correcting factor involving the mean free path of phonons in the bulk material and the pore radius via Knudsen's number Kn, i.e.

$$\lambda_{pSi}^{eff,Lys} = \frac{\lambda_m^0}{1+\frac{4}{3}Kn}(1-\varepsilon)^3 . \tag{3.44}$$

In Eq. (3.44), the porosity dependence is given by the term $(1-\varepsilon)^3$, which means that the mean free path does not depend on the porosity as is the case in our model.

In another model, based on a formalism rather similar to the model by Lysenko et al. [81], Sumirat et al. [82] write the effective thermal conductivity in the form:

$$\lambda_{pSi}^{eff,Sum} = \lambda_m^0 \frac{(1-\varepsilon)}{1+\varepsilon^{1/3}Kn}. \tag{3.45}$$

Comparison between our model and the five other ones expressed by relations (3.41) to (3.45) is shown in Figure 3.10, wherein the effective thermal conductivity is plotted as a function of the porosity for a pore radius of 10 nm. Irrespective of the selected model, the thermal conductivity is decreasing with the porosity, and this reduction is easily understood owing to the weak thermal conductivity of air enclosed in the pores.

Our model is shown to be in satisfactory agreement with the hydrodynamic one, at values of porosity smaller than 0.2. In contrast, our results disagree with Eucken's approximation and the simplified formula (3.43), which predict overestimated results, particularly

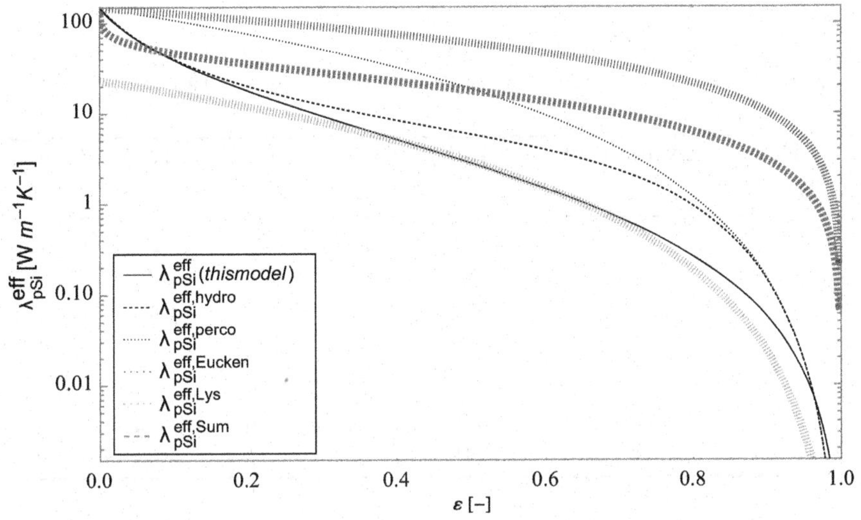

FIGURE 3.10
Effective thermal conductivity as a function of porosity for a pore radius of 10 nm. Comparison between our model and five other ones. (Modified from: Machrafi, H., Lebon, G. 2015. Size and porosity effects on thermal conductivity of nanoporous material with an extension to nanoporous particles embedded in a host matrix. *Physics Letters, Section A* 379:968–973.)

Eucken's approach. This is a clear indication that pore sizes play a decisive role in the calculations of thermal conductivity in nanoporous materials (an observation already well known for nanocomposites [54]). The hydrodynamic formulation does consider pore size and the ballistic property of the heat transport which is not the case of the simplified formula which express the thermal conductivity exclusively as a function of the porosity. In the limit of very large porosity, the size is no longer relevant for any of the models, and the thermal conductivity tends uniformly to that of air, at the exception of Eucken's. The results from Figure 3.10 show that for small porosities, there is a rather good agreement between our model and the hydrodynamic one. This is explained by the fact that at small porosities, the collision mean free path in the bulk material [see Eq. (6)] is rather large so that the phonons travelling in the bulk are hardly influenced by the presence of the pores. However, at larger porosities, a clear discrepancy is visible. This can be interpreted by the particular treatment reserved in our model to the influence of the pore size and the mean free path, which enter into the description of heat transport in both the bulk material and the nanopores. At intermediate porosity range ($0.3 < \varepsilon < 0.7$), a rather good agreement is found between our model and the Lysenko model [81]. This indicates that the dependence of the mean free path on nanopore size and porosity (only considered in our model) becomes relevant at porosities larger than 0.7. At small porosities, however, the Lysenko model shows an important discrepancy with respect to ours, and, more disturbing, when porosity tends to zero, the thermal conductivity does not approach the bulk value of silicon given by $\lambda_m^0 = 150\,\mathrm{W\,m^{-1}\,K^{-1}}$. The formula (3.45) proposed by Sumirat et al. [82] leads to results that are intermediate between Eucken's formula and the Lysenko model. The question that finally arises is, what is the best approach?

Complementary information is provided by Table 3.6 wherein the various models are compared to experimental data.

TABLE 3.6
Thermal Conductivities (in W m^{-1} K^{-1}) for Different Porosities and Pore Radii (in nm) and Comparison Between the Five Models Discussed in this Section with Experimental Data.

Porosity	Pore Radius	Experiments	(3.38)	(3.41)	(3.42)	(3.43)	(3.44)	(3.45)
0.40[a]	1–5	1.2	0.6–2.7	75	1.0–4.6	32.4	0.6–2.7	3–13
0.40[b]	100	31.2	32.0	75	29.1	32.4	20.9	69
0.50[b]	10	3.9	2.8	60	5.9	18.7	2.9	17.6
0.60[b]	10	2–5	1.5	46	4.0	9.6	1.5	13.5
0.64[c]	2	0.2	0.23	41	0.96	7.0	0.25	2.9
0.71[c]	2	0.14	0.13	32	0.77	3.7	0.13	2.3
0.79[c]	3	0.06	0.09	23	0.67	1.4	0.07	2.3
0.89[c]	5	0.04	0.04	11	0.18	0.2	0.02	1.9

[a]Ref. [83].
[b]Ref. [71].
[c]Ref. [77].

Table 3.6 indicates that irrespective of the model, the thermal conductivity decreases with the porosity and that at fixed porosity, it is considerably reduced by passing from macro- to nanopores. It is worth to stress that our model fits rather well experiments for all the values of porosities considered. The hydrodynamic model predicts satisfactory the results at small porosities, but not at larger porosities. The Lysenko model is in agreement with experiments at intermediate porosities, except for a 100-nm pore radius. It is confirmed that the results provided by the Eucken, Sumirat and percolation models are rather far from the experimental data for all the ranges considered.

A further comparison between our model, the hydrodynamic model and experimental data [71,77,81] for several values of the pore radii (a_{po} = 2, 5, 10 and 100 nm) is shown in Figure 3.11a. The same comparison is shown in Figure 3.11b, but instead of the hydrodynamic model, we compare with the Lysenko model.

A good agreement between our model and the experimental data is observed. There is only one point, at radius 10 nm and porosity 0.6 that seems to be in less satisfactory accord. However, it should be stressed that this point is experimentally not well defined. Indeed, the experimental values of the thermal conductivity at these values is found between 2 and 5 W m^{-1} K^{-1}, while our model predicts 1.5 W m^{-1} K^{-1}. Recalling that the curve starts with a value of the thermal conductivity of 150 W m^{-1} K^{-1} (at zero porosity), the agreement may be considered as rather fair. Anyway, for all the other reported experimental points, the agreement with our model is better than with the five other discussed models. For porosities smaller than 0.6 and pore sizes of 10 and 100 nm, (see Figure 3.11a) a good agreement with experimental data is reached between our model (solid line) and the hydrodynamic one. Contrary to our model, the hydrodynamic model does not compare well experimentally at pore sizes of 2 and 5 nm at porosities larger than 0.6. The poor correspondence at larger pore sizes has been discussed earlier. The good correspondence observed at higher pore sizes is due to the fact that at larger pore sizes (macroporous media), the ballistic effects become less important and classical descriptions remain satisfactory. At intermediate porosities, the Lysenko model agrees well with our model and the experimental data, except at a pore radius of 100 nm (see Figure 3.11b). It is concluded that only our model seems to be in accord with the experiments in the whole range of porosities and pore sizes considered in this work.

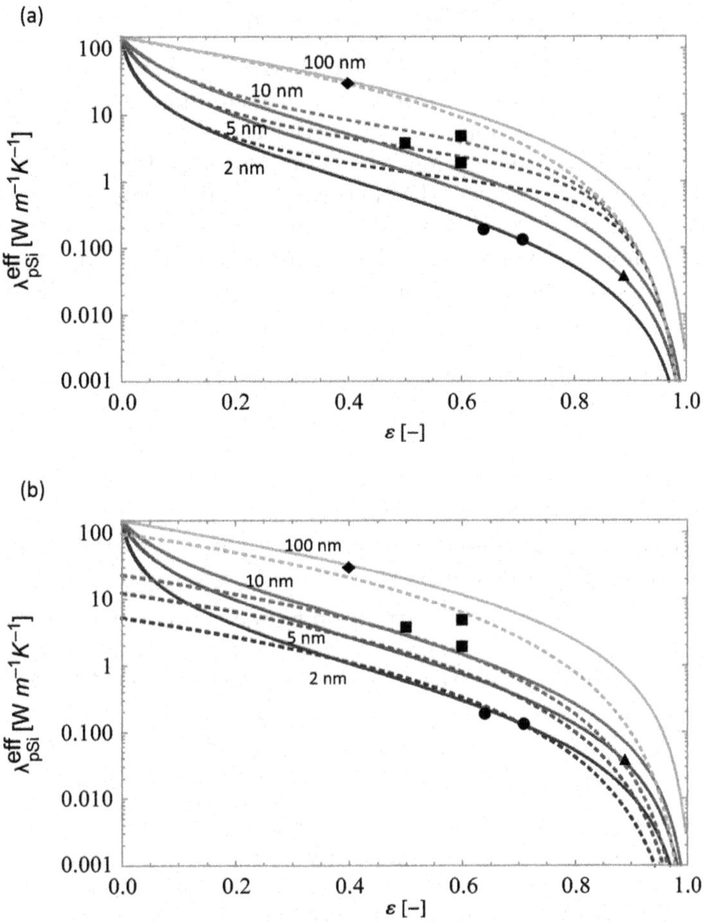

FIGURE 3.11
Effective thermal conductivity (W m^{-1} K^{-1}) as a function of porosity for pore radii of 2 (•), 5 (▲), 10 (■) and 100 (♦) nm. Comparison of the results of our model (solid lines) with reported experimental data [71,77] and, in dashed lines, the hydrodynamic model [78] (a) and the Lysenko model [23] (b). (Modified from: Machrafi, H., Lebon, G. 2015. Size and porosity effects on thermal conductivity of nanoporous material with an extension to nanoporous particles embedded in a host matrix. *Physics Letters, Section A* 379:968–973.)

We also observe that generally, at a fixed value of the porosity, the thermal conductivity decreases under a reduction in the radius of the pores. This is easily understandable as smaller pore radii correspond to an increase in the total pore–matrix interface, at a given value of porosity. This results in an increase in the scattering of matrix phonons at the pore–matrix interface and consequently a reduction in the effective thermal conductivity. However, Eq. (3.38) shows interestingly that this is not the only reason for the reduction in the effective thermal conductivity. Indeed, smaller pore radii also result into larger values of the Knudsen number, whence a smaller λ_p-value is obtained according to Eq. (3.38). This means that the increase in phonon collisions inside the pores (not considered up to now) contributes further to the reduction in the effective thermal conductivity [84].

3.4.2 Nanoporous Particles in a Composite

In Section 3.3, we have calculated the effective thermal conductivity of Si/Ge composites with silicon spherical nanoparticle embedded in a germanium matrix. Here, instead of bulk Si particles, we will consider nanoporous Si particles (p-Si) by focusing on the effects of porosity [84]. The idea is to evaluate how the combination of two types of nano-inclusions (nanopores in Si nanoparticles) affects the effective thermal conductivity. This calls for some modifications and adaptations to the model developed in Section 3.4.1.

Let us first determine the expression of the thermal conductivity of the p-Si nanoparticles of radius a_p assumed to be dispersed (with a volume fraction φ) in a Ge matrix. In virtue of Eq. (3.40), it is given by

$$\lambda_{p-Si} = \frac{3\lambda^0_{pSi}}{4\pi^2 Kn_p^2}\left[\frac{2\pi Kn_p}{\arctan\left(2\pi Kn_p\right)} - 1\right], \tag{3.46}$$

The next step is to state precisely what represent the quantities Kn_p ($\equiv \ell_{p,b}/a_p$) and λ^0_{pSi}. The latter designates the thermal conductivity of nanoporous silicon (p-Si) particles (with porosity ε and pore radius a_p) by taking into account the size effects of the nanopores but not those of the nanoparticles themselves, its expression is given by Eq. (3.38) λ^{eff}_{pSi} or, to be explicit, $\lambda^0_{pSi} \equiv \lambda^{eff}_{pSi}$. We now turn our attention to the expression of the Knudsen number Kn_p of the p-Si nanoparticles; it is defined as $Kn_p = \ell_{p-Si}/a_p$, with ℓ_{p-Si} the mean free path of phonons in the p-Si nanoparticles approximated by (using Matthiesen's rule)

$$\frac{1}{\ell_{p-Si}} = \frac{1}{\ell_p} + \frac{1}{\ell_{air}}, \tag{3.47}$$

where the quantities ℓ_{air} and ℓ_p are the mean free paths in air and bulk silicon, respectively. It remains to determine the thermal conductivity of the Ge host matrix, λ_{mGe}. It is given by the equivalent of relation (3.7), with ℓ_m expressed by Eqs. (3.8) and (3.9). The overall effective thermal conductivity is still given by relation (3.1) wherein the volume fraction φ stands for that of p-Si nanoparticles in Ge, so that

$$\lambda^{eff}_{p-Si,Ge} = \lambda_{mGe}\frac{2\lambda_{mGe} + (1+2\alpha_{i,*})\,\lambda_{p-Si} + 2\varphi\left[(1-\alpha_{i,*})\,\lambda_{p-Si} - \lambda_{mGe}\right]}{2\lambda_{mGe} + (1+2\alpha_{i,*})\,\lambda_{p-Si} - \varphi\left[(1-\alpha_{i,*})\,\lambda_{p-Si} - \lambda_{mGe}\right]}. \tag{3.48}$$

The quantity $\alpha_{i,*}$ in Eq. (3.48) is given by Eq. (3.2), but where the thermal boundary resistance coefficient R_* is given by

$$R_* = \frac{4}{c_{mGe}v_{mGe}} + \frac{4}{c_{p-Si}v_{p-Si}}, \tag{3.49}$$

where c_{mGe} and v_{mGe} are, respectively, the specific heat capacity and phonon group velocity corresponding to germanium. The specific heat capacity c_{p-Si} and phonon group velocity v_{p-Si} are supposed to be given by the following mean values:

$$c_{p-Si} = \varepsilon c_{air} + (1-\varepsilon)\,c_p, \tag{3.50}$$

$$v_{p-Si} = \varepsilon v_{air} + (1-\varepsilon)\,v_p, \tag{3.51}$$

where the sub-index "p" refers to bulk silicon. Figure 3.12 depicts the dependence of the effective thermal conductivities of the nanocomposite p-Si/Ge versus the volume fraction φ of the p-Si particles for several porosities ε, pore radii a_{po} and particle radii a_p. Note that $\varepsilon = 0$ corresponds to bulk silicon nanoparticles embedded in germanium as studied in [54], whereas $\varepsilon = 1$ describes nanoporous Ge samples [76].

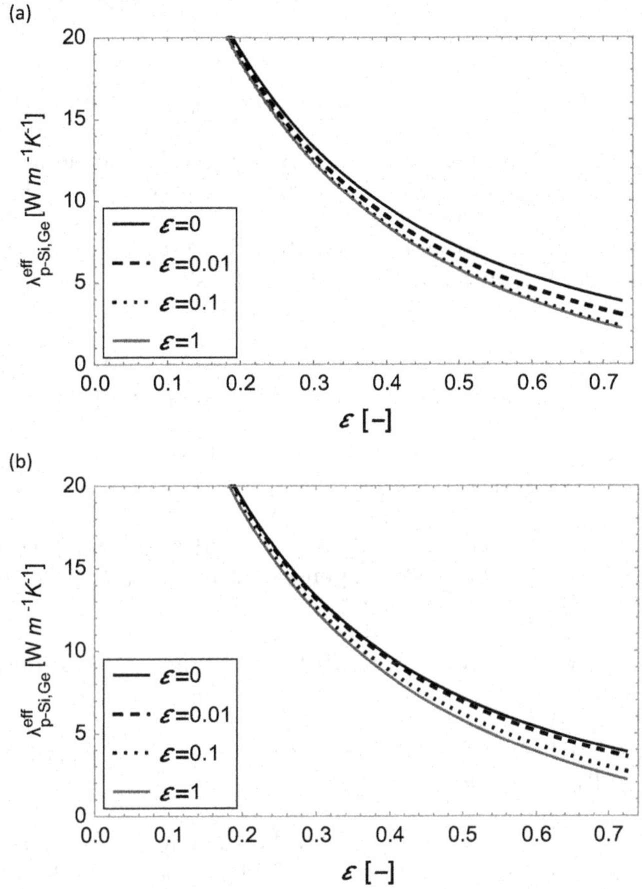

FIGURE 3.12
Effective thermal conductivity as a function of the p-Si nanoparticle volume fraction, with (a) $a_{po} = 1$ nm and $a_p = 25$ nm, and (b) $a_{po} = 5$ nm and $a_p = 25$ nm, for several porosities. (Modified from: Machrafi, H., Lebon, G. 2015. Size and porosity effects on thermal conductivity of nanoporous material with an extension to nanoporous particles embedded in a host matrix. *Physics Letters, Section A* 379:968–973.)

Figure 3.12a indicates that thermal conductivity is decreasing with increasing volume fraction of the porous p-Si nanoparticles. The effective thermal conductivity is rather insensitive to the porosity of the p-Si particles; indeed, a porosity of 0.1 leads to results practically equivalent to fully porous particles with $\varepsilon = 1$, which corresponds to the maximum isolation of the sample. By increasing the radius of the pores (from $r_p = 1$–5 nm) (see Figure 3.12b), a value of ε larger than 0.1 is necessary to reach the same result, though the thermal conductivity still remains rather insensitive to the porosity. It is interesting to note that for large radii of the p-Si nanoparticles ($a_p = 100$ nm) the reduction of the thermal conductivity is much larger when ε is increased. The reduction is still reinforced by diminishing the size a_{po} of the nanopores inside the p-Si particles. We can conclude that the behaviour of the effective thermal conductivity is governed by the following four parameters: porosity, pore size, particle size and particle volume fraction. It appears that for small particle sizes ($a_p \leq 25$ nm), porosity hardly influences the value of the thermal conductivity, while for large particle sizes ($a_p \geq 100$ nm), porosity plays a more influential role.

References

[1] Maxwell, J.C. 1881. *Treatise on Electricity and Magnetism.* Oxford: Clarendon.

[2] Hasselman, D.P.H., Johnson, L.F. 1987. Effective thermal conductivity of composites with interfacial thermal barrier resistance. *Journal of Composite Materials* 21:508–515.

[3] Nan, C.W., Birringer, R., Clarke, D.R., Gleiter, H. 1997. Effective thermal conductivity of particulate composites with interfacial thermal resistance. *Journal of Applied Physics* 81:6692.

[4] Lebon, G., Machrafi, H. 2015. Thermal conductivity of tubular nanowire composites based on a thermodynamical model. *Physica E* 71:117–122.

[5] Hamilton, R.I., Crosser, O.K. 1962. Thermal conductivity of heterogeneous two component systems. *Industrial and Engineering Chemistry Fundamentals* 1:187–191.

[6] Wang, B.X., Zhou, L.P., Peng, X.F. 2003. A fractal model for predicting the effective thermal conductivity of liquid with suspension of nanoparticles. *International Journal of Heat and Mass Transfer* 46:2665–2672.

[7] Barnes, H.A., Hutton, J.F., Walters, K. 1989. *An Introduction to Rheology.* Amsterdam: Elsevier.

[8] Goodwin, J.W., Hughes, R.W. 2000. *Rheology for Chemists—An Introduction.* Cambridge: Baker & Taylor Books.

[9] Prasher, R., Evans, W., Meakin, P., Fish, J., Phelan, P., Keblinski, P. 2006. Effect of aggregation on thermal conduction in colloidal nanofluids. *Applied Physics Letters* 89:143119.

[10] Hui, P.M., Zhang, X., Markworth, A.J., Stroud, D. 1999. Thermal conductivity of graded composites: Numerical simulations and an effective medium approximation. *Journal of Materials Science* 34:5497–5503.

[11] Mingo, N., Yang, L., Li, D., Majumdar, A. 2003. Predicting the thermal conductivity of Si and Ge nanowires. *Nano Letters* 3:1713–1716.

[12] Callaway, J. 1959. Model for lattice thermal conductivity at low temperature. *Physical Review* 113:1046–1051.

[13] Glassbrenner, C.J., Slack, G.A. 1964. Thermal conductivity of silicon and germanium from 3°K melting point. *Physical Review A* 134:1058–1069.

[14] Xie, S.H., Zhu, B.K., Li, J.B., Wei, X.Z., Xu, Z.K. 2004. Preparation and properties of polyimide/aluminum nitride composites. *Polymer Testing* 23:797–804.

[15] Xu, Y., Chung, D.D.L. 2000. Increasing the thermal conductivity of boron nitride and aluminum nitride particle epoxy-matrix composites by particle surface treatments. *Composite Interfaces* 7:243–256.

[16] Agarwal, S., Khan, M.M.K., Gupta, R.K. 2008. Thermal conductivity of polymer nanocomposites made with carbon nanofibers. *Polymer Engineering and Science* 48:2474–2481.

[17] Wong, C.P., Bollampally, R.S. 1999. Comparative study of thermally conductive fillers for use in liquid encapsulants for electronic packaging. *IEEE Transactions on Advanced Packaging* 22:54–59.

[18] Han, Z., Wood, J.W., Herman, H., Zhang, C., Stevens, G.C. 2008. Thermal properties of composites filled with different fillers. *International Symposium Electrical Insulation*, pp. 497–501.

[19] Fukushima, K., Takahashi, H., Takezawa, Y., Hattori, M., Itoh, M., Yonekura, M. 2004. High thermal conductive epoxy resins with controlled high-order structure (electrical insulation applications), *Conference Electrical Insulation and Dielectric Phenomena*, pp. 340–343.

[20] Miyazaki, Y., Nishiyama, T., Takahashi, H., Katagiri, J.I., Takezawa, Y. 2009. Development of highly thermoconductive epoxy composites, *Conference Electrical Insulation and Dielectric Phenomena*, pp. 638–641.

[21] Ordonez-Miranda, J., Alvarado-Gil, J.J. 2012. Thermal conductivity of nanocomposites with high volume fractions of particles. *Composites Science and Technology* 72:853–857.

[22] Tian, W.X., Yang, R.G. 2008. Phonon transport and thermal conductivity percolation in random nanoparticle composites. *Computer Modeling in Engineering and Sciences* 24:123–141.

[23] Tian, W.X., Yang, R.G. 2007. Effect of interface scattering on phonon thermal conductivity percolation in random nanowire composites. *Applied Physics Letters* 90:263105–263108.

[24] Yang, R.G., Chen, G., Dresselhaus, S.M. 2005. Thermal conductivity of simple and tubular nanowire composites in the longitudinal direction. *Physical Review B* 72:125418–125424.

[25] Prasher, R. 2005. Thermal boundary resistance of nanocomposites. *International Journal of Heat and Mass Transfer* 48:4942–4952.

[26] Jou, D., Casas-Vazquez, J., Lebon, G. 2010. *Extended Irreversible Thermodynamics*, fourth ed. New York: Springer.

[27] Chen, H., Witharana, S., Jin, Y., Kim, C., Ding, Y. 2009. Predicting thermal conductivity of liquid suspensions of nanoparticles (nanofluids) based on rheology. *Particuology* 7:151–157.

[28] Chen, H., Ding, Y., Tan, C. 2007. Rheological behaviour of nanofluids. *New Journal of Physics* 9:367–390.

[29] Kochetov, R., Korobko, A.V., Andritsch, T., Morshuis, P.H.F., Picken, S.J., Smit, J.J. 2011. Modelling of the thermal conductivity in polymer nanocomposites and the impact of the interface between filler and matrix. *Journal of Physics D* 44:395401.

[30] Timofeeva, E.V., Gavrilov, A.N., McCloskey, J.M., Tolmachev, V.V. 2007. Thermal conductivity and particle agglomeration in alumina nano-fluids: Experiment and theory. *Physical Review* 76:061203.

[31] Anoop, K.B., Kabelac, S., Sundararajan, T., Das, S.K. 2009. Rheological and flow characteristics of nanofluids: Influence of electroviscous effects and particle agglomeration. *Journal of Applied Physics* 106:034909.

[32] Machrafi, H., Lebon, G., Iorio, C.S. 2016. Effect of volume-fraction dependent agglomeration of nanoparticles on the thermal conductivity of nanocomposites: Applications to epoxy resins, filled by SiO_2, AlN and MgO nanoparticles. *Composites Science and Technology* 130:78–87.

[33] Šupová, M., Martynková, G.S., Barabaszová, K. 2011. Effect of nanofillers dispersion in polymer matrices: A review. *Science of Advanced Materials* 3:1–25.

[34] Ammala, A., Bell, C., Dean, K. 2008. Poly(ethylene terephthalate) clay nanocomposites: Improved dispersion based on an aqueous ionomer. *Composites Science and Technology* 68:1328–1337.

[35] Nolte, H., Schilde, C., Kwade, A. 2012. Determination of particle size distributions and the degree of dispersion in nanocomposites. *Composites Science and Technology* 72:948–958.

[36] Finney, E.E., Shields, S.P., Buhro, W.F., Finke, R.G. 2012. Gold nanocluster agglomeration kinetic studies: Evidence for parallel bimolecular plus autocatalytic agglomeration pathways as a mechanism-based alternative to an Avrami-based analysis. *Chemistry of Materials* 24:1718–1725.

[37] Yon, J., Bescond, A., Ouf, F.X. 2015. A simple semi-empirical model for effective density measurements of fractal aggregates. *Journal of the Aerospace Sciences* 87:28–37.

[38] Krishnan, A., Xu, L.R. 2011. A simple effective flaw model on analyzing the nanofiller agglomeration effect of nanocomposite materials. *Journal of Nanomaterials* 2012:483093.

[39] Fu, Y.X., He, Z.X., Mo, D.C., Lu, S.S. 2014. Thermal conductivity enhancement of epoxy adhesive using graphene sheets as additives. *International Journal of Thermal Science* 86:276–283.

[40] Fu, Y.X., He, Z.X., Mo, D.C., Lu, S.S. 2014. Thermal conductivity enhancement with different fillers for epoxy resin adhesives. *Applied Thermal Engineering* 66:493–498.

[41] Wang, Z., Qi, R., Wang, J., Qi, B. 2015. Thermal conductivity improvement of epoxy composite filled with expanded graphite. *Ceramics International* 41:13541–13546.

[42] Zhou, T., Wang, X., Cheng, P., Wang, T., Xiong, D., Wang, X. 2013. Improving the thermal conductivity of epoxy resin by the addition of a mixture of graphite nanoplatelets and silicon carbide microparticles. *eXpress Polymer Letters* 7:585–594.

[43] Hong, J.P., Sun, S.W., Hwang, T., Oh, J.S., Hong, C., Nam, J.D. 2012. High thermal conductivity epoxy composites with bimodal distribution of aluminum nitride and boron nitride fillers. *Thermochimica Acta* 537:70–75.

[44] Yu, W., Xie, H., Yin, L., Zhao, J., Xia, L., Chen, L. 2015. Exceptionally high thermal conductivity of thermal grease: Synergistic effects of graphene and alumina. *International Journal of Thermal Science* 91:76–82.

[45] Prasher, R., Phelan, P.E., Bhattacharya, P. 2006. Effect of aggregation kinetics on the thermal conductivity of nanoscale colloidal solutions (nanofluid). *Nano Letters* 6:1529–1534.

[46] Lebon, G., Machrafi, H., Grmela, M. 2015. An extended irreversible thermodynamic modelling of size-dependent thermal conductivity of spherical nanoparticles dispersed in homogeneous media. *Proceedings of the Royal Society A* 471:20150144.

[47] Moreira, D.C., Braga, N.R., Sphaier, L.A., Nunes, L.C.S. 2016. Size effect on the thermal intensification of alumina-filled nanocomposites. *Journal of Composite Materials* 50:3699–3707.

[48] Duong, H.M., Yamamoto, N., Bui, K., Papavassiliou, D.V., Maruyama, S., Wardle, B.I. 2010. Morphology effects on nonisotropic thermal conduction of aligned single-walled and multi-walled carbon nanotubes in polymer nanocomposites. *Journal of Physical Chemistry C* 114:8851–8860.

[49] Zeng, T., Chen, G. 2001. Phonon heat conduction in thin films: Impacts of thermal boundary resistance and internal heat generation. *Journal of Heat Transfer* 123:340–347.

[50] Zhao, Y., Zhu, C., Wang, S., Tian, J.Z., Yang, D.J., Chen, C.K., Cheng, H., Hing, P. 2004. Pulsed photothermal reflectance measurement of the thermal conductivity of sputtered aluminum nitride thin films. *Journal of Applied Physics* 96:4563.

[51] Harri, D.C., Cambrea, L.R., Johnson, I.F., Seaver, R.T., Baronowski, M., Gentilman, R., Nordahl, C.S., Gattuso, T., Silberstein, S., Rogan, P., Hartnett, T., Zelinski, B., Sunne, W., Fest, E., Pois, W.H., Willingham, C.B., Turri, G., Warren, C., Bass, M., Zelmon, D.E., Goodrich, S.M. 2013. Properties of an infrared-transparent $MgO:Y_2O_3$ nanocomposite. *Journal of the American Ceramic Society* 96:3828–3835.

[52] Carter, C.B., Norton, M.G. 2007. *Ceramic Materials: Science and Engineering.* New York: Springer.

[53] Chen, G. 1998. Thermal conductivity and ballistic-phonon transport in the cross-plane direction of superlattices. *Physical Review B* 57:14958.

[54] Machrafi, H., Lebon, G. 2014. Effective thermal conductivity of spherical particulate nanocomposites: Comparison with theoretical models, Monte Carlo simulations and experiments. *Journal of International Nanoscience* 13:1450022.

[55] Choudhury, M., Mohanty, S., Nayak, S.K., Aphale, R. 2012. Preparation and characterization of electrically and thermally conductive polymeric nanocomposites. *Journal of Minerals and Materials Characterization and Engineering* 11:744–756.

[56] Kochetov, R., Andritsch, T., Lafont, U., Morshuis, P.H.F., Picken, S.J., Smit, J.J. 2009. Thermal behaviour of epoxy resin filled with high thermal conductivity nanopowders. *IEEE Electrical Insulation Conference*, pp. 524–528.

[57] Gómez-Graña, S., Hubert, F., Testard, F., Guerrero-Martínez, A., Grillo, I., Liz-Marzán, L.M., Spalla, O. 2012. Surfactant (bi)layers on gold nanorods. *Langmuir* 28:1453–1459.

[58] Jones, O.G., Mezzenga, R. 2012. Inhibiting, promoting, and preserving stability of functional protein fibrils. *Soft Matter* 8:876–895.

[59] Choi, S.U.S., Eastman, J.A. 1995. *Enhancing Thermal conductivity of Fluids with Nanoparticles.* Lemont, IL: Argonne Press.

[60] De Tomas, C., Cantarero, A., Lopeandia, A.F., Alvarez, F.X. 2014. Thermal conductivity of group-IV semiconductors from a kinetic-collective model. *Proceedings of the Royal Society A* 470:2169.

[61] Kakak, S., Pramuanjaroenkij, A. 2009. Review of heat transfer enhancement with nanofluids. *International Journal of Heat and Mass Transfer* 52:3187–3169.

[62] Tesfai, W., Singh, P.K., Masharga, S., Souier, T., Chiesa, M., Shatilla, Y. 2012. Investigating the effects of suspensions nano-structure on the thermophysical properties of nanofluids. *Journal of Applied Physics* 112:114315.

[63] Sellitto, A., Cimmelli, V.A., Jou, D. 2013. Thermoelectric effects and size dependency of the figure of merit in cylindrical nanowires. *International Journal of Heat and Mass Transfer* 57:109–116.

[64] Venkatasubramanian, R., Siivola, E., Colpitts, T., O'Quinnet, B. 2001. Thin-film thermoelectric devices with high room-temperature figures of merit. *Nature* 413:597–602.

[65] Zhao, L.D., Lo, S.H., Zhang, Y., Sun, H., Tan, G., Uher, C., Wolverton, C., Dravid, V.P., Kanatzidis, M.G. 2014. Ultralow thermal conductivity and high thermoelectric figure of merit in SnSe crystals. *Nature* 508:373–377.

[66] Wang, Z., Mingo, N. 2010. Diameter dependence of SiGe nanowire thermal conductivity. *Applied Physics Letters* 97:101903.

[67] Kim, H., Kim, I., Choi, H., Kim, W. 2010. Thermal conductivities of Si1-xGex nanowires with different germanium concentrations and diameters. *Applied Physics Letters* 96:233106.

[68] Lebon, G. 2014. Heat conduction at micro and nanoscales; A review through the prism of extended irreversible thermodynamics. *Journal of Non-Equilibrium Thermodynamics* 39:35–58.

[69] Cimmelli, V.A., Sellitto, A., Jou, D. 2014. A nonlinear thermodynamic model for a breakdown of the Onsager symmetry and the efficiency of thermoelectric conversion in nanowires. *Proceedings of the Royal Society A* 470:2170.

[70] Michaelides, E.E. 2013. Transport properties of nanofluids. *A Critical Review. Journal of Non-Equilibrium Thermodynamics* 38:1–31.

[71] Benedetto, G., Boarino, L., Spagnolo, R. 1997. Evaluation of thermal conductivity of porous silicon layers by a photoacoustic method. *Applied Physics A* 64:155–159.

[72] Bernini, U., Lettieri, S., Maddalena, P. Vitiello, R., Francia, G.D. 2001. Evaluation of the thermal conductivity of porous silicon layers by an optical pump-probe method. *Journal of Physics C* 13:1141–1150.

[73] Criado-Sancho, M., del Castillo, L.F., Casas-Vázquez, J., Jou, D. 2012. Theoretical analysis of thermal rectification in a bulk Si/nanoporous Si device. *Physics Letters A* 376:1641–1644.

[74] Criado-Sancho, M., Jou, D. 2013. Heat transport in bulk/nanoporous/bulk silicon devices. *Physics Letters A* 377:486–490.

[75] Lee, J.H., Grossman, J.C., Reed, J., Galli, G. 2007. Lattice thermal conductivity of nanoporous Si: Molecular dynamics study. *Applied Physics Letters* 91:223110.

[76] Jean, V., Fumeron, S., Termentzidis, K., Tutashkonko, S., Lacroix, D. 2014. Monte Carlo simulations of phonon transport in nanoporous silicon and germanium. *Journal of Applied Physics* 115:024304.

[77] Gesele, G., Linsmeier, J., Drach, V., Fricke, J., Arens-Fischer, R. 1997. Temperature-dependent thermal conductivity of porous silicon. *Journal of Physics D* 30:2911–2916.

[78] Alvarez, F.X., Jou, D., Sellitto, A. 2010. Pore-size dependence of the thermal conductivity of porous silicon: A phonon hydrodynamic approach. *Applied Physics Letters* 97:033103.

[79] Green, D.W., Perry, R.H. 2008. *Perry's Chemical Engineers' Handbook*, eighth ed. New York: McGraw-Hill.

[80] Bruggeman, D.A.G. 1935. Berechnung verschiedener physikalischer Konstanten von heterogenen Substanzen. I. Dielektrizitätskonstanten und Leitfähigkeiten der Mischkörper aus isotropen Substanzen. *Annalen der Physik* 24:636–664.

[81] Lysenko, V., Roussel, P., Remaki, B., Delhomme, G., Dittmar, A., Barbier, D., Strikha, V., Martelet, C. 2000. Study of nano-porous silicon with low thermal conductivity as thermal insulating material. *Journal of Porous Materials* 7:177–182.

[82] Sumirat, I., Ando, Y., Shimamura, S. 2006. Theoretical consideration of the effect of porosity on thermal conductivity of porous materials. *Journal of Porous Materials* 13:439–443.

[83] Drost, A., Steiner, P., Moser, H., Lang, W. 1995. Thermal conductivity of porous silicon. *Sensors and Materials* 7:111–120.

[84] Machrafi, H., Lebon, G. 2015. Size and porosity effects on thermal conductivity of nanoporous material with an extension to nanoporous particles embedded in a host matrix. *Physics Letters, Section A* 379:968–973.

Part II

Selected Applications

4

Thermal Rectifier Efficiency of Various Bulk–Nanoporous Silicon Devices

4.1 Principles of Thermal Rectifiers

Thermal rectifiers have attracted an increased interest not only in fundamental but also in applied research, [1–6] presently framed within the relatively new topic of *phononics* [7–9]. The concept of thermal rectification can be traced back to Starr in 1936 [10]. By analogy with electronic diodes used for the control of electric current, thermal rectifiers are devices allowing heat to flow easily in one direction but offering strong resistance in the reverse direction. Search for suitable materials and configurations enhancing heat rectification has been intensely activated principally during the two last decades [11–14]. Among the current and widely used techniques, let us mention these based on adsorption/desorption properties [15], phase change [16], photon radiation [17] and thermal expansion–contraction [6] heat transfer between materials with different temperature-dependent heat conductivities [18,19], the main objective is the search for thermal diodes with the highest performance. The subject opens the way to specific applications, especially in the field of energy-saving structures, solar energy conversion, nano-electronic cooling, aero-spatial industry and cryogenics.

In the present theoretical approach, we consider successively discontinuous devices constituted by the junction of bulk and nanoporous Si and porous non-homogenous Si. The purpose of this work is the search of suitable material configurations leading to the most efficient heat rectification. Although the actual analysis is limited to silicon, there will be no difficulty to extend it to other materials.

The underlying property responsible for heat rectification is the behaviour of the thermal conductivity of nanoporous materials [1,20]. When the dimensions of the pores are comparable or smaller than the mean free path of the phonons, the effective thermal conductivity of the porous material is generally much lower than that of the corresponding bulk crystal.

Thermal rectification in Si devices has been recently analysed theoretically by Criado-Sancho et al. [1]. These authors base their approach on an expression of Fourier's law using for the effective thermal conductivity an expression derived in the framework of phonon hydrodynamics. Our aim is to complement and compare with Criado-Sancho et al. results, proposing novel configurations with better results. The present analysis uses instead for the thermal conductivity a relation proposed by Sumirat et al. [21], which emphasizes more particularly the effect of phonon scattering by pores. The Sumirat model has been validated with results derived by Machrafi and Lebon [22] in the framework of extended thermodynamics [23], it is mathematically simpler than the hydrodynamic model used by Criado et al. and allows for a more exhaustive exploration of several aspects of the problem.

In this chapter, we investigate which configuration among several bulk–nanoporous lattices will yield the most efficient rectification. We will examine successively several different devices consisting of juxtaposed homogeneous bulk and nanoporous Si samples and will end with non-homogeneous porous materials wherein both porosity and particles' size are varying in space.

The important quantity characterizing rectification is the thermal rectifying coefficient R (also called diodicity by some researchers) defined by the ratio of the heat flows in the direct and reverse directions:

$$R \equiv \frac{|q_d|}{|q_r|} \tag{4.1}$$

with q_d the heat current in the direct "conducting" direction and q_r in the reverse "insulating" one. To get an amplification effect, one must have $R \neq 1$; when $|q_d|$ is larger than $|q_r|$ (which is the situation investigated in the present work), one has $R > 1$. Of course, if the conducting and insulating directions would be switched, one should have $R < 1$. In two phase situations involving bulk and porous Si samples, it was found [1] that R is of the order of 1.5, and it is also noted that the effect of porosity on heat conduction is particularly relevant at low temperatures not exceeding 200 K. Similar values for R were obtained with two polycrystalline cobalt oxides segments [19]; more recently, a solid-state structure with shape memory alloys was built by Tso et al. [4] who recorded an R-value of about 90.

The mechanism underlying thermal rectification has also been interpreted as a consequence of a negative differential thermal resistance defined as $R_H = (\partial q/\partial T_H)_{T_L}$ and an equivalent definition for R_L obtained by permuting the different indices H and L. Classically, the heat flow increases with increasing thermal gradients, and therefore, both R_H and R_L are positive so that by defining R as $R = |R_L| / (|R_H| + |R_L|)$, it is found that $R < 1$ and the device will not work as a rectifier for which it is imperative that either R_L or R_H be negative. In the following, we will not follow this route. Instead, we will focus on the definition (4.1) of diodicity. This choice is not unique, though. For instance, some authors (e.g. [4,18]) use the following quantities to define the level of rectification, either

$$R^* = \frac{|q_d| - |q_r|}{|q_d| + |q_r|} = \frac{R-1}{R+1} \tag{4.2}$$

or

$$R^{**} = \frac{|q_d| - |q_r|}{|q_r|} = R - 1 \tag{4.3}$$

By selecting one or another of these definitions, one obtains values of the diodicity that may differ from several orders of magnitude.

4.2 Thermal Conductivity of Bulk and Porous Silicon

4.2.1 Thermal Conductivity

The expression of the thermal conductivity of porous materials is a central quantity in the determination of the rectifying coefficient (see for instance [24] for a review). In the present analysis, we use for the effective thermal conductivity of porous Si the analytical approximate result derived by Sumirat et al. [21] on the basis of the kinetic theory of phonons in solids (taking into account non-local effects at non-equilibrium state), i.e. Eq. (3.45), which is rewritten here for completeness:

$$\lambda^{\mathit{eff}}(\varphi, T, Kn) = \lambda^0(T)\frac{(1-\varepsilon)}{1+\sqrt[3]{\varepsilon}Kn(T)}. \tag{4.4}$$

This formula will be used as a mathematically simple efficient tool in order to perform our investigation. The quantity $\lambda^0(T)$ is the bulk thermal conductivity, which depends generally on the temperature T. For Si, it is of the order 3,700 W m^{-1} K^{-1} at 40 K, 2,200 W m^{-1} K^{-1} at 58 K, 1,350 W m^{-1} K^{-1} at 79 K, 800 W m^{-1} K^{-1} at 105 K, 560 W m^{-1} K^{-1} at 127 K, and 475 W m^{-1} K^{-1} at 153 K [10]. The behaviour of the silicon bulk thermal conductivity versus the temperature in the range 40–150 K is given by the solid black line of Figure 4.1. In relation (4.2), ε designates the porosity, i.e. the ratio of the volume occupied by the pores and the total volume, and Kn is the Knudsen number defined by $Kn\,(T) \equiv \ell\,(T)/a_p$, with a_p the pore radius and $\ell\,(T)$ the temperature-dependent mean free path of the phonons. The mean free path is calculated from the classical expression of the thermal conductivity derived from Debye's model, i.e. $\lambda\,(T) = \frac{1}{3}c\,(T)\,v\ell\,(T)$. From the experimental values of the thermal conductivity $\lambda\,(T)$ (e.g. [25]), the knowledge of the specific heat $c\,(T)$ [25] and the assumption that the phonon group velocity v is temperature independent [25–27], one obtains directly the value of the mean free path $\ell\,(T)$. It is equal to 6,680 nm at 50 K, 1,430 nm at 80 K, 700 nm at 100 K, 340 nm at 130 K and 180 nm at 150 K. For large Knudsen numbers (i.e. when the phonon mean free path is much longer than the radius of the pores), the heat transfer is mainly described by ballistic contributions of phonon collisions against pores. For low Knudsen numbers, the diffusive regime is dominant. Making an interpolation of the mean free path as a function of the temperature, we can plot from Sumirat's expression (4.1) the effective thermal conductivity of the nanoporous medium as a function of the temperature for several pore radii (see Figure 4.1). In [21], the dependence of λ^{eff} on the temperature is not made explicit as only the dependence on the porosity is investigated. However, it is clear from the above discussion that expression (4.1) is strongly temperature dependent.

It is seen that the conductivities exhibit opposite trends as a function of the temperature: in the case of bulk Si, it is decreasing, whereas for porous systems, it is increasing; note however that at relatively large pores sizes ($a_p > 100$ m), it is decreasing after reaching a maximum. This property is the key for obtaining thermal rectification. Since the bulk phonon mean free path is relatively large and strongly temperature dependent at low temperatures, the contributions of the pores become especially relevant. As the phonon mean free path increases when temperature is lowered, the increase in the denominator of relation (4.2) will be partially compensated by the increase in the numerator, hence the occurrence of a maximum. At higher temperatures, around 140 K, the contribution of the temperature

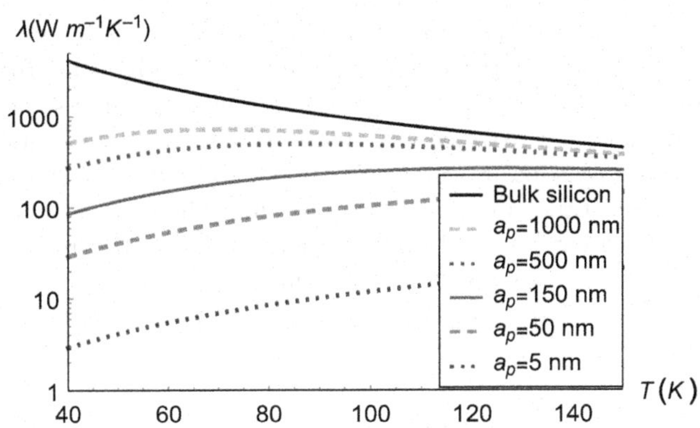

FIGURE 4.1
Thermal conductivity versus temperature for bulk [$\lambda^0(T)$] and nanoporous Si [Eq. (4.2)]: porosity $\varepsilon = 0.1$ and pore sizes a_p are varying from 1,000 to 5 nm. (Modified from: Machrafi, H., Lebon, G., Jou, D. 2016. Thermal rectifier efficiency of various bulk–nanoporous silicon devices. *International Journal of Heat and Mass Transfer* 97:603–610.)

effect on the mean free path of the phonons becomes less important and, at large pores ($a_p > 500$ m), the conductivity of nanoporous Si is close to that of the bulk one. This is understandable since at larger pore sizes, the ballistic contribution becomes negligible with respect to the diffusive one ($Kn \ll 1$) and Fourier's heat transfer is dominating. The effect of the porosity on the thermal conductivity is not to be investigated further, since it has been thoroughly studied in a previous work [22]. It is generally admitted that porosity reduces the thermal conductivity. As for the pore sizes, its role is also experimentally well known. For instance, in [28,29] it is observed that at room temperature, for a porosity $\varepsilon = 0.4$, the effective thermal conductivity of porous Si is 31.2 W m^{-1} K^{-1} for $a_p = 100$ nm [28] and 1.2 W m^{-1} K^{-1} for $a_p = 5$ nm [29]. These results are much lower than the thermal conductivity of bulk Si which is 148 W m^{-1} K^{-1} under the same temperature conditions. We will no longer comment about the parametric dependence of the thermal conductivity as the purpose of this work is rather to investigate the performances of the coupled bulk–nanoporous materials from the point of view of thermal rectification. We choose therefore in the forthcoming one single value for the porosity, say $\varepsilon = 0.1$, and one pore size, say $a_p = 150$ nm.

4.2.2 Notions on the Thermal Boundary Resistance

In principle, interfacial effects between bulk and nanoporous Si are expected. A way to account for them is to evaluate the dimensionless coefficient α [30,31] defined by

$$\alpha_i = \frac{R_{th}\lambda^0}{L}, \tag{4.5}$$

in analogy with (3.2), where R_{th} is a measure of the thermal boundary resistance, λ^0 the bulk thermal conductivity and L a reference length, say the mean length of the segments. The product $R_{th}\lambda^0 = \alpha_K$ is the so-called Kapitza length. Values of $\alpha_K \ll 1$ mean weak interfacial effects. If $R_{th} = 0$, one has $\alpha_K = 0$ and the interface is said to be perfect. Generally, R_{th} is non-zero and is given by Eq. (3.3). To give an idea of the order of magnitude of the coefficient α_i, we have calculated it at an intermediate temperature $T = 100$ K. At this value, the specific heats of Si and air (supposed to fill the pores) are known to be 0.6*10^6 J m^{-3} K^{-1} and 2.9*10^3 J m^{-3} K^{-1}, respectively. The velocity of phonons at 100 K in air is about 200 m s^{-1} and that of silicon is equal to 6,400 m s^{-1}. The latter result is derived from Debye's model, $\lambda^0 = \frac{1}{3}cv\ell$, with $\lambda^0 = 900$ W m^{-1} K^{-1} as obtained from Figure 4.1. The corresponding values in the porous segment are obtained from the volumetric average of the silicon and air data [22]. Making use of these numerical values, it is found that the thermal boundary resistance coefficient and the Kapitza length are given by $R_{th} = 2.2 * 10^{-9}$ m^2 K W^{-1} and $\alpha_K \approx 2 * 10^{-6}$ m. The latter result is significant as it justifies that the effect of interfacial thermal boundary resistance can be neglected for device lengths L longer than 2 μm, which represents a reasonable value within the context of the present work.

Thermal boundary resistance may become dominant when the number of device layers is increased as in superlattice structures. In principle, the role of a thermal boundary resistance could be explicitly studied by introducing between the bulk and porous segments, a third one undergoing a temperature drop, in analogy with the analysis presented in Section 4.3.2.

4.3 Configurations for Thermal Rectifiers

4.3.1 Homogeneous Two- and Three-Phase Systems

The simplest configuration is represented in Figure 4.2 consisting of a homogeneous two-phase system: bulk Si of length L_1 in series with nanoporous Si with length L_2 with the

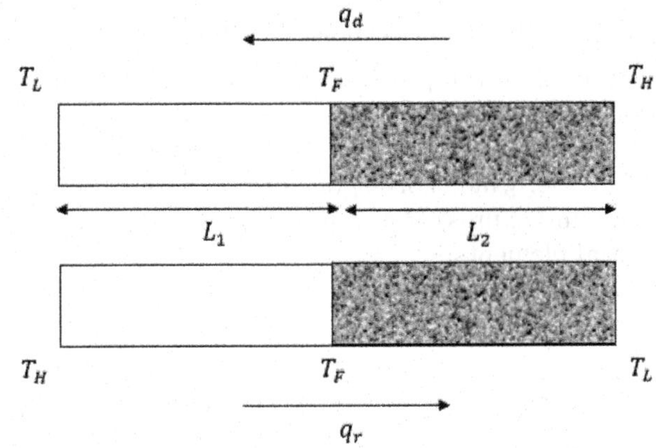

FIGURE 4.2
Schematic picture of the thermal transistor: (a) direct heat flow, (b) reverse heat flow. Bulk material is represented by a virgin surface and the porous phase by a shaded one. The temperatures at the ends of the device are T_H (high) and T_L (low), respectively, whereas at the bulk–porous interface, the temperature is T_F. The lengths of crystalline and porous Si are L_1 and L_2, respectively. (From: Machrafi, H., Lebon, G., Jou, D. 2016. Thermal rectifier efficiency of various bulk–nanoporous silicon devices. *International Journal of Heat and Mass Transfer* 97:603–610.)

hot face, respectively, at the porous and the bulk side. The first situation is referred to as the direct direction (with heat flux q_d) and the second one as the reverse one (with heat flux q_r). The rectifying effect is shown through the difference of values of the direct (q_d) and the reverse (q_r) values of the heat flux.

We assume that the assembly is made out of two homogeneous materials taking the form of rods or cylinders of constant cross section, wide enough in comparison with the value of the bulk phonon mean free path, in order that the collisions with the walls do not contribute to the thermal conductivity. The hot head of the device is kept at $T_H = 150$ K, and the cold side T_L varies from 40 to 150 K. The heat flow in each element is governed by Fourier's law:

$$q = -\lambda^{eff}(T, \varepsilon, Kn)\,(dT/dx)\,, \tag{4.6}$$

where λ^{eff} is given by expression (4.4) in the case of the nanoporous element and by $\lambda^0(T)$ for the bulk segment. The hypothesis of stationarity implies that the heat flux remains uniform throughout the sample, its expressions in the direct and reverse directions read, respectively, as [18–20]

$$q_d \equiv -\int_{T_H}^{T_F} \frac{\lambda^{eff}}{L_2} dT = -\int_{T_F}^{T_L} \frac{\lambda^0}{L_1} dT, \tag{4.7}$$

$$q_r \equiv -\int_{T_H}^{T_F} \frac{\lambda^0}{L_1} dT = -\int_{T_F}^{T_L} \frac{\lambda^{eff}}{L_2} dT. \tag{4.8}$$

Equalizing the second and third terms of Eqs. (4.7) and (4.8), respectively, leads to an algebraic relation of the form $f(L_1/L_2, T_H, T_L, T_F) = 0$, allowing to determine the intermediate temperature T_F in terms of L_1/L_2, T_H and T_L. After the temperature junction T_F is known, the heat fluxes q_d and q_r can be calculated and finally the value of $R = |q_d| / |q_r|$ can be determined.

4.3.2 Bulk–Porous–Bulk and Porous–Bulk–Porous Si Configurations

We perform the same study as in the previous subsection, but with systems composed out of three elements: porous–bulk–porous on one side (Figure 4.3) and bulk–porous–bulk on the other side (Figure 4.4). Since we already have considered the influence of the L_1/L_2 ratio, we will select $L_1/L_2 = 4$, which corresponds to the largest rectifying coefficient found in the previous subsection and express R as a function of T_L for various values of the aspect ratio L_1/L_3 of the external elements.

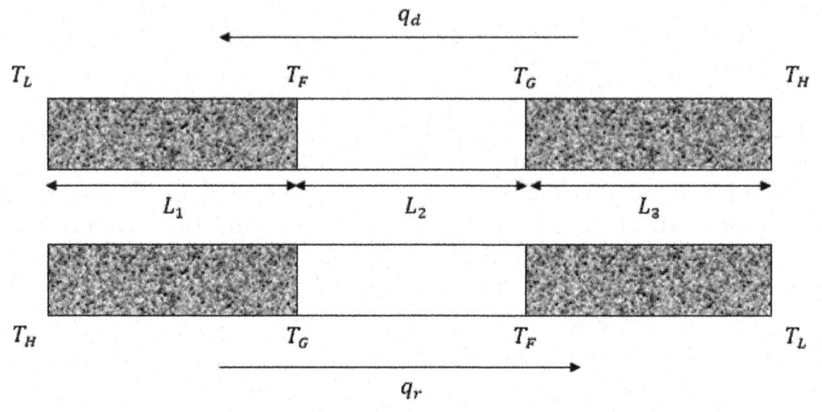

FIGURE 4.3
The device is composed of three homogeneous regions of porous Si (left side), bulk Si (middle side), and porous Si (right side). The system is subject to a temperature difference $\Delta T = T_H(\text{high}) - T_L(\text{low})$ at its boundaries. T_F and T_G denote temperatures at the interfaces. The porous elements of lengths L_1 and L_3 are characterized by the same porosity. (From: Machrafi, H., Lebon, G., Jou, D. 2016. Thermal rectifier efficiency of various bulk–nanoporous silicon devices. *International Journal of Heat and Mass Transfer* 97: 603–610.)

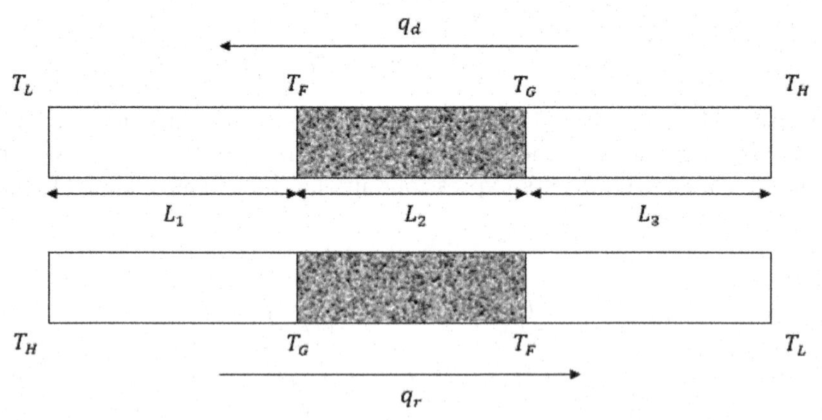

FIGURE 4.4
The device is composed of three regions of bulk Si (left side), porous Si (middle side) and bulk Si (right side). (From: Machrafi, H., Lebon, G., Jou, D. 2016. Thermal rectifier efficiency of various bulk–nanoporous silicon devices. *International Journal of Heat and Mass Transfer* 97:603–610.)

The heat fluxes for the porous–bulk–porous device are given by

$$q_d \equiv -\int_{T_H}^{T_G} \frac{\lambda^{eff}}{L_3} dT = -\int_{T_G}^{T_F} \frac{\lambda^0}{L_2} dT = -\int_{T_F}^{T_L} \frac{\lambda^{eff}}{L_1} dT, \tag{4.9}$$

$$q_r \equiv -\int_{T_H}^{T_G} \frac{\lambda^{eff}}{L_1} dT = -\int_{T_G}^{T_F} \frac{\lambda^0}{L_2} dT = -\int_{T_F}^{T_L} \frac{\lambda^{eff}}{L_3} dT, \tag{4.10}$$

and heat fluxes for the bulk–porous–bulk device are given by

$$q_d \equiv -\int_{T_H}^{T_G} \frac{\lambda^0}{L_3} dT = -\int_{T_G}^{T_F} \frac{\lambda^{eff}}{L_2} dT = -\int_{T_F}^{T_L} \frac{\lambda^0}{L_1} dT, \tag{4.11}$$

$$q_r \equiv -\int_{T_H}^{T_G} \frac{\lambda^0}{L_1} dT = -\int_{T_G}^{T_F} \frac{\lambda^{eff}}{L_2} dT = -\int_{T_F}^{T_L} \frac{\lambda^0}{L_3} dT. \tag{4.12}$$

To determine the values of the intermediate temperatures T_F and T_G, for a fixed T_H and a varying T_L, we equalize on one side the second and third terms of Eqs. (4.11) and (4.12) and the second and fourth terms on the other side. After calculating the direct and reverse heat fluxes, one obtains directly the corresponding rectifying coefficient. The bulk–porous–bulk thermal rectification device seems to be promising. In the case of the porous–bulk–porous configuration, the rectifying coefficient remains close to unity for a large range of the L_1/L_3 ratio, meaning that this kind of device is not suitable for thermal rectification. This can be understood by noticing that in this configuration, we have two porous media with a relatively large thermal resistance. So, the thermal properties of the bulk phase in between them will not undergo much modifications, whether one porous phase is larger than the other or not. Therefore, this configuration (porous–bulk–porous thermal rectification) is of little interest within the perspective of heat rectifiers.

4.3.3 Graded Porosity

Instead of discontinuous configurations, we consider a sample with a gradually changing porosity along the dimensionless x-axis (see Figure 4.5), in other words $\varepsilon = \varepsilon(x)$ and thus $\lambda^{eff} = \lambda^{eff}(T, x)$. In terms of non-dimensional quantities, Fourier's law can be written as follows:

$$\frac{dT}{dx} = -\frac{q}{\lambda^{eff}(T, x)}, \tag{4.13}$$

where the temperature is scaled by the highest temperature T_H, the space coordinate by the length L of the rod, the effective thermal conductivity by the bulk conductivity λ^0 and the heat flux by $\lambda^0 T_H/L$. By selecting a graded structure, one avoids of course the problems raised by the presence of the aforementioned thermal boundary resistances.

It is assumed that in the graded porosity thermal rectification device, the porosity varies along the sample according to the two following distributions:

$$\varepsilon = \varepsilon_0 x^n \tag{4.14}$$

$$\varepsilon = \varepsilon_0 \left(ax^2 + bx + c\right), \tag{4.15}$$

where ε_0 is a constant and n, a, b and c are non-dimensional quantities. The above dependencies of ε with respect to x are expected to be reproducible in experiments that are actually still lacking. In Eq. (4.14), the space dependency is monotonic so that $\varepsilon(x + dx) > \varepsilon(x)$. For $n = 1$, we have a linear relation of $\varepsilon(x)$, for $n = 2$, a quadratic one, etc. The coefficients a, b and c in Eq. (4.15) will be determined in Section 4.4.3 by imposing that the porosity takes a maximum value at a given position in the sample. The direct and reverse heat fluxes can now be calculated by solving Eq. (4.13) in both directions. We keep the hot side temperature at $T_H = 150$ K and the cold side temperature at $T_L = 40$ K.

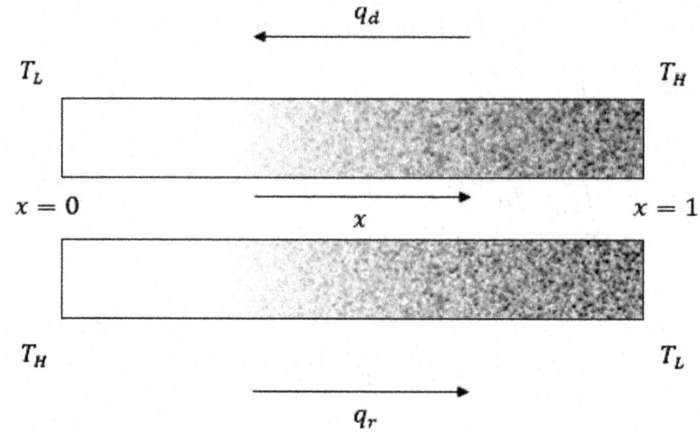

FIGURE 4.5
Sketch of the graded porosity distribution, with, at the left of the figure a zero porosity increasing gradually from zero to a maximum at the right. The direct flow (q_d) refers to the hottest temperature T_H imposed on the highest porosity side and the reverse (q_r) to the same side subject to the lowest temperature T_L. (From: Machrafi, H., Lebon, G., Jou, D. 2016. Thermal rectifier efficiency of various bulk–nanoporous silicon devices. *International Journal of Heat and Mass Transfer* 97:603–610.)

4.3.4 Graded Pore Size

We consider the same system as in the previous subsection (illustrated by Figure 4.5), but instead of the porosity, it is the pore size (graded-pore-size thermal rectification device) that is assumed to vary continuously along the system. So, we still have $\lambda^{eff} = \lambda^{eff}(T, x)$ in Eq. (4.13), but now $\varepsilon = \varepsilon_0 =$ constant and $a_p = a_p(x)$. It is assumed that the pore size varies linearly between a fixed minimum pore size $a_{p,min} = 10$ nm at $x = 1$, and a variable maximum $a_{p,max}$ at $x = 0$. The dependence of the size with x is thus given by

$$a_p(x) = a_{p,max} - (a_{p,max} - a_{p,min})\, x. \tag{4.16}$$

4.4 Analysis of Thermal Rectification

4.4.1 Homogeneous Two- and Three-Phase Systems

In Figure 4.6 are plotted the values of R as a function of T_L for several values of L_1/L_2 and with fixed values of ε (= 0.1) and a_p (=150 nm).

Several comments are in form. First, Figure 4.6 shows that $R > 1$, for all the L_1/L_2 ratios considered in this work. This can be understood by recalling that the porous region is characterized by the largest thermal resistance, so that a higher temperature (T_H in the direct case) at the porous side will contribute to a larger heat flow than by prescribing a lower temperature (T_L in the reverse case).

Second, by increasing the L_1/L_2 ratio from 0.1, the rectifying coefficient increases gradually and attains a maximum value at $L_1/L_2 = 4$, but afterwards, it decreases with larger L_1/L_2 values. This can be interpreted as follows. For $L_1 \gg L_2$, the porous region of dimension L_2 is so tiny that it will not contribute to the heat flux, and things happen as there were only one single region, the direction of the heat flux is of slight influence, and the rectifying

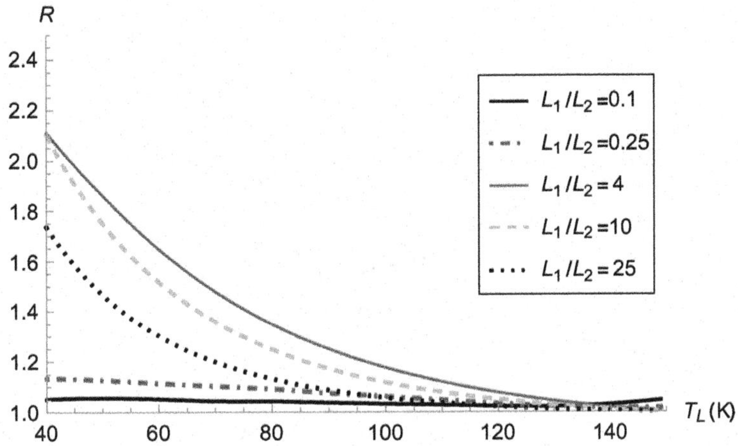

FIGURE 4.6
Rectifying coefficient R as a function of T_L for several values of the ratio L_1/L_2 for the device in Figure 4.2, $\varepsilon = 0.1$, $a_p = 150$ nm and $T_H = 150$ K. (From: Machrafi, H., Lebon, G., Jou, D. 2016. Thermal rectifier efficiency of various bulk–nanoporous silicon devices. *International Journal of Heat and Mass Transfer* 97:603–610.)

coefficient tends to unity. By increasing L_2, we have again a two-phase situation; the heat flux in the direct case becomes larger and larger, so that an increase in L_2 is tantamount to an increase in the rectifying coefficient. Finally, for $L_2 \gg L_1$, we recover a quasi-one-phase situation with R decreasing once more to unity. It is therefore not surprising to observe an extremum value of R at intermediate values of L_1 and L_2.

4.4.2 Bulk–Porous–Bulk and Porous–Bulk–Porous Si Configurations

As shown in Figure 4.7, the results of the bulk–porous–bulk device are more promising with the rectifying coefficient becoming larger than one with increasing L_1/L_3 ratio. This is easily explained as for $L_1 > L_3$, the temperature at the bulk–porous interface (T_G in the direct case) is higher than that in the case $L_1 < L_3$, which causes a larger direct heat flux than the reverse one. It is interesting to observe that, depending on the value of L_1/L_3-, the rectifying coefficient is smaller than one for $L_1/L_3 < 1$ and larger than one for $L_1/L_3 > 1$. Such a configuration presents a kind of reversibility in that the preferred direction of the heat will flow can be controlled through the geometrical configuration. Note finally that for the three-element system with $L_1/L_3 = 100$ (recall that $L_1/L_2 = 4$), the corresponding curve is akin to that of the two-element system with the same value $L_1/L_2 = 4$. This is understandable since for $L_1/L_3 = 100$, the system is comparable to a bulk–porous one, with the right-side bulk element becoming negligible. It can be concluded that, concerning thermal rectification, the performance of the three-element system is not superior to that of the two-element system. This is the motivation to discuss different systems in Sections 4.4.3 and 4.4.4.

4.4.3 Graded Porosity

Figure 4.8 shows the rectifying coefficient R as a function of the exponent n of Eq. (4.14). R is increasing with n and attains a maximum value around $n = 20$. It should be noted.

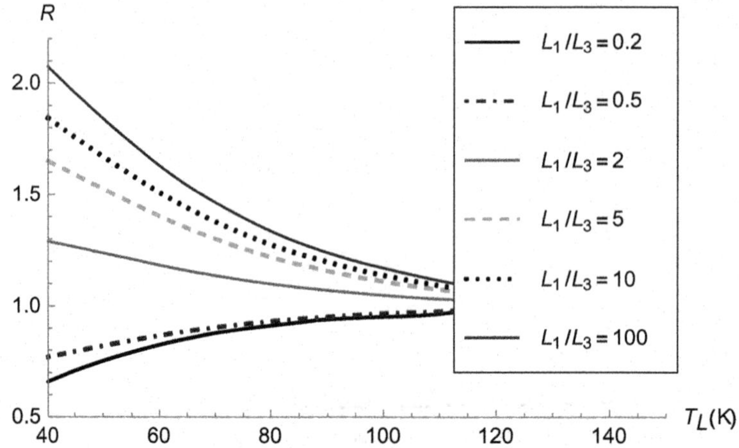

FIGURE 4.7
Rectifying coefficient as a function of T_L for several values of the ratio L_1/L_3 for the device of Figure 4.3, at L_1/L_2, $\varepsilon = 0.1$, $a_p = 150$ nm and $T_H = 150$ K. (From: Machrafi, H., Lebon, G., Jou, D. 2016. Thermal rectifier efficiency of various bulk–nanoporous silicon devices. *International Journal of Heat and Mass Transfer* 97:603–610.)

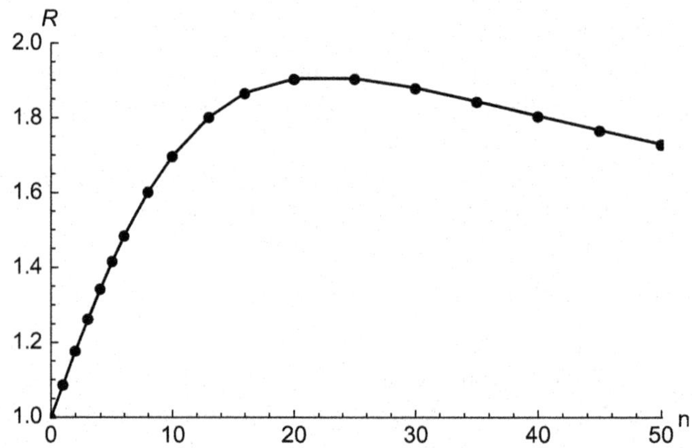

FIGURE 4.8
Rectifying coefficient as a function of the porosity degree n from Eq. (4.14) with $\varepsilon_0 = 0.1$, $a_p = 150$ nm, $T_H = 150$ K and $T_L = 40$ K. (From: Machrafi, H., Lebon, G., Jou, D. 2016. Thermal rectifier efficiency of various bulk–nanoporous silicon devices. *International Journal of Heat and Mass Transfer* 97:603–610.)

However, such high n-values are from a technical point of view rather difficult to manage. Therefore, for the sake of simplicity, let n be bounded to $n = 4$. It will be shown later that choosing higher values of n will not influence the value R of significantly. Nonetheless, from a theoretical point of view, it is interesting to observe that R will not grow indefinitely but is bounded by a finite maximum value.

Going now to Eq. (4.15), we are left with the determination of the unknown quantities a, b and c. We first impose $\varepsilon(0) = 0$, $\varepsilon(1) \neq 0$ and $0 < \frac{\varepsilon(1)}{\varepsilon_0} < 1$. Therefore, ε will reach a maximum in the region $0.5 < x < 1$. From $\varepsilon(0) = 0$, follows that $c = 0$. Requiring that

ε is maximum at x_{max} leads to $a = -\left(\frac{1}{x_{max}}\right)^2$ and $b = \frac{2}{x_{max}}$. In Figure 4.9 is plotted the rectifying coefficient as a function of x_{max}.

The dashed line in Figure 4.9 represents the value of R for $n = 1$ in Eq. (4.14). The results of Figure 4.9 indicate that the parabolic dependence (4.15) of the porosity with respect to x does not provide better results than the linear one.

4.4.4 Graded Pore Size

The rectifying coefficient as a function of $a_{p,max}$ is represented in Figure 4.10.

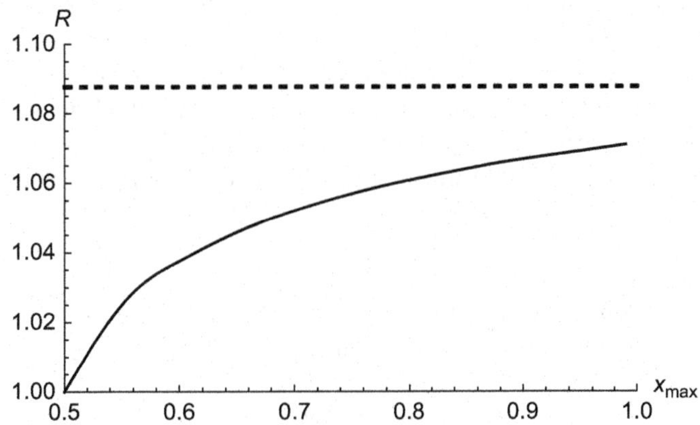

FIGURE 4.9
Rectifying coefficient as a function of the position x_{max} corresponding to the maximum value of the porosity given by Eq. (4.15) with $\varepsilon_0 = 0.1$, $a_p = 150$ nm, $T_H = 150$ K and $T_L = 40$ K. (From: Machrafi, H., Lebon, G., Jou, D. 2016. Thermal rectifier efficiency of various bulk–nanoporous silicon devices. *International Journal of Heat and Mass Transfer* 97:603–610.)

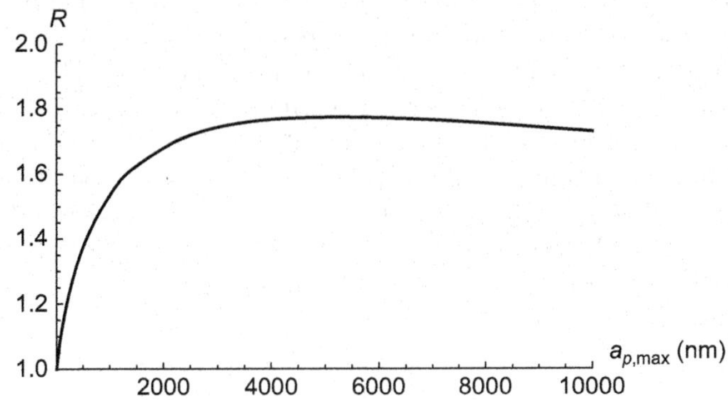

FIGURE 4.10
Rectifying coefficient as a function of the maximum pore size $a_{p,max}$, with $a_p(x)$ given by Eq. (14) at $\varepsilon = 0.1$, $a_{p,min} = 10$ nm, $T_H = 40$ K and $T_L = 40$ K. (From: Machrafi, H., Lebon, G., Jou, D. 2016. Thermal rectifier efficiency of various bulk–nanoporous silicon devices. *International Journal of Heat and Mass Transfer* 97:603–610.)

It is shown that the rectifying coefficient is of the same order of magnitude as for the two-phase bulk/porous medium, the best result is obtained with rather large pore sizes at $x = 1$, around $a_{p,max} = 5000$ nm. The results of Figures 4.9 and 4.10 suggest to combine both graded pore size and graded porosity.

4.5 Combining Graded Porosity and Pore Size

Comparing Figures 4.9 and 4.10 leads to the conclusion that a non-homogeneous porosity [with $n = 20$ in Eq. (4.14)] leads to a rectifying coefficient R around 1.8 which is approximately the same value obtained for non-homogeneous pore sizes with $a_{p,max} = 5,000$ nm. The question may therefore be raised whether a combination of both effects can further enhance the rectifying coefficient. By taking $n = 20$ in Eq. (4.14) and $a_{p,max} = 5,000$ nm in Eq. (4.16), the rectifying coefficient is raised up to $R = 2.24$, which is quite higher than for each individual case. Even if we take a more realistic value for n, say $n = 4$ in Eq. (4.14), in combination with $a_{p,max} = 5,000$ nm in Eq. (4.16), the rectifying coefficient is found to be given by

$$R = 2.12. \tag{4.17}$$

Based on the different definitions used by some researchers and expressed by relations (4.2) and (4.3), one would have obtained the following:

$$R^* = 0.36, \tag{4.18}$$

$$R^{**} = 1.12. \tag{4.19}$$

As mentioned earlier, we can see indeed that the selection of the value of n has only a slight influence on the value of R. Summing up, we obtain the highest thermal rectifying coefficient by combining non-homogeneous porosity and variable pore size along the device. The most favourable configuration is the side with the highest porosity being associated with the smallest pore size. It has been checked that if the side with the highest porosity contains the largest pores, the rectifying coefficient falls down.

To give an idea of the order of magnitudes of the R-values recorded in other studies, the values $R = 1.35$, 1.43 and even 90 were obtained, respectively, in the case of materials with asymmetric shapes [32], LaCo oxides [19] and shape memory alloys [4]. In [33], a maximum value of $R^* = 0.39$ is found. A higher thermal rectification $R^* = 0.67$ is obtained by pressurizing an aluminium–stainless steel combination [34]. More recently, molecular dynamics simulations of asymmetric graphene ribbons predict $R^* = 0.33 - 0.64$ [35] depending on the length of the device, and $R^* = 0.13$ at 25 K for argon–krypton devices [36]. It appears that our R-values are among the highest ones, which suggests opening the way to the building-up of a new class of rectifiers based on nanoporous materials [37].

References

[1] Criado-Sancho, M., del Castillo, L.F., Casas-Vázquez, J., Jou, D. 2012. Theoretical analysis of thermal rectification in a bulk Si/nanoporous Si device. *Physics Letters A* 376:1641–1644.

[2] Alvarez, F.X., Jou, D., Sellitto, A. 2010. Pore-size dependence of the thermal conductivity of porous silicon: A phonon hydrodynamic approach. *Applied Physics Letters* 97:033103.

[3] Lebon, G., Jou, D., Casas-Vázquez, J. 2008. *Understanding Non-Equilibrium Thermodynamics*. Berlin: Springer.

[4] Tso, C.Y., Chao, C.Y.H. 2016. Solid-state thermal diode with shape memory alloys. *International Journal of Heat and Mass Transfer* 93:605–611.

[5] Somers, R.R., Fletcher, L.S., Flack, R.D. 1987. Explanation of thermal rectification. *AIAA Journal* 25:620–621.

[6] Dos Santos Bernardes, M.A. 2014. Experimental evidence of the working principle of thermal diodes based on thermal stress and thermal contact conductance–thermal semiconductors. *International Journal of Heat and Mass Transfer* 73:354–357.

[7] Wang, L., Li, B. 2008. Phononics gets hot. *Physics World* 21:27–29.

[8] Li, N., Ren, J., Wang, L., Zhang, G., Hänggi, P., Li, B. 2012. Colloquium: Phononics: Manipulating heat flow with electronic analogs and beyond. *Reviews of Modern Physics* 84:1045.

[9] Maldovan, M. 2013. Sound and heat revolutions in phononics. *Nature* 503:209–217.

[10] Starr, C. 1936. The copper oxide rectifier. *Journal of Applied Physics* 7:15–19.

[11] Go, D.B., Sen, M. 2010. On the condition of thermal rectification using bulk materials. *Journal of Heat Transfer* 132:124502.

[12] Hu, J., Ruan, X., Chen, Y.P. 2009. Thermal conductivity and thermal rectification in graphene nanoribbons: A molecular dynamics study. *Nano Letters* 9:2730–2735.

[13] Chang, C.W., Okawa, D., Majumdar, A., Zettl, A. 2006. Solid-state thermal rectifier. *Science* 314:1121–1124.

[14] Li, B., Wang, L., Casati, G. 2004. Thermal diode: Rectification of heat flux. *Physical Review Letters* 93:184301.

[15] Catarino, I., Bonfait, G., Duband, L. 2008. Neon gas-gap heat switch. *Cryogenics* 48:17–25.

[16] Boryeko, J.B., Zhao, Y., Chen, C.H. 2011. Planar jumping drop thermal diodes. *Applied Physics Letters* 99:234105.

[17] Ben-Abdallah, P., Biehs, S.A. 2013. Phase-change radiative thermal diode. *Applied Physics Letters* 103:191907.

[18] Dames, C. 2009. Solid-state thermal rectification with existing bulk materials. *Journal of Heat Transfer* 131:061301.

[19] Kobayashi, W., Teraoka, Y., Terasaki, I. 2009. An oxide thermal rectifier. *Applied Physics Letters* 95:171905.

[20] Criado-Sancho, M., Jou, D. 2013. Heat transport in bulk/nanoporous/bulk silicon devices. *Physics Letters A* 377:486–490.

[21] Sumirat, I., Ando, Y., Shimamura, S. 2006. Theoretical consideration of the effect of porosity on thermal conductivity of porous materials. *Journal of Porous Materials* 13:439–443.

[22] Machrafi, H., Lebon, G. 2015. Size and porosity effects on thermal conductivity of nanoporous material with an extension to nanoporous particles embedded in a host matrix. *Physics Letters A* 379:968–973.

[23] Jou, D., Casas-Vazquez, J., Lebon, G. 2010. *Extended Irreversible Thermodynamics*, fourth ed. Berlin: Springer.

[24] Wang, M., Pan, N. 2008. Predictions of effective physical properties of complex multiphase materials. *Materials Science and Engineering Reports* 63:1–30.

[25] Lebon, G., Machrafi, H., Grmela, M. 2015. An extended irreversible thermodynamic modelling of size dependent thermal conductivity of spherical nanoparticles dispersed in semiconductors. *Proceedings of the Royal Society A* 471:20150144.

[26] Callaway, J. 1959. Model for lattice thermal conductivity at low temperature. *Physical Review* 113:1046–1051.

[27] Mingo, N., Yang, L., Li, D., Majumdar, A. 2003. Predicting the thermal conductivity of Si and Ge nanowires. *Nano Letters* 3:1713–1716.

[28] Gesele, G., Linsmeier, J., Drach, V., Fricke, J., Arens-Fischer, R. 1997. Temperature-dependent thermal conductivity of porous silicon. *Journal of Physics D* 30:2911–2916.

[29] Benedetto, G., Boarino, L., Spagnolo, R. 1997. Evaluation of thermal conductivity of porous silicon layers by a photoacoustic method. *Applied Physics A* 64:155–159.

[30] Swartz, E.T., Pohl, R.O. 1969. Thermal boundary resistance. *Reviews of Modern Physics* 61:605–664.

[31] Chen, G. 1998. Thermal conductivity and ballistic-phonon transport in the cross-plane direction of superlattices. *Physical Review B* 57:14958–14973.

[32] Sawaki, D., Kobayashi, W., Moritomo, Y., Terasaki, I. 2011. Thermal rectification in bulk materials with asymmetric shape. *Applied Physics Letters* 98:081915.

[33] Walker, D.G. 2006. Thermal rectification mechanisms including noncontinuum effects, in: *proceedings of the Joint ASME-ISHMT Heat Transfer Conference*, Guwahati, India.

[34] Barzelay, M.H., Tong, K.N., Holloway, G.F. 1955. Effects of pressure on thermal conductance of contact joints, Technical Report 3295, NACA.

[35] Yang, N., Zhang, G., Li, B. 2009. Thermal rectification in asymmetric graphene ribbons. *Applied Physics Letters* 95:033107.

[36] Roberts, N., Walker, D. 2008. Molecular dynamics simulation of thermal transport with asymmetric and rough interfaces, in: *Proceedings of International Symposium on Transport Phenomena*, Reykjavik, Iceland.

[37] Machrafi, H., Lebon, G., Jou, D. 2016. Thermal rectifier efficiency of various bulk-nanoporous silicon devices. *International Journal of Heat and Mass Transfer* 97:603–610.

5

Thermoelectric Devices

5.1 Thermodynamics Behind Thermoelectric Devices

Transport properties play a central role in materials sciences. One of the current interests lies in the optimal application of thermoelectricity to convert (waste) heat to usable electricity (the Seebeck effect) [1]. This requires materials with high Seebeck coefficient, high electrical conductivity and low thermal conductivity. This is not an easy task because usually a high electrical conductivity implies also a high thermal conductivity, the two being linked as seen in section 5.3.1. Other effects include the Peltier effect, where it is an electric current that heats the material, and the Thomson effect, which is a combination of the former two [2]. Recently, research on thermoelectric materials has known an extraordinary burst [3]. A combination of the Seebeck coefficient, and the electrical and thermal conductivities is found to be directly related to the efficiency of thermoelectric energy conversion. The combination of the aforementioned parameters is often resumed into a so-called figure of merit Z, or the dimensionless version $ZT = TS^2\sigma_e/\left(\lambda_e + \lambda_{ph}\right)$, where T is the temperature, S the Seebeck coefficient, σ_e the electric conductivity, λ_e the electron thermal conductivity (the contribution of the electrons to the heat transfer) and λ_{ph} the phonon thermal conductivity (the contribution of the phonons to the heat transfer) [4–6]. The figure of merit has increased by more than one order of magnitude since the end of the twentieth century. This was in particular achieved in superlattices in which one dimension is smaller than the mean free path of phonons but larger than that of electrons. Many aim at increasing further this figure of merit by the use of nanostructured materials [7]. Therefore, we investigate in this work the particular properties of thermal and electric conduction at nanoscales and apply this to nanostructured materials. Such materials know a huge variety of applications, such as heat conduction enhancement in polyesters [8] or energy storage systems [9], to mention a few. As far as goes nanostructured materials for thermoelectric systems [10–13], a decrease in the thermal conductivity is often sought for in order to increase the figure of merit ZT [5,6,14]. The figure of merit is inversely dependent not only on the thermal conductivity but also on the electric conductivity, which are both size dependent [7]. The figure of merit also depends on other material-related properties, such as the hole and electron mobility's and carrier concentrations [11,12].

Nanostructured materials can be used in different forms. We are interested here in nanofilms and nanocomposites. Nanofilms can be between two other bulk materials, and nanocomposites are generally made out of a homogeneous matrix in which nanoparticles are dispersed. For nanofilms and particles with characteristic lengths of the same order of magnitude or smaller than the phonon and electron mean free paths, the Fourier theory, based on the classical approach of thermodynamics, is not able to predict the heat flux thermal interactions between the nanofilm or nanoparticle and the bulk material next to it. Therefore, we propose to investigate the problem of heat conduction in nanostructured thermoelectric materials by the more sophisticated thermodynamic formalism, presented in the preceding chapters and confirmed in the literature [15,16]. In this approach, the heat flux is elevated to the status of independent variable at the same footing as the temperature. The total heat flux $\boldsymbol{q}$ is assumed to be the sum of a phonon contribution $\boldsymbol{q_{ph}}$ and an electric one $\boldsymbol{q_e}$ [7], which are due to the phonon and electron motion through the material:

$$\boldsymbol{q} = \boldsymbol{q_e} + \boldsymbol{q_{ph}}. \tag{5.1}$$

In the presence of an electric field and without local heat supply, the partial energy balances for the phonons and the electrons are

$$\frac{\partial u_{ph}}{\partial t} + \nabla \cdot \boldsymbol{q_{ph}} = 0, \tag{5.2a}$$

$$\frac{\partial u_e}{\partial t} + \nabla \cdot \boldsymbol{q_e} = \boldsymbol{E} \cdot \boldsymbol{I}, \tag{5.2b}$$

where u_{ph}, u_e, $\boldsymbol{E}$ and $\boldsymbol{I}$ are the phonon internal energy, electron internal energy, the electric field and the electric current density. The total energy balance is then, in the absence of a magnetic field, given by

$$\frac{\partial u}{\partial t} + \nabla \cdot \boldsymbol{q} = \boldsymbol{E} \cdot \boldsymbol{I}, \tag{5.3}$$

with

$$u = u_{ph} + u_e, \tag{5.4}$$

which also implies $c = c_{ph} + c_e$, used later on, c being the specific (at constant volume) heat capacity. The continuity law for electric charge is

$$\frac{\partial \varrho_e}{\partial t} + \nabla \cdot \boldsymbol{I} = 0, \tag{5.5}$$

where ϱ_e is the density of electric charge. These equations give us now the evolution of u_{ph}, u_e and ϱ_e, with the corresponding fluxes $\boldsymbol{q_{ph}}$, $\boldsymbol{q_e}$ and $\boldsymbol{I}$. The basic principles allow postulating additional evolution equations for these fluxes (which are now considered as independent variables) as [15]

$$\tau_{ph}\frac{\partial \boldsymbol{q_{ph}}}{\partial t} + \boldsymbol{q_{ph}} = -\lambda_{ph}\nabla T + \ell_{ph}^2\left(\nabla^2 \boldsymbol{q_{ph}} + 2\nabla\nabla \cdot \boldsymbol{q_{ph}}\right), \tag{5.6a}$$

$$\tau_e\frac{\partial \boldsymbol{q_e}}{\partial t} + \boldsymbol{q_e} = -\left(\lambda_e + \mathrm{S}\Pi\sigma_e\right)\nabla T + \ell_e^2\left(\nabla^2 \boldsymbol{q_e} + 2\nabla\nabla \cdot \boldsymbol{q_e}\right) + \Pi\sigma_e \boldsymbol{E}, \tag{5.6b}$$

$$\tau_I\frac{\partial \boldsymbol{I}}{\partial t} + \boldsymbol{I} = \sigma_e\left(\boldsymbol{E} - \mathrm{S}\nabla T\right) + \ell_I^2\left(\nabla^2 \boldsymbol{I} + 2\nabla\nabla \cdot \boldsymbol{I}\right), \tag{5.6c}$$

where τ, ℓ and Π are the relaxation time, mean free path and the Peltier coefficient, respectively, and the subscripts *ph* and *e* stand for phonon and electron, respectively. The last term in Eq. (5.6a) denotes the Guyer–Krumhansl contribution to denote the non-locality of the phonon heat flux, while the equivalents are also present in Eqs. (5.6b)–(5.6c) for the electron heat flux and electric current density, respectively. When the Guyer–Krumhansl contribution is neglected, we obtain the Cattaneo equation [17], and if furthermore the time dependency is neglected, the classical Fourier law is obtained. Eq. (5.6b) is obtained with the hypothesis that the electronic heat flux is governed by the same principles as the phonon one in Eq. (5.6a) with additional terms for the Peltier effect in Eq. (5.6b). Neglecting again the Guyer–Krumhansl and time dependency terms in Eq. (5.6b) gives the classical equation describing the Peltier effect [7]. Since the electric current also passes through the same material, we assume also that it follows the same thermodynamic principles as Eq. (5.6a) and (5.6b). Neglecting Guyer–Krumhansl and the time dependency term in Eq. (5.6c) gives $\boldsymbol{I} = \sigma_e\left(\boldsymbol{E} - \mathrm{S}\nabla T\right)$, and omitting also an electrical field gives $\boldsymbol{I} = -\mathrm{S}\sigma_e\nabla T$, which shows that an electric current can be obtained by a temperature gradient, which is called the Seebeck effect. In the absence of an imposed electric field (where we do not consider the Peltier effect), Eq. (5.6) reduces to

$$\tau_{ph}\frac{\partial \boldsymbol{q_{ph}}}{\partial t} + \boldsymbol{q_{ph}} = -\lambda_{ph}\nabla T + \ell_{ph}^2\left(\nabla^2\boldsymbol{q_{ph}} + 2\nabla\nabla\cdot\boldsymbol{q_{ph}}\right), \tag{5.7a}$$

$$\tau_{e}\frac{\partial \boldsymbol{q_{e}}}{\partial t} + \boldsymbol{q_{e}} = -\lambda_{e}\nabla T + \ell_{e}^2\left(\nabla^2\boldsymbol{q_{e}} + 2\nabla\nabla\cdot\boldsymbol{q_{e}}\right), \tag{5.7b}$$

$$\tau_{I}\frac{\partial \boldsymbol{I}}{\partial t} + \boldsymbol{I} = -\mathrm{S}\sigma_{e}\nabla T + \ell_{I}^2\left(\nabla^2\boldsymbol{I} + 2\nabla\nabla\cdot\boldsymbol{I}\right). \tag{5.7c}$$

This development suggests that in order to account for non-local effects, higher order fluxes should be taken into account. Also, it shows nicely the relation between on one side an imposed temperature gradient (for instance, the heat from sunrays) and on the other side a heat flux and electric current. The system (5.1)–(5.6) (of which (5.7) is a special case) can be applied to calculate the efficiency of thermoelectric systems [3]. This is, however, not the purpose of this work, but is placed here for information and should be put in context with respect to the development from our theory in Chapter 8. Before calculating the efficiency of thermoelectric systems, it is important to first investigate the optimal conditions that allow such an efficiency to be improved. Such conditions are set by an optimization of the physical parameters that are present in the Eqs. (5.1)–(5.6): the phonon and electron contributions of the thermal conductivity, the Seebeck coefficient and the electric conductivity. As will be of use later on, the system (5.1)–(5.6) shows also the analogy between these different physical parameters in that they all appear before the temperature gradient.

The way the heat flux and the electric current pass through the thermoelectric material is crucial for its performance and depends directly on the phonon thermal conductivity, electron thermal conductivity and electric conductivity as is shown in system (5.7). The figure of merit has for bulk materials a typical non-dimensional value around 1 and is a complex interaction between the thermal conductivity, electric conductivity and the Seebeck coefficient, which depends on the carrier concentration. While it is already known that the figure of merit can be increased by using nanoscaled materials, it is the purpose of this chapter to quantitatively model the phenomena of the several thermoelectric components of the figure of merit at nanoscale. In doing this, it is aimed to propose a concrete easy-to-use formulation that still captures the several nanoscale features that influence the figure of merit. Non-local effects of heat transfer by phonons and electrons become important at nanoscale, and Fourier's law is no longer applicable. Cattaneo's adaptation of Fourier's law shows the improvement for high-frequency systems, but still is not accurate at nanoscale. Via a new thermodynamic formulation, taking into account the aforementioned non-local effects, we show that size effects at nanoscale become important.

5.2 Basics in Nanoscale Heat and Electric Transfer

We will derive here the thermal conductivity valid at nanoscales. We will perform this by deriving the phonon contribution of the thermal conductivity $\lambda_{ph,\mathcal{N}}$ at nanoscales, with the subscript $\mathcal{N}$ designating the nanoscale. The electron contribution of the thermal conductivity $\lambda_{e,\mathcal{N}}$ is exactly the same, by replacing the phonon specific heat c_{ph}, phonon lattice velocity v_{ph} and phonon mean free path $\ell_{ph,b}$ by the electron specific heat c_e, electron thermal velocity v_e and electron mean free path $\ell_{e,b}$, respectively. The analogy of the electron contribution with that of the phonon one is generally proposed throughout the chapter.

From a physical point of view, we can consider the phonons and the electrons as gas-like constituents that have both gas-like behaviour, flowing through the material lattice. Since we consider that both the phonons and the electrons behave like a gas, we assume that they also follow the same thermodynamic principles. Therefore, the non-local effects that are introduced for the phonon contribution apply for the electron contribution of the thermal conductivity. As they also apply for the electron contribution, the electrical conductivity can be treated in the same way in the same framework. The Seebeck coefficient that relates a temperature gradient with an electric current can then also be treated that way, albeit not directly. Indeed, the systems (5.6) and (5.7) show nicely that the thermal and electric conductivities precede the temperature gradient, whereas the Seebeck coefficient only precedes the temperature gradient in the form of a product with the electric conductivity. The phonon thermal conductivity at nanoscale is given by

$$\lambda_{ph,\mathcal{N}} = \lambda_{ph}^0 \, f\left(Kn, s\right), \tag{5.8}$$

where λ_{ph}^0 is the value of the phonon thermal conductivity for the bulk material at macroscale (λ_e^0 being the equivalent for the electron contribution of the thermal conductivity) and given by the classical Boltzmann phonon expression:

$$\lambda_{ph}^0 = \frac{1}{3}\left(c_{ph} v_{ph} \ell_{ph,b}\right)|_{T_{ref}}, \tag{5.9}$$

the quantity $f\left(Kn, s\right)$ being a correction factor, taking into account the dimension of the nanoparticles, their shape and their specularity. T_{ref} is the reference temperature, say the room temperature. Moreover, Kn is the Knudsen number defined as the ratio of the mean free path of the phonons inside the nanomaterial $\ell_{ph,b}$ (the subscript b stands for the bulk material), and the "specular" characteristic length $\mathcal{Y}_s$:

$$Kn_{ph} = \ell_{ph,b}/\mathcal{Y}_s. \tag{5.10}$$

The specular characteristic length $\mathcal{Y}_s$ is defined by

$$\mathcal{Y}_s = \mathcal{Y}\left(1+s\right)/\left(1-s\right), \tag{5.11a}$$

$$\mathcal{Y} \equiv \delta_{\mathcal{N}} \text{ (nanofilm)}, \tag{5.11b}$$

$$\mathcal{Y} \equiv a_p \text{ (nanocomposite)}, \tag{5.11c}$$

where $\delta_{\mathcal{N}}$ is the nanofilm thickness (used in this subsection) and a_p the radius of a nanoparticle embedded in a nanocomposite (used in the next subsection). In expression (5.11a), the symbol s $(0 \leq s \leq 1)$ denotes the surface specularity on the nanoscale material, expressing the probability of specular scattering of phonons on the boundary (whether it be that of the nanofilm or the nanoparticle). For $s = 0$, the surface is called diffuse, meaning that the direction of phonons after impact is independent of the direction of the impacting phonons, in which case y_s is simply y. For $s \to 1$, we have a surface on which the impacting phonons influence the direction of the out coming phonons and the surface is said to be perfectly specular. Another interesting interpretation of $s \to 1$, which should be kept in mind, is that the dimension of the nanoscale material becomes macroscale, i.e. $\mathcal{Y} \gg \ell_{ph,b}$ or $Kn_{ph} \to 0$.

It should also be noticed that the contribution of the phonon collisions is present in the correction factor $f\left(Kn\right)$ as well. This correction factor is already developed in Chapter 3, leading to Eq. (3.14). The equivalent is the following expression for $\lambda_{\mathcal{N},ph}$:

$$\lambda_{ph,\mathcal{N}} = \frac{3\lambda_{ph}^0}{4\pi^2 Kn_{ph}^2}\left[\frac{2\pi Kn_{ph}}{\arctan\left(2\pi Kn_{ph}\right)} - 1\right], \tag{5.12}$$

with Kn_{ph} given by Eq. (5.10).

5.3 Nanofilm Thermoelectric Devices

5.3.1 Theory

In this subsection, we continue with the model presented in Section 5.2 in order to apply it to nanofilm thermoelectric systems. In analogy with Eq. (5.12) (and the proved analogy between thermal and electrical flux in Chapter 1), we also obtain the following expression for the electron contribution of the thermal conductivity at nanoscales:

$$\lambda_{e,\mathcal{N}} = \frac{3\lambda_e^0}{4\pi^2 Kn_e^2}\left[\frac{2\pi Kn_e}{\arctan(2\pi Kn_e)} - 1\right], \tag{5.13}$$

where $Kn_e = \Lambda_{e,b}/\ell_s$ in analogy with Eq. (5.10). The total thermal conductivity at nanoscales is then defined as

$$\lambda_{tot,\mathcal{N}} = \lambda_{ph,\mathcal{N}} + \lambda_{e,\mathcal{N}}.$$

The electric conductivity at nanoscale is defined in analogy to the electron contribution of the thermal conductivity as

$$\sigma_{\mathcal{N}} = \frac{\lambda_{e,\mathcal{N}}}{L_e T}, \tag{5.14}$$

or

$$\sigma_{\mathcal{N}} = \frac{3\lambda_e^0}{4\pi^2 Kn_e^2 L_e T}\left[\frac{2\pi Kn_e}{\arctan(2\pi Kn_e)} - 1\right], \tag{5.15}$$

with $\lambda_e^0 \equiv \sigma^0 L_e T$, where L_e is the Lorentz number and T the absolute temperature. The Lorentz number is determined by

$$L_e = \frac{\pi^2}{3}\left(\frac{k_B}{e_c}\right)^2, \tag{5.16}$$

where k_B is Boltzmann's constant and e_c the elementary charge. Furthermore, the electric conductivity of a bulk material is dependent not only on the electron contribution of the thermal conductivity but also on the electron n_n and hole n_p carriers concentration as well as electron μ_n and hole μ_p mobility [18]:

$$\sigma^0 = n_n e_c \mu_n + n_p e_c \mu_p. \tag{5.17}$$

Here, the subscripts n and p stand for the denomination of n-type and p-type carriers, typically used for electrons and holes, respectively. The thermoelectric performance is typically assessed by the figure of merit $Z_{\mathcal{N}}$, where the non-dimensional version $ZT_{\mathcal{N}}$ is given by

$$ZT_{\mathcal{N}} = T\frac{S_{\mathcal{N}}^2 \sigma_{\mathcal{N}}}{\lambda_{ph,\mathcal{N}} + \lambda_{e,\mathcal{N}}} = \frac{S_{\mathcal{N}}^2/L_e}{\frac{\lambda_{ph,\mathcal{N}}}{\lambda_{e,\mathcal{N}}} + 1}. \tag{5.18}$$

Here, $S_{\mathcal{N}}$ is the Seebeck coefficient (the electric conductivity in a material is determined via the mobility of both electrons and holes) given by [19]

$$S_{\mathcal{N}} = \frac{n_n \mu_n S_n + n_p \mu_p S_p}{n_n \mu_n + n_p \mu_p}. \tag{5.19}$$

with

$$S_x = (-1)^z \frac{8\pi^2 k_B^2}{3e_c h^2} m_x^* T \left(\frac{\pi}{3n_x}\right)^{2/3}, \tag{5.20}$$

where h is Planck's constant, m_x^* is the effective mass of electrons or holes (expressed in free electron mass m^0), and the subscript x denotes whether it concerns the Seebeck coefficient of electrons ($x = n$ and $z = 1$) or that of holes ($x = p$ and $z = 2$). If the material is only of the n-type, then $S_\mathcal{N} = S_n\,(< 0)$, and if the material is only of the p-type, then $S_\mathcal{N} = S_p\,(> 0)$. With the Seebeck coefficient and the electric conductivity, a power factor can be defined as

$$PF_\mathcal{N} = S_\mathcal{N}^2 \sigma_\mathcal{N}. \tag{5.21}$$

There is still a note to be made. The carrier concentration of the holes and electrons cannot be assumed to stay constant with respect to the film's size. Nanoscaling causes the scattering of the carriers [20] (in the direction of the film thickness) at the film's boundary (ballistic effect). Since the carrier concentrations appear in the definition of the electric conductivity in Eq. (5.17), we assume that the variation of the carrier concentrations (and its scattering behaviour) also follows that of the electric conductivity (which follows that of the thermal conductivity) at nanoscale [see Eq. (5.15)]. Also, we should notice that the mobility will increase for decreasing carrier concentration, governed by an inverse power law [21]. From Eqs. (5.15) and (5.17), we have

$$n_x \mu_x \sim \sigma_\mathcal{N} = \frac{3\sigma^0}{4\pi^2 Kn_e^2}\left[\frac{2\pi Kn_e}{\arctan(2\pi Kn_e)} - 1\right], \tag{5.22}$$

Knowing that $S_x \sim n_x^{-2/3}$ from Eq. (5.20) and that $\sigma \sim n_x$ and noting that from [12] the power factor $S_x^2\sigma \sim \mu_x$ (neglecting any variations of the effective mass), it can be easily deduced that $\mu_x \sim n_x^{-1/3}$ (also observed by [22]), which leads to

$$n_x^{2/3} \sim \sigma_\mathcal{N}, \tag{5.23}$$

or

$$n_x = n_{x,0}\left(\frac{\sigma_\mathcal{N}}{\sigma^0}\right)^{3/2} = n_{x,0}\left(\frac{\lambda_{e,\mathcal{N}}}{\lambda_e^0}\right)^{3/2}, \tag{5.24}$$

where $n_{x,0}$ is the carrier concentration of the bulk material (without taking into account the effect of electron scattering due to the nanoparticles). In this subsection, we have obtained a mathematical model for describing the thermal conductivity (with its phonon and electron contributions), the electric conductivity, the Seebeck coefficient (taking into account the scattering of the carriers and the change in the mobility) and the resulting figure of merit, all at nanoscale. This model is suitable for nanofilms, and in the following section, we will apply this model in a case study. Interestingly, it should be noted that this model is also valid for nanoparticles, with some slight modifications, presented in Section 5.4. Furthermore, since nanoparticles are embedded in a matrix (nanocomposites), the latter must also be taken into account.

5.3.2 Case Study: Thin Films of Bi and Bi_2Te_3

5.3.2.1 Material Properties

Historically, bismuth was the first material showing substantial thermoelectric coefficients [23], and bismuth telluride is one of the best performing thermoelectric materials at room temperature [24]. As such, these materials lend themselves for a case study. In order to give a realistic quantitative character to this case study, the material properties have to be obtained as accurately as possible. Bismuth is a semimetal where the electric conductivity passes via both hole and electron mobility. We can obtain the electric conductivity via Eq. (5.17). The hole (μ_p) and electron (μ_n) mobilities are 1 and 0.4 $\mathrm{m^2\ V^{-1}\ s^{-1}}$, respectively [25]. The carrier concentrations of the holes and electrons are each 3.5*10^{24} m^{-3} [25]. This gives an electric conductivity σ^0 of 7.85*10^5 Ω^{-1} m^{-1}. The Lorentz number can be obtained by Eq. (5.16) to be 2.44*10^{-8} W Ω K^{-2}. The electron thermal conductivity is then given by $\lambda_e^0 \equiv \sigma^0 L_e T$, obtaining 5.75 W K^{-1} m^{-1}. The electron thermal velocity is given by $\sqrt{3k_BT/m^*}$ [26]. With $m^* = 0.16m^0$ [27] for the electrons (the same holds for the holes), we find an electron thermal velocity of 2.92*10^5 m s^{-1}. The electron mean free path is 0.67 nm [28,29]. This gives with the electron version of Eq. (5.9) an electron specific heat capacity of 0.088 MJ m^{-3} K. Assuming that the total specific heat capacity (1.21 MJ m^{-3} K^{-1} [30]) is the sum of the electron and phonon ones, we can find the phonon specific heat capacity of 1.12 MJ m^{-3} K^{-1}. The phonon mean free path is 3.0 nm [31]. Via Eq. (5.9), we can find a phonon group velocity of 1980 m s^{-1}. This value corresponds nicely with the value of 1790 m s^{-1}, proposed by [28]. This gives finally all the electron and phonon material properties needed for the calculation of the effective thermal conductivities for bismuth. In order to calculate the figure of merit, we still need to calculate the Seebeck coefficient. Using Eqs. (5.19) and (5.20), we can find a bulk Seebeck coefficient of 139 μV K^{-1} for bismuth. A resume of these properties are given in Table 5.1.

As for p-type (the electric conductivity passes only via the mobility of holes) bismuth telluride the electron (c_e, v_e and $\ell_{e,b}$) and phonon (c_{ph}, v_{ph} and $\ell_{ph,b}$) material properties are given by [32]. As for the Seebeck coefficient, we use Eq. (5.20) with $x = p$ and $m^* = 1.25m^0$ (using the six-valley model) [27] for the holes (p-type) and obtain finally a bulk value of 188 μV K^{-1} for Bi_2Te_3. The mobility is 420*10^{-4} m^2 V^{-1} s^{-1} [33], and the electric conductivity can be found by $\sigma^0 = \lambda_e^0/L_eT$, with $\lambda_e^0 = \frac{1}{3}c_e v_e \ell_{e,b}$, so that the electric conductivity is found to be 3.28*10^5 Ω^{-1} m^{-1}. With this, we can find the hole carrier concentration from Eq. (5.17) to be 4.87*10^{25} m^{-3}. A resume of these properties are given in Table 5.1. Table 5.2 shows the values of the used physical constants.

5.3.2.2 Discussion

Figure 5.1 presents the phonon, electron and total thermal conductivities (also for several specularities), electric conductivity and the dimensionless figure of merit for Bi and Bi_2Te_3.

We start the discussion first for $s = 0$. From Figure 5.1(Ia,Ib,IIa,IIb), we can clearly see that the thermal conductivities for both considered nanofilms (Bi and Bi_2Te_3) decrease considerably for thicknesses below 50 nm. This is due to the scattering effect of the phonons and electrons. As the film thickness decreases, the ballistic effects (phonons colliding with boundaries of which the dimensions approach that of the mean free paths) become more important. The scattering effect has clear influences on the scattering of the carriers translated into a sharp increase in the Seebeck coefficient [see Eq. (5.24)] in Figure 5.1(Ic,IIc). The electric conductivity follows the behaviour of the electron thermal conductivity [see Figure 5.1(Id,IId)] for the same reasons, i.e. the scattering of the electrons. Figure 5.1(Ie,IIe) shows the power factor, which increases as the nanofilm thickness decreases. Finally, Figure 5.1(If,IIf) shows the figure of merit, which shows us that the decreasing thermal conductivity and the increasing Seebeck coefficient result into an increased figure of merit. Note the relatively high value for the figure of merit in Figure 5.1(IIf). It should be mentioned that a value of $ZT_{\mathcal{N}} = 8$ (for $s = 0$ and 0.2) is quite on the high side. It is still shown here as an extrapolation for the purposes of mathematical interest, showing to what extent nanofilms can increase the figure of merit. We can also see that the electron thermal conductivity decreases less than the phonon one, which also contributes to increased figure of merit. As far as concerns the difference between Bi and Bi_2Te_3, it can be said that they show the same behaviour. The results only differ quantitatively. Bi_2Te_3 presents a higher Seebeck coefficient and power factor than Bi, resulting into higher figure of merits. It is interesting to note that the values at $s \to 1$ are constant. This can be interpreted in two ways. Firstly,

TABLE 5.1
Electron and Phonon Material Properties of Bi and Bi_2Te_3 at $T = 300$ K.

Material	Electron			Phonon			Carrier		
	c_e [MJ m^{-3} K^{-1}]	v_e [km s^{-1}]	$\ell_{e,b}$ [nm]	c_{ph} [MJ m^{-3} K^{-1}]	v_{ph} [km s^{-1}]	$\ell_{ph,b}$ [nm]	$n_{p,m}$ [10^{24} m^{-3}]	$n_{n,m}$	S_m [μV K^{-1}]
Bi	0.088	292	0.67	1.12	1.98	3.0	3.5	3.5	59
Bi_2Te_3	1.01	7.83	0.91	0.19	8.43	3.0	48.7	–	188

TABLE 5.2
Physical Constants at T = 300 K.

Physical Constant	k_B [J K^{-1}]	h [J s]	e_c [C]	L_e [W Ω K^{-2}]	m^0 [kg]
Value	1.38*10^{-23}	6.626*10^{-34}	1.602*10^{-19}	2.44*10^{-8}	9.11*10^{-31}

for $s \to 1$, the electron and phonon scattering occurs on a so-called smooth surface so that the incoming phonons and electrons leave the boundary of the nanofilm at the same angle. In other words, the boundaries are specular. This results into zero scattering, and the heat transfer occurs fully, so that the thermal conductivities, electric conductivity and carriers do not undergo the effect of scattering and preserve their bulk values. As a consequence, also the Seebeck coefficient and power factor remain the same, resulting altogether in a constant figure of merit. Another interpretation is to say that for $s \to 1$, Eq. (5.11a) shows that $\mathcal{Y}_s \to \infty$. This means that the nanofilm behaves as if it were a macrofilm with bulk properties. As far as goes the results for $0 < s < 1$, they show intermediate tendencies between those of $s = 0$ and $s \to 1$, as expected.

5.4 Nanocomposite Thermoelectric Devices

5.4.1 Theory

In this section, we use the models from Sections 5.2 and 5.3.1 in order to apply it for nanocomposites, making some modifications. We begin with a remark about the thermal (with both phonon and electron contributions) and electrical conductivities: using the definition of the Knudsen number (5.10) and the corresponding equations (5.11), we can easily notice that Eq. (5.13) can, next to nanofilms, also be used for nanoparticles. The thermal conductivity (both phonon ($\lambda_{ph,\mathcal{N}}$) and electron ($\lambda_{e,\mathcal{N}}$) contributions) and electric conductivity ($\sigma_{\mathcal{N}}$) of the nanoparticles are then given keeping in mind that now $\mathcal{Y} \equiv a_p$. We will have to add the contribution of the bulk matrix ($\lambda_{ph,m}$, $\lambda_{e,m}$ and σ_m), respectively, within which the nanoparticles are embedded. As in the previous section, we will extend our model for the phonon contribution. The electron contribution follows exactly the same procedure, replacing the phonon specific heat $c_{ph,m}$, lattice velocity $v_{ph,m}$ and phonon mean free path $\ell_{ph,b,m}$ by the electron specific heat $c_{e,m}$, electron thermal velocity $v_{e,m}$ and electron mean free path $\ell_{e,b,m}$, respectively, m the subscript denoting that the property concerns the bulk matrix. The phonon thermal conductivity of the bulk matrix is given by the classical Boltzmann phonon expression:

$$\lambda_{ph,m} = \frac{1}{3}\left(c_{ph,m} v_{ph,m} \ell_{ph,m}\right)|_{T_{ref}}. \tag{5.25}$$

Within the matrix, the phonons experience phonon–phonon interactions and the mean free path is given by the Matthiessen rule:

$$\frac{1}{\ell_{ph,m}} = \frac{1}{\ell_{ph,b,m}} + \frac{1}{\ell_{ph,coll,m}}. \tag{5.26}$$

where $\ell_{ph,b,m}$ denotes the mean free path in the bulk matrix, and $\ell_{coll,m}$, the supplementary contribution due to the interactions at the particle–matrix interface, is given by

$$\ell_{coll,m} = \frac{4a_{p,s}}{3\varphi}, \tag{5.27}$$

where φ is the volume fraction of the nanoparticles in the matrix. This is a heterogeneous medium, which means that we cannot just use Eq. (5.13). We have said earlier that Eq. (5.13) can also be used for nanoparticles. Still, Eq. (5.13) is only valid for materials that are nanoscaled and homogeneous, and the latter of which is obviously not the case for nanocomposites. Therefore, in order to calculate the thermal conductivity for

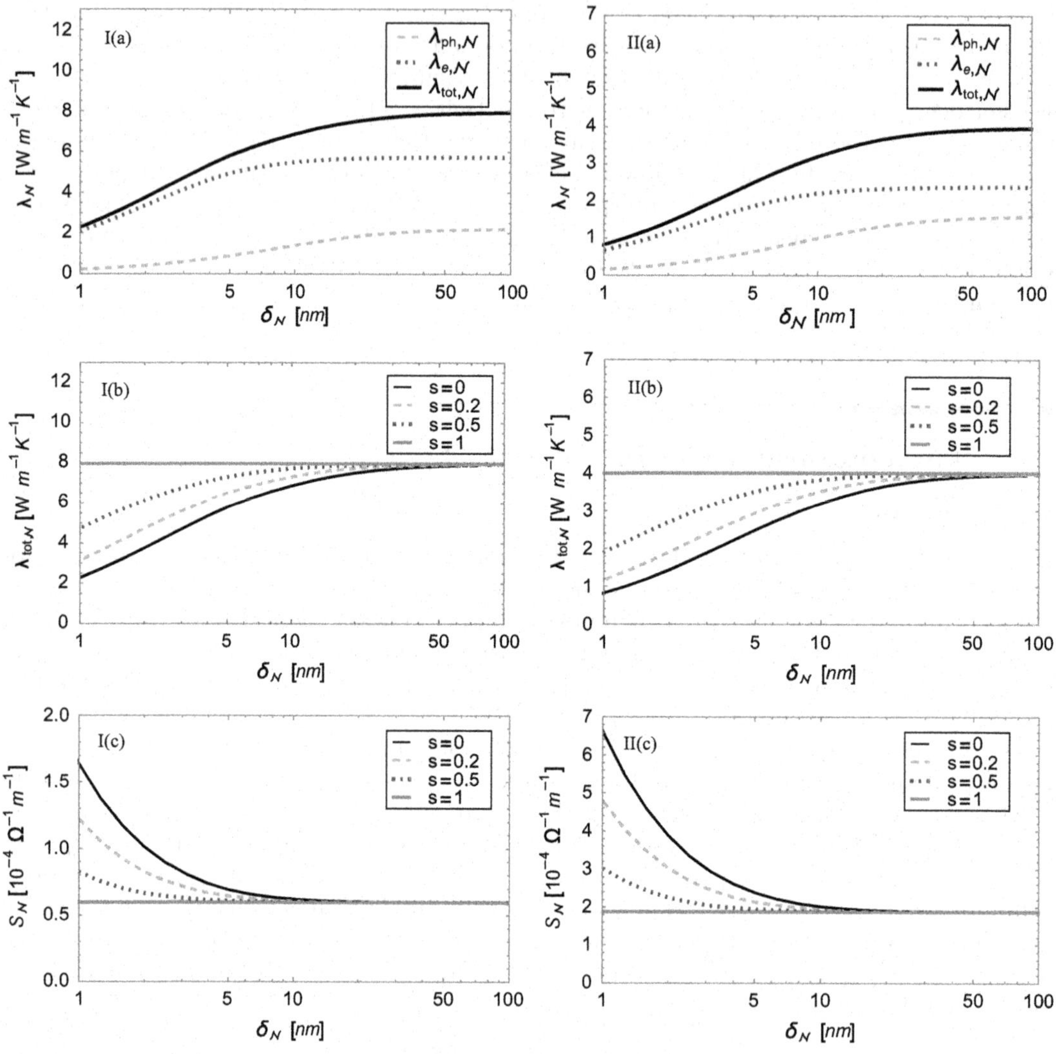

FIGURE 5.1
Dependence on the film thickness $\delta_{\mathcal{N}}$ of Bi (left, indicated by I) and Bi_2Te_3 (right, indicated by II) of (a) the phonon, electric and total thermal conductivity ($\lambda_{ph,\mathcal{N}}$, $\lambda_{e,\mathcal{N}}$ and $\lambda_{tot,\mathcal{N}}$) for $s = 0$, and (b) the total thermal conductivity $\lambda_{tot,\mathcal{N}}$, (c) the Seebeck coefficient $S_{\mathcal{N}}$, (d) the electric conductivity $\sigma_{\mathcal{N}}$, (e) the power factor $PF_{\mathcal{N}}$ and (f) the dimensionless figure of merit $ZT_{\mathcal{N}}$, for $s = 0$, 0.2, 0.5 and 1. (Modified from: Machrafi, H. 2016. An extended thermodynamic model for size-dependent thermoelectric properties at nanometric scales: Application to nanofilms, nanocomposites and thin nanocomposite films. *Applied Mathematical Modelling* 40:2143–2160.)

(Continued)

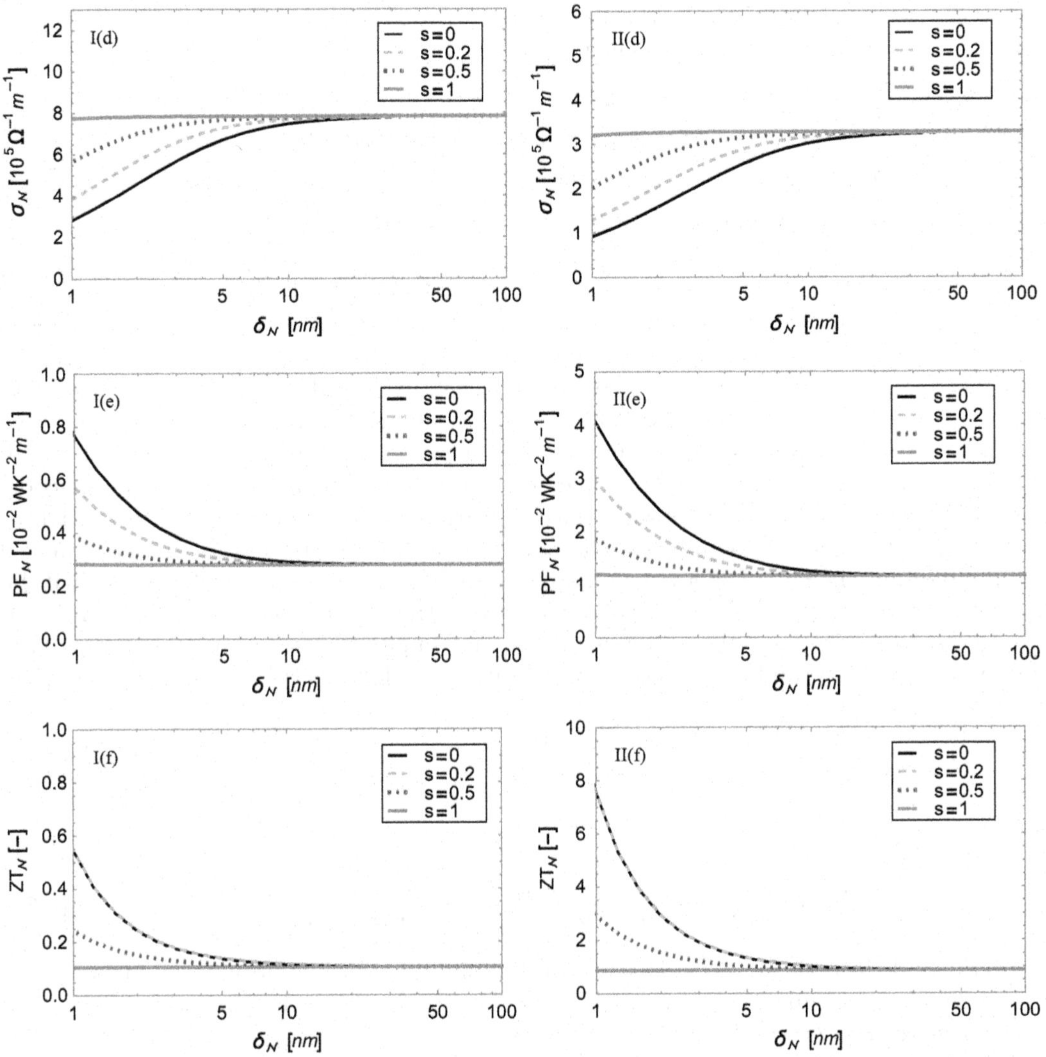

FIGURE 5.1 (CONTINUED)
Dependence on the film thickness $\delta_{\mathcal{N}}$ of Bi (left, indicated by I) and Bi_2Te_3 (right, indicated by II) of (a) the phonon, electric and total thermal conductivity ($\lambda_{ph,\mathcal{N}}$, $\lambda_{e,\mathcal{N}}$ and $\lambda_{tot,\mathcal{N}}$) for $s = 0$, and (b) the total thermal conductivity $\lambda_{tot,\mathcal{N}}$, (c) the Seebeck coefficient $S_{\mathcal{N}}$, (d) the electric conductivity $\sigma_{\mathcal{N}}$, (e) the power factor $PF_{\mathcal{N}}$ and (f) the dimensionless figure of merit $ZT_{\mathcal{N}}$, for $s = 0,\ 0.2,\ 0.5$ and 1. (Modified from: Machrafi, H. 2016. An extended thermodynamic model for size-dependent thermoelectric properties at nanometric scales: Application to nanofilms, nanocomposites and thin nanocomposite films. *Applied Mathematical Modelling* 40:2143–2160.)

nanocomposites, we have to relate the one in the nanoparticle [given by Eq. (5.13)] to the one in the bulk matrix [given by Eq. (5.25)]. For this purpose, we will make use of the effective medium approach [34,35] which provides a process of homogenization of the heterogeneous medium formed by the matrix and the particles. The effective medium theory calculates effective properties for media with located symmetric inclusions. The basic formula for the effective phonon thermal conductivity coefficient λ_{ph}^{eff} is the equivalent of Eq. (3.1), adapted here for the present purposes:

$$\lambda_{ph}^{eff} = \lambda_{ph,m} \frac{2\lambda_{ph,m} + (1 + 2\alpha_{ph})\lambda_{ph,\mathcal{N}} + 2\varphi\left[(1 - \alpha_{ph})\lambda_{ph,\mathcal{N}} - \lambda_{ph,m}\right]}{2\lambda_{ph,m} + (1 + 2\alpha_{ph})\lambda_{ph,\mathcal{N}} - \varphi\left[(1 - \alpha_{ph})\lambda_{ph,\mathcal{N}} - \lambda_{ph,m}\right]}. \tag{5.28}$$

$$\alpha_{ph} = R_{ph}\lambda_{ph,m}/a_{p,s}. \tag{5.29}$$

The quantity R_{ph} is the equivalent of Eq. (3.3) and is here given by

$$R_{ph} = 4/C_{ph,m}v_{ph,m} + 4/C_{ph}v_{ph}, \tag{5.30}$$

Note that the result (5.30) was established in the case of diffusive surfaces. The subscript "ph, m" stands for the phonon contribution in the bulk matrix, whereas the subscript "ph" denotes the phonon contribution in the nanoparticle (also equal to the one in the nanofilm in the previous section). It is important to note that equation (5.28) stands for the effective thermal conductivity that would be measured in a nanocomposite composed out of a bulk matrix and a volume fraction of nanoparticles; it is thus a combination of macroscopic and microscopic heat transfer phenomena. In thermoelectric devices, the combination of macroscopic and microscopic phenomena can lead to different results as the nanoparticle volume fraction changes. For low volume fractions, the macroscopic heat transfer will be more important, whereas for higher volume fractions, the microscopic one will take the lead. The electron contribution follows the same path, defining the electron thermal conductivity, electric conductivity and the electron contribution of the effective thermal conductivity which are given by,

$$\lambda_{e,m} = \frac{1}{3}\left(c_{e,m}v_{e,m}\ell_{e,m}\right)|_{T_{ref}}, \tag{5.31}$$

$$\sigma_m = \frac{\lambda_{e,m}}{L_e T}, \tag{5.32}$$

$$\lambda_e^{eff} = \lambda_{e,m}\frac{2\lambda_{e,m} + (1 + 2\alpha_e)\lambda_{e,\mathcal{N}} + 2\varphi\left[(1 - \alpha_e)\lambda_{e,\mathcal{N}} - \lambda_{e,m}\right]}{2\lambda_{e,m} + (1 + 2\alpha_e)\lambda_{e,\mathcal{N}} - \varphi\left[(1 - \alpha_e)\lambda_{e,\mathcal{N}} - \lambda_{e,m}\right]}, \tag{5.33}$$

respectively. In Eq. (5.33), $\lambda_{e,\mathcal{N}}$ is given by Eq. (5.13) and α_e is obtained from α_{ph} from Eq. (5.29), by replacing $\lambda_{ph,m}$ by $\lambda_{e,m}$ (5.31) and R_{ph} by $R_e = 4/c_{e,m}v_{e,m} + 4/c_e v_e$. The total effective thermal conductivity is given by

$$\lambda_{tot}^{eff} = \lambda_{ph}^{eff} + \lambda_e^{eff}, \tag{5.34}$$

whereas the effective electric conductivity is expressed as

$$\sigma^{eff} = \frac{\lambda_e^{eff}}{L_e T}. \tag{5.35}$$

As for the Seebeck coefficient, we cannot take a volumetric average since this means neglecting the effect of size. Also, we need to redefine the Seebeck coefficient for the nanocomposite:

$$S_m = \frac{n_{n,m}\mu_n S_{n,m}^* + n_{p,m}\mu_p S_{p,m}^*}{n_{n,m}\mu_n + n_{p,m}\mu_p}, \tag{5.36a}$$

$$S_{\mathcal{N}} = \frac{n_{n,\mathcal{N}}\mu_n S_{n,\mathcal{N}}^* + n_{p,\mathcal{N}}\mu_p S_{p,\mathcal{N}}^*}{n_{n,\mathcal{N}}\mu_n + n_{p,\mathcal{N}}\mu_p}, \tag{5.36b}$$

where S_m is the Seebeck coefficient of the matrix (the bulk material) and $S_{\mathcal{N}}$ the Seebeck coefficient of the nanoparticle. Furthermore, we define

$$S_{x,y}^* = (-1)^z \frac{8\pi^2 k_B^2}{3e_c h^2} m_x^* T \left(\frac{\pi}{3n_{x,y}}\right)^{2/3}, \tag{5.37}$$

where the subscript x denotes whether it concerns the Seebeck coefficient of electrons ($x = n$ and $z = 1$) or that of holes ($x = p$ and $z = 2$), and the subscript y denotes whether it concerns the bulk material ($y = m$) or the nanoparticle ($y = \mathcal{N}$). If the material is only of the n-type, then $S_{\mathcal{N}} = S^*_{n,\mathcal{N}}$ (for the nanoparticle) and $S_m = S^*_{n,m}$ (for the bulk matrix), and if the material is only of the p-type , then $S_{\mathcal{N}} = S^*_{p,\mathcal{N}}$ (for the nanoparticle) and $S_m = S^*_{p,m}$ (for the bulk matrix). In this definition of the Seebeck coefficient, we need again to pay attention to the carrier concentration of the holes and electrons, but this time because of two reasons: the nanoparticle size is at nanoscale, and they are embedded in another material. This causes the scattering [20] of the carriers not only within the nanoparticles but also at the nanoparticle–matrix boundary, which influences the carrier concentration within the matrix as well. We assume again, as in Section 5.3.1, that the variation of the carrier concentrations (and its scattering behaviour) also follows that of the electric conductivity (which follows that of the thermal conductivity) for both the nanoparticles [see Eq. (5.15)] and the bulk matrix. From Eqs. (5.15), (5.17), (5.31) and (5.32), we have

$$n_{x,\mathcal{N}}\mu_x \sim \sigma_{\mathcal{N}} = \frac{3\sigma^0}{4\pi^2 Kn_e^2}\left[\frac{2\pi Kn_e}{\arctan(2\pi Kn_e)} - 1\right], \tag{5.38}$$

$$n_{x,m}\mu_x \sim \sigma_m = \frac{1}{3L_eT}\left(c_{e,m}v_{e,m}\ell_{e,m}\right)|_{T_{ref}}. \tag{5.39}$$

As noticed earlier, we take the mobility to be dependent on the carrier concentration as $\mu_x \sim n_x^{-1/3}$, which leads to

$$n_{x,\mathcal{N}}^{2/3} \sim \sigma_{\mathcal{N}}, \tag{5.40a}$$

$$n_{x,m}^{2/3} \sim \sigma_m. \tag{5.40b}$$

This leads finally to

$$n_{x,\mathcal{N}} = n_{x,0}\left(\frac{\sigma_{\mathcal{N}}}{\sigma^0}\right)^{3/2} = n_{x,0}\left(\frac{\lambda_{e,\mathcal{N}}}{\lambda_e^0}\right)^{3/2}, \tag{5.41a}$$

$$n_{x,m} = n_{x,0}\left(\frac{\sigma_m}{\sigma_m^0}\right)^{3/2} = n_{x,0}\left(\frac{\lambda_{e,m}}{\lambda_{e,m}^0}\right)^{3/2} = n_{x,0}\left(\frac{\ell_{e,m}}{\ell_{e,b,m}}\right)^{3/2}, \tag{5.41b}$$

where σ_m^0 and $\lambda_{e,m}^0$ are the electric conductivity and the electron thermal conductivity, respectively, of the matrix bulk material (without taking into account the effect of electron scattering due to the nanoparticles, i.e. $\ell_{e,m} \equiv \ell_{e,b,m}$ and $\varphi = 0$).

The size dependency of the Seebeck coefficient for nanofilms has been taken into account via the carrier concentration. For nanocomposites, however, it is quite different, since the two materials are present, one of which is at nanoscales. This will lead to an effective Seebeck coefficient. However, the Seebeck coefficient is not to be taken size dependent in the same way as the thermal conductivity, since Eqs. (5.6c) and (5.7c) suggest that S and σ are to be evaluated together when it comes to the extended thermodynamic approach of nanocomposite thermoelectric devices in this work (assuming $\ell_I \equiv \ell_e$). Therefore, with Eq. (5.14) [keeping in mind Eq. (5.15)], we have

$$(S\sigma)_{\mathcal{N}} = \frac{S_{\mathcal{N}}}{L_eT}\lambda_{e,\mathcal{N}}. \tag{5.42}$$

For the matrix, an analogous expression can be obtained:

$$(S\sigma)_m = \frac{S_m}{L_eT}\lambda_{e,m}, \tag{5.43}$$

Then, we have in analogy with Eq. (5.33)

$$(S\sigma)^{eff} = (S\sigma)_m \frac{2(S\sigma)_m + (1+2\alpha_e)(S\sigma)_{\mathcal{N}} + 2\varphi\left[(1-\alpha_e)(S\sigma)_{\mathcal{N}} - (S\sigma)_m\right]}{2(S\sigma)_m + (1+2\alpha_e)(S\sigma)_{\mathcal{N}} - \varphi\left[(1-\alpha_e)(S\sigma)_{\mathcal{N}} - (S\sigma)_m\right]}. \tag{5.44}$$

The effective Seebeck coefficient S^{eff} and the effective power factor PF^{eff} can then be easily obtained a posteriori, by defining

$$S^{eff} = (S\sigma)^{eff}/\sigma^{eff}, \tag{5.45}$$

$$PF^{eff} = S^{eff^2}\sigma^{eff} = (S\sigma)^{eff^2}/\sigma^{eff} \tag{5.46}$$

Finally, with the preceding equations, we can obtain the effective non-dimensional figure of merit for nanocomposites:

$$ZT^{eff} = T\frac{\left(S^{eff}\right)^2\sigma^{eff}}{\lambda_{ph}^{eff} + \lambda_e^{eff}} = T\frac{(S\sigma)^{eff^2}/\sigma^{eff}}{\lambda_{ph}^{eff} + \lambda_e^{eff}} = L_e T^2 \frac{(S\sigma)^{eff^2}}{\lambda_e^{eff}\left(\lambda_{ph}^{eff} + \lambda_e^{eff}\right)}. \tag{5.47}$$

The model presented in this section will be applied in two case studies in the next section.

5.4.2 Two Case Studies: Nanocomposites of Bi Nanoparticles in Bi_2Te_3 and of Bi_2Te_3 Nanoparticles in Bi

Figure 5.2 presents the phonon, electron and total thermal conductivities (also for several specularities), electric conductivity and the dimensionless figure of merit as a function of the volume fraction of nanoparticles for two systems: left, Bi nanoparticles embedded in Bi_2Te_3, and, right, Bi_2Te_3 nanoparticles embedded in Bi. The size of the nanoparticles is 1 nm.

Again, we start with discussing the results for $s = 0$. From Figure 5.2(Ia,Ib,IIa,IIb), we can see that the effective thermal conductivities decrease as the volume fraction of the nanoparticles increase. The reason of this decrease is twofold. Firstly, due to an increase in the nanoparticles, the mean free path of phonons and electrons in the host matrix decreases [see Eqs. (5.27) and (5.31)] and so does the overall mean free path of phonons and electrons [Eq. (5.26)]. This causes the effective thermal conductivity of the matrix to decrease (Eq. (5.25) for the phonon contribution and Eq. (5.31) for the electron one). Secondly, due to the small size (1 nm) of the nanoparticles, the phonon and electron scattering cause a considerable decrease in the nanoparticle effective phonon thermal conductivity [Eq. (5.12) and for the electron contribution, see Eq. (5.13)]. The effective electric conductivity follows, for the same reasons, the behaviour of the thermal one [see Figure 5.2(Id,IId)]. As for the effective Seebeck coefficient in Figure 5.2(Ic,IIc), it increases as the volume fraction increases. However, this increase is not as strong as the decrease in the thermal and electric conductivities. This can be explained by considering that the scattering effect on the effective Seebeck coefficient has two sources. The first can be understood via the principles in Section 5.1 [see Eq. (5.7c)], which is explicitly expressed in Eq. (5.44) for the effective Seebeck coefficient. Following Eq. (5.44), the effective Seebeck coefficient should decrease due to the scattering effect, in analogy with the effective thermal conductivity [Eqs. (5.28) and (5.33)]. The second, however, shows via Eq. (5.37) that the effective Seebeck coefficient is inversely proportional to the carrier concentration by a factor 2/3. The scattering behaviour of the carriers is proportional to the decrease in the electric conductivity by a factor 3/2 [see Eqs. (5.40a) and (5.40b)]. Therefore, the proportionality of the decreasing and increasing effects of the scattering process on the effective Seebeck coefficient competes with each other, so that the scattering effect is of more importance to the effective thermal conductivities. It appears, by the way, that the scattering process of the carriers is stronger than that expressed by Eq. (5.40), which explains the eventual increase in the effective Seebeck coefficient. Figure 5.2(Ie,IIe) shows that the power factor decreases. This is due to the stronger decrease in the effective electric conductivity with respect to the increase in the effective Seebeck coefficient. This decreasing power factor results into a less strong increase

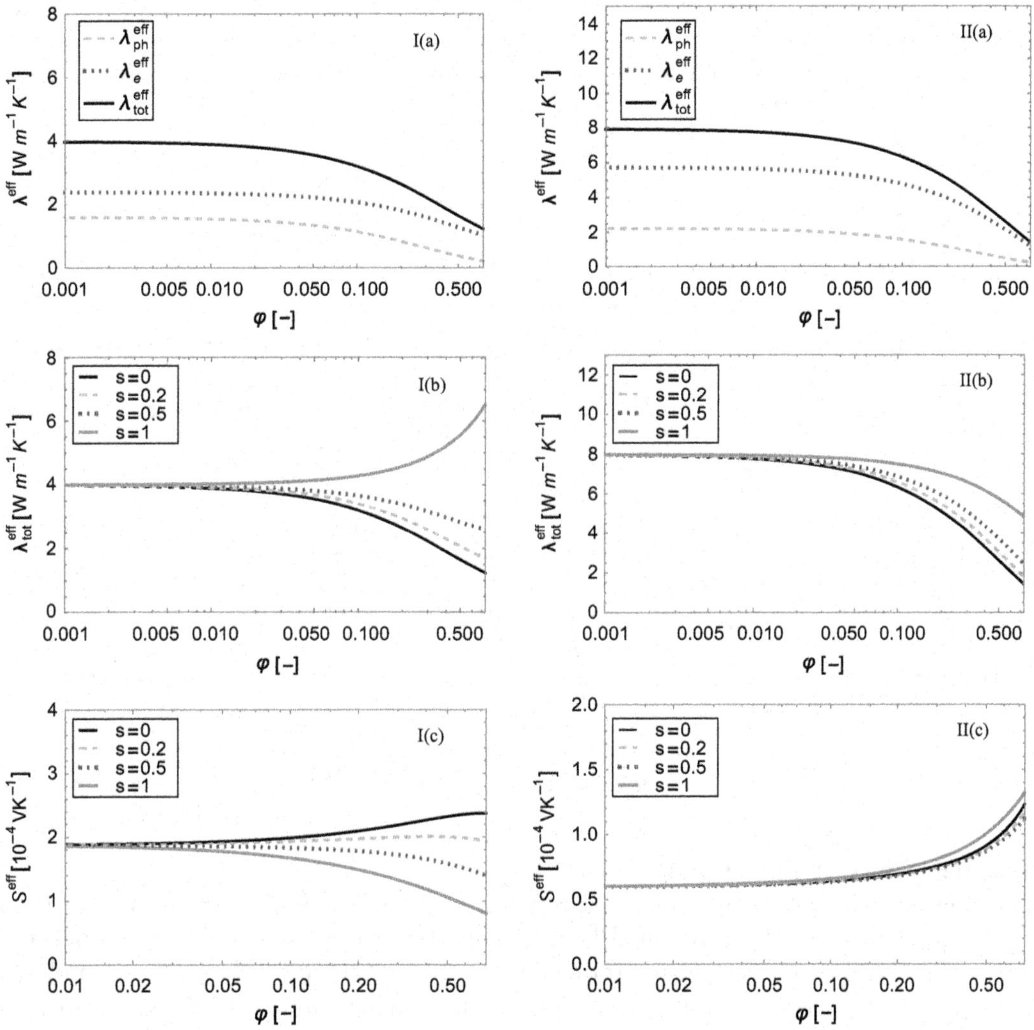

FIGURE 5.2
Dependence on the volume fraction φ of, respectively, 1-nm Bi nanoparticles in a matrix of Bi_2Te_3 (left, indicated by I) and 1-nm Bi_2Te_3 nanoparticles in a matrix of Bi (right, indicated by II) of (a) the phonon, electric and total effective thermal conductivity (λ_{ph}^{eff}, λ_e^{eff} and λ_{tot}^{eff}, respectively) for $s = 0$, and (b) the total effective thermal conductivity λ_{tot}^{eff}, (c) the effective Seebeck coefficient S^{eff}, (d) the effective electric conductivity σ^{eff}, (e) the effective power factor PF^{eff} and (f) the dimensionless effective figure of merit ZT^{eff}, for $s = 0,\ 0.2,\ 0.5$ and 1. (Modified from: Machrafi, H. 2016. An extended thermodynamic model for size-dependent thermoelectric properties at nanometric scales: Application to nanofilms, nanocomposites and thin nanocomposite films. *Applied Mathematical Modelling* 40:2143–2160.)

(*Continued*)

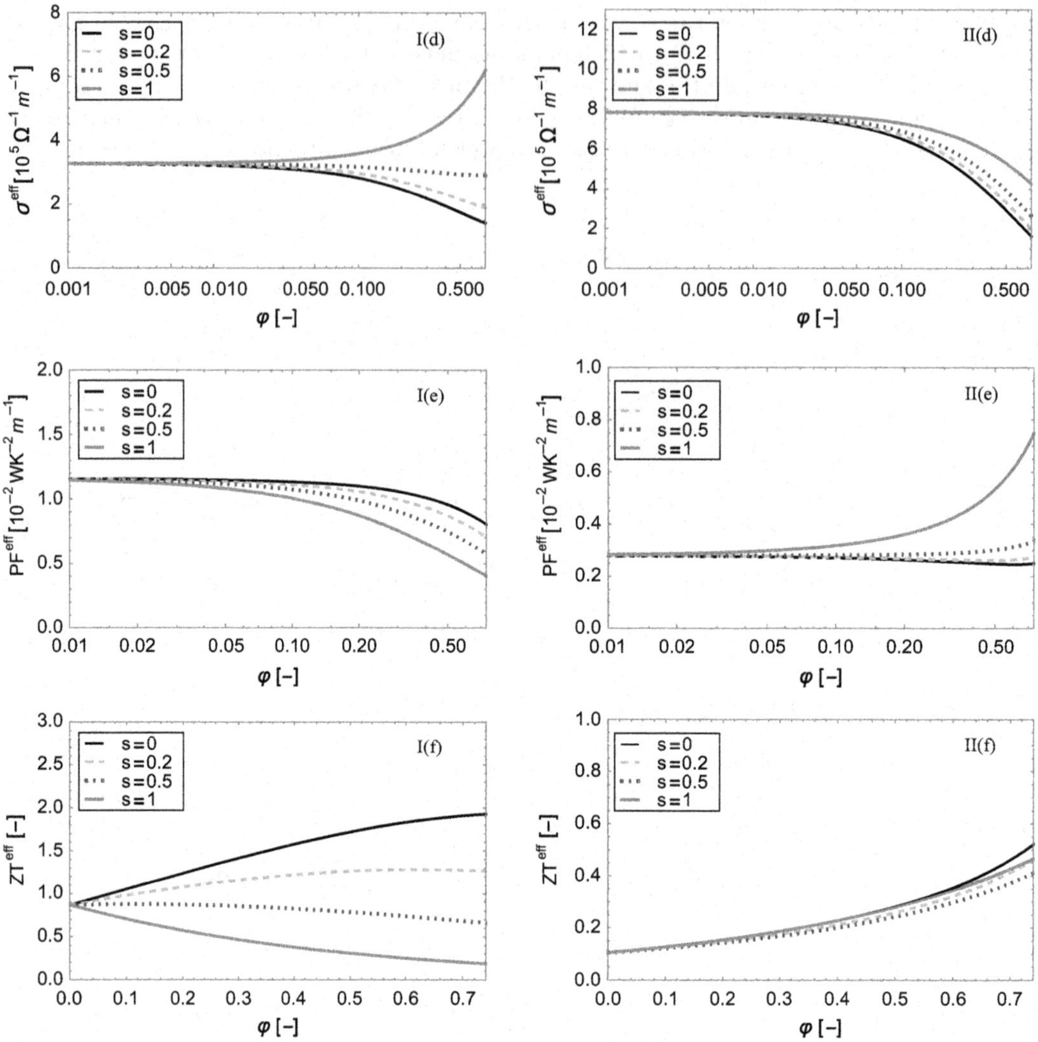

FIGURE 5.2 (CONTINUED)
Dependence on the volume fraction φ of, respectively, 1-nm Bi nanoparticles in a matrix of Bi_2Te_3 (left, indicated by I) and 1-nm Bi_2Te_3 nanoparticles in a matrix of Bi (right, indicated by II) of (a) the phonon, electric and total effective thermal conductivity (λ_{ph}^{eff}, λ_e^{eff} and λ_{tot}^{eff}, respectively) for $s = 0$, and (b) the total effective thermal conductivity λ_{tot}^{eff}, (c) the effective Seebeck coefficient S^{eff}, (d) the effective electric conductivity σ^{eff}, (e) the effective power factor PF^{eff} and (f) the dimensionless effective figure of merit ZT^{eff}, for $s = 0,\ 0.2,\ 0.5$ and 1. (Modified from: Machrafi, H. 2016. An extended thermodynamic model for size-dependent thermoelectric properties at nanometric scales: Application to nanofilms, nanocomposites and thin nanocomposite films. *Applied Mathematical Modelling* 40:2143–2160.)

in the figure of merit [see Figure 5.2(If,IIf)] than was the case for the nanofilm in Section 5.3.2, but increase nonetheless.

Now, comparing the two nanocomposites, we observe that the tendencies of all the parameters behave in the same way as a function of the volume fraction of nanoparticles. However, the dependency on the s-parameter is not the same at all. As far as goes the Bi nanoparticles in Bi_2Te_3, we obtain expected results. A higher s-value results into a smoother surface, causing less scattering. It also can be interpreted by a higher particle radius [see Eqs. (5.11a) and (5.11c)], which at a given volume fraction decreases the particle–matrix interface, creating less obstacles for the phonons and electrons to transfer heat, increasing the thermal conductivity. However, in the case we have Bi_2Te_3 nanoparticles in a Bi matrix, we obtain surprising results. Of course, the lesser scattering effect of the nanoparticles and the particle–matrix interface play an important role, but there seems to be another effect that counteracts it. Figure 5.2(IIb) shows that, even though for $s \to 1$, the effective thermal conductivity increases, this increase is less pronounced compared to that in Figure 5.2(Ib). The same goes for the effective thermal conductivity shown in Figure 5.2(Id,IId), respectively. Moreover, being the most pronounced difference, the effective Seebeck coefficient at $s \to 1$ decreases in Figure 5.2(Ic) and increases in Figure 5.2(IIc). Table 5.1 shows that bulk p-type Bi_2Te_3 has a much higher Seebeck coefficient than Bi. We have already said earlier that at $s \to 1$, bulk material values become important. Therefore, as the volume fraction increases at $s \to 1$, the effective Seebeck coefficient increases as well. In the fictive limit of $\varphi \to 1$, which is not possible here, since we assume that the nanoparticles are undeformable spheres that do not merge with other ones (at maximum packing $\varphi_{max} = \pi/\sqrt{18}$), we obtain the bulk value for Bi_2Te_3, whereas at $\varphi = 0$, we obtain that of Bi. The sum of all these effects gives eventually the result that the figure of merit in Figure 5.2(IIf) is hardly influence by the s-value.

We have seen that the s-parameter can influence greatly the thermoelectric properties, and we said that one of the interpretations is higher nanoparticle radii for higher s-values. The nanoparticle radius used for the results in Figure 5.2 is 1 nm. We are therefore interested to assess the influence of the nanoparticle radius on the figure of merit. For this purpose, Figure 5.3 presents the dimensionless effective figure of merit as a function of the inverse phonon Knudsen number $Kn_{ph}^{-1} \sim a_p$ at $s = 0$ and for volume fractions $\varphi = 0,\ 0.2,\ 0.4$ and 0.7. Note that doing the same as a function of the inverse electron Knudsen number mounts to the same tendency since $Kn_e = Kn_{ph}\ell_e/\ell_{ph}$. It should be mentioned that, regarding the values of the mean free paths in Table 5.2, the phonon Knudsen number cannot realistically be of order 10 or larger (the nanoparticle radii should be significantly larger

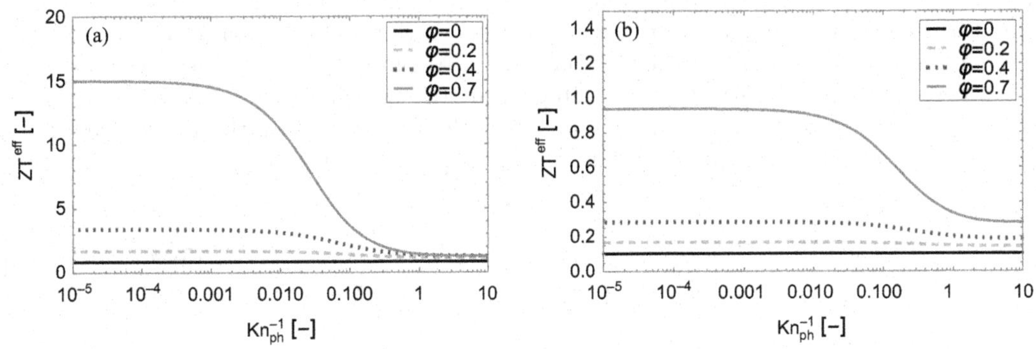

FIGURE 5.3
Dependence on the inverse phonon Knudsen number Kn_{ph}^{-1}, for several volume fractions, of the effective dimensionless figure of merit ZT^{eff}, for (a) Bi nanoparticles in a matrix of Bi_2Te_3 and (b) Bi_2Te_3 nanoparticles in a matrix of Bi. (From: Machrafi, H. 2016. An extended thermodynamic model for size-dependent thermoelectric properties at nanometric scales: Application to nanofilms, nanocomposites and thin nanocomposite films. *Applied Mathematical Modelling* 40:2143–2160.)

than the atom sizes). This means that in real-case situations and for the materials in this work, $Kn_{ph} \ll 10$, or $Kn_{ph}^{-1} \gg 0.1$. Nonetheless, we would like to perform a dimensionless analysis and investigate the behaviour of the figure of merit with respect to nanoparticle size. Therefore, we used a much larger range of Kn_{ph}^{-1}, so that we capture all the phenomena for mathematical interest. This point is related to the results of both Figure 5.1(IIf) ($ZT_{\mathcal{N}} = 8$) and Figure 5.3a ($ZT^{\text{eff}} = 15$ for $\varphi = 0.7$). As such, the results will indeed be qualitatively valid for other materials, which have much higher mean free paths.

Figure 5.3 shows that the nanoparticle size greatly influences the figure of merit (and thus the thermoelectric properties). The value at $\varphi = 0$ is obviously the figure of merit value at $\delta_{\mathcal{N}} \to \infty$ or $s \to 1$ shown in Figure 5.1(If,IIf). We can see that for increasing nanoparticle radius (decreasing Kn_{ph}), the figure of merit decreases and for decreasing nanoparticle radius (increasing Kn_{ph}), the figure of merit increases. This shows clearly the positive effect of nanocomposites on the figure of merit with respect to nanofilms for very large Knudsen numbers. It should be noted that the value of the figure of merit is higher in case of Figure 5.3a than in case of Figure 5.3b. This is easily understood by noticing that Figure 5.2f shows that Bi nanoparticles in a matrix of Bi_2Te_3 show already a higher figure of merit than Bi_2Te_3 nanoparticles in a matrix of Bi. For an even smaller nanoparticle (Kn_{ph}^{-1} decreases), this difference is more accentuated. Finally, it appears that with nanofilms, one may obtain even higher figure of merits as described in the previous section. However, from a practical point of view, it can give problems in the robustness or production of the nanofilms if the thickness is too small. One solution could be to use a combination of nanofilms and nanocomposites. This could increase the thermoelectric performances of nanocomposites without decreasing too much the thickness of the nanocomposite material.

5.5 Thin-Film Nanocomposite Thermoelectric Devices

5.5.1 Theory

In this section, we extend the model from Sections 5.3.1 and 5.4.1 to thermoelectric systems in the form of nanofilms composed out of nanocomposites (of course, the film thickness should be larger than the particle diameters). The description of the conductivities and the components of the figure of merit for the nanoparticles will remain the same as the previous. The differences are the parts that describe the matrix which is now also at nanoscale in contrast to the previous section. The new nanoscale matrix phonon thermal conductivity, $\lambda_{ph,m}^{\mathcal{N}}$, will be given by Eq. (5.12), where λ_{ph}^{0} should now be given by Eq. (5.25):

$$\lambda_{ph,m}^{\mathcal{N}} = \frac{3\lambda_{ph,m}}{4\pi^2 Kn_{ph}^{\mathcal{N}2}}\left[\frac{2\pi Kn_{ph}^{\mathcal{N}}}{\arctan\left(2\pi Kn_{ph}^{\mathcal{N}}\right)} - 1\right] = \frac{C_{ph,m}v_{ph,m}\Lambda_{ph,m}}{4\pi^2 Kn_{ph}^{\mathcal{N}2}}\left[\frac{2\pi Kn_{ph}^{\mathcal{N}}}{\arctan\left(2\pi Kn_{ph}^{\mathcal{N}}\right)} - 1\right] \tag{5.48}$$

where $Kn_{ph}^{\mathcal{N}} = \ell_{ph,b,m}/\delta_{m,s}^{\mathcal{N}}$ is the phonon Knudsen number of the nanoscale matrix, where $\delta_{m,s}^{\mathcal{N}} = \delta_{m}^{\mathcal{N}}(1+s)/(1-s)$, with $\delta_{m}^{\mathcal{N}}$ the film thickness of the nanocomposite. The nanoscale effective phonon thermal conductivity, $\lambda_{ph}^{\mathcal{N},\text{eff}}$, is then, in analogy with Eq. (5.28), given by

$$\lambda_{ph}^{\mathcal{N},\text{eff}} = \lambda_{ph,m}^{\mathcal{N}} \frac{2\lambda_{ph,m}^{\mathcal{N}} + \left(1+2\alpha_{ph}^{\mathcal{N}}\right)\lambda_{ph,\mathcal{N}} + 2\varphi\left[\left(1-\alpha_{ph}^{\mathcal{N}}\right)\lambda_{ph,\mathcal{N}} - \lambda_{ph,m}^{\mathcal{N}}\right]}{2\lambda_{ph,m}^{\mathcal{N}} + \left(1+2\alpha_{ph}^{\mathcal{N}}\right)\lambda_{ph,\mathcal{N}} - \varphi\left[\left(1-\alpha_{ph}^{\mathcal{N}}\right)\lambda_{ph,\mathcal{N}} - \lambda_{ph,m}^{\mathcal{N}}\right]}, \tag{5.49}$$

with $\lambda_{ph,\mathcal{N}}$ given by Eq. (5.12). In expression (5.49), $\alpha_{ph}^{\mathcal{N}}$ is given by Eq. (5.29), but replacing $\lambda_{ph,m}$ by $\lambda_{ph,m}^{\mathcal{N}}$. In analogy with Eq. (5.49), the electron contribution of the thermal conductivity is given by

$$\lambda_e^{\mathcal{N},\mathit{eff}} = \lambda_{e,m}^{\mathcal{N}} \frac{2\lambda_{e,m}^{\mathcal{N}} + \left(1 + 2\alpha_e^{\mathcal{N}}\right)\lambda_{e,\mathcal{N}} + 2\varphi\left[\left(1 - \alpha_e^{\mathcal{N}}\right)\lambda_{e,\mathcal{N}} - \lambda_{e,m}^{\mathcal{N}}\right]}{2\lambda_{e,m}^{\mathcal{N}} + \left(1 + 2\alpha_e^{\mathcal{N}}\right)\lambda_{e,\mathcal{N}} - \varphi\left[\left(1 - \alpha_e^{\mathcal{N}}\right)\lambda_{e,\mathcal{N}} - \lambda_{e,m}^{\mathcal{N}}\right]}, \tag{5.50}$$

with $\lambda_{e,\mathcal{N}}$ given by Eq. (5.13). In expression (5.50), $\alpha_e^{\mathcal{N}}$ is obtained from Eq. (5.29), by replacing R_{ph} by $R_e = 4/c_{e,m}v_{e,m} + 4/c_e v_e$ and $\lambda_{e,m}$ by $\lambda_{e,m}^{\mathcal{N}}$, and the latter is given, in analogy with Eq. (5.48), by

$$\lambda_{e,m}^{\mathcal{N}} = \frac{c_{e,m}v_{e,m}\ell_{e,m}}{4\pi^2 Kn_e^{\mathcal{N}2}}\left[\frac{2\pi Kn_e^{\mathcal{N}}}{\arctan\left(2\pi Kn_e^{\mathcal{N}}\right)} - 1\right], \tag{5.51}$$

with $Kn_e^{\mathcal{N}} = \ell_{e,b,m}/\delta_{m,s}^{\mathcal{N}}$ the electron Knudsen number of the nanoscale matrix, where $\delta_{m,s}^{\mathcal{N}}$ is here again the speculate film thickness. The total nanoscale effective thermal conductivity is given by

$$\lambda_{tot}^{\mathcal{N},\mathit{eff}} = \lambda_{ph}^{\mathcal{N},\mathit{eff}} + \lambda_e^{\mathcal{N},\mathit{eff}}, \tag{5.52}$$

whereas the nanoscale effective electric conductivity is expressed as follows [in analogy with Eq. (5.35)]:

$$\sigma^{\mathcal{N},\mathit{eff}} = \frac{\lambda_e^{\mathcal{N},\mathit{eff}}}{L_e T}. \tag{5.53}$$

As for the Seebeck coefficient, we have in analogy with Eq. (5.44)

$$(S\sigma)^{\mathcal{N},\mathit{eff}} = (S\sigma)_m^{\mathcal{N}} \frac{2(S\sigma)_m^{\mathcal{N}} + \left(1 + 2\alpha_e^{\mathcal{N}}\right)(S\sigma)_{\mathcal{N}} + 2\varphi\left[\left(1 - \alpha_e^{\mathcal{N}}\right)(S\sigma)_{\mathcal{N}} - (S\sigma)_m^{\mathcal{N}}\right]}{2(S\sigma)_m^{\mathcal{N}} + \left(1 + 2\alpha_e^{\mathcal{N}}\right)(S\sigma)_{\mathcal{N}} - \varphi\left[\left(1 - \alpha_e^{\mathcal{N}}\right)(S\sigma)_{\mathcal{N}} - (S\sigma)_m^{\mathcal{N}}\right]}, \tag{5.54}$$

with $(S\sigma)_{\mathcal{N}}$ given by Eq. (5.42), $\alpha_e^{\mathcal{N}}$ defined under Eq. (5.50) and $(S\sigma)_m^{\mathcal{N}}$ derived from $(S\sigma)_m$ [see Eq. (5.43)] by replacing $\lambda_{e,m}$ by $\lambda_{e,m}^{\mathcal{N}}$ [see Eq. (5.51)] and S_m by $S_m^{\mathcal{N}}$. Note that $(S\sigma)_m^{\mathcal{N}} \equiv S_m^{\mathcal{N}}\sigma_m^{\mathcal{N}}$, with $\sigma_m^{\mathcal{N}}$ and $S_m^{\mathcal{N}}$ the nanoscale electric conductivity and Seebeck coefficient of the matrix, which are given by, respectively,

$$\sigma_m^{\mathcal{N}} = \frac{\lambda_{e,m}^{\mathcal{N}}}{L_e T}. \tag{5.55}$$

$$S_m^{\mathcal{N}} = \frac{n_{n,m}\mu_n S_{n,m}^{*,\mathcal{N}} + n_{p,m}\mu_p S_{p,m}^{*,\mathcal{N}}}{n_{n,m}\mu_n + n_{p,m}\mu_p}, \tag{5.56}$$

where

$$S_{x,y}^{*,\mathcal{N}} = (-1)^z \frac{8\pi^2 k_B^2}{3e_c h^2} m_x^* T \left(\frac{\pi}{3n_{x,y}^{\mathcal{N}}}\right)^{2/3}, \tag{5.57}$$

with $n_{x,y}^{\mathcal{N}}$ defined [in analogy with Eqs. (5.41)]. In the case of the nanoparticles $n_{x,y}^{\mathcal{N}} = n_{x,\mathcal{N}}$ [Eq. (5.41a)]. In the case of the nanoscale matrix $n_{x,y}^{\mathcal{N}} = n_{x,m}^{\mathcal{N}}$ [keeping in mind the definitions of x, y and z and the explanation of using Eq. (5.56), all to be consulted in the text under Eq. (5.37)] given by

$$n_{x,m}^{\mathcal{N}} = n_{x,0}\left(\frac{\sigma_m^{\mathcal{N}}}{\sigma_m^0}\right)^{3/2} = n_{x,0}\left(\frac{\lambda_{e,m}^{\mathcal{N}}}{\lambda_{e,m}^0}\right)^{3/2}. \tag{5.58}$$

The effective nanoscale Seebeck coefficient $S^{\mathcal{N},\text{eff}}$ and the effective nanoscale power factor $PF^{\mathcal{N},\text{eff}}$ can then be easily obtained a posteriori, by defining

$$S^{\mathcal{N},\text{eff}} = (S\sigma)^{\mathcal{N},\text{eff}} / \sigma^{\mathcal{N},\text{eff}}, \tag{5.59}$$

$$PF^{\mathcal{N},\text{eff}} = S^{\mathcal{N},\text{eff}^2} \sigma^{\mathcal{N},\text{eff}} = (S\sigma)^{\mathcal{N},\text{eff}^2} / \sigma^{\mathcal{N},\text{eff}}. \tag{5.60}$$

Finally, the nanoscale effective non-dimensional figure of merit for nanofilms composed out of nanocomposites is given by

$$ZT^{\mathcal{N},\text{eff}} = T\frac{\left(S^{\mathcal{N},\text{eff}}\right)^2 \sigma^{\mathcal{N},\text{eff}}}{\lambda_{ph}^{\mathcal{N},\text{eff}} + \lambda_e^{\mathcal{N},\text{eff}}} = T\frac{(S\sigma)^{\mathcal{N},\text{eff}^2} / \sigma^{\mathcal{N},\text{eff}}}{\lambda_{ph}^{\mathcal{N},\text{eff}} + \lambda_e^{\mathcal{N},\text{eff}}} = L_e T^2 \frac{(S\sigma)^{\mathcal{N},\text{eff}^2}}{\lambda_e^{\mathcal{N},\text{eff}} \left(\lambda_{ph}^{\mathcal{N},\text{eff}} + \lambda_e^{\mathcal{N},\text{eff}}\right)}. \tag{5.61}$$

5.5.2 Discussion on a Gedankenexperiment

Figure 5.4 shows the effect of the film thickness of the nanocomposites considered in the previous section. Here, the nanoparticles have a radius of 1 nm, $s = 0$, and the results are presented for volume fractions $\varphi = 0,\ 0.2,\ 0.4$ and 0.7.

From the results of Figure 5.4, we can see indeed that decreasing the film thickness of the nanocomposites, we can improve furthermore the figure of merit. The improvement is not large but still significant. This weak improvement can be understood due to the minimum film thickness considered, i.e. $\delta_{m,min}^{\mathcal{N}} = 5$ nm, which is larger than both the phonon and electron mean free paths (though still in the same order of magnitude). Therefore, the Knudsen number is smaller than one. As shown in Figure 5.3, this leads to poorer thermoelectric properties. We did not consider smaller thicknesses, since the nanoparticle size is 2 nm (radius of 1 nm). We consider that the thin film should be at least twice the particle size (four times the radius), so that the thin film is still considered as a composite. Let us imagine a smaller particle. We consider here only the case of a Bi nanoparticle embedded in a Bi_2Te_3 host matrix, as an example. A Bi atom has a van der Waals radius of approximately 0.2 nm. Let us consider, for the purposes of this Gedankenexperiment,

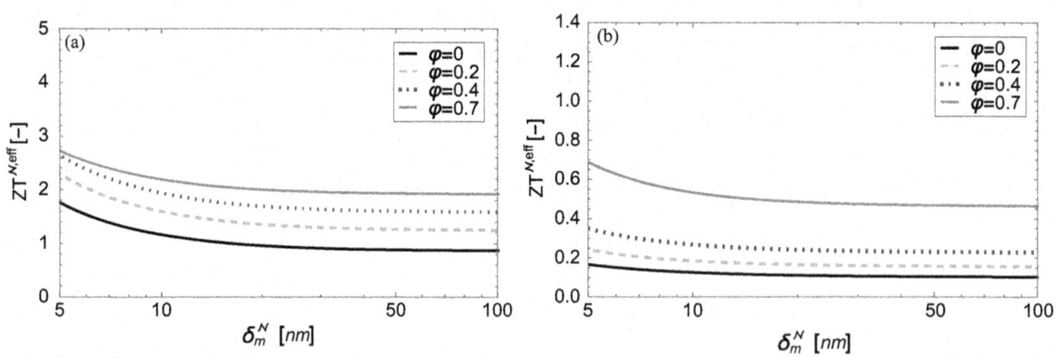

FIGURE 5.4
Dependence of the dimensionless effective figure of merit $ZT^{\mathcal{N},\text{eff}}$ on the nanocomposite film thickness $\delta_m^{\mathcal{N}}$ of, respectively, (a) 1-nm Bi nanoparticles in a matrix of Bi_2Te_3 and (b) 1-nm Bi_2Te_3 nanoparticles in a matrix of Bi , for volume fractions $\varphi = 0,\ 0.2,\ 0.4$ and 0.7. (From: Machrafi, H. 2016. An extended thermodynamic model for size-dependent thermoelectric properties at nanometric scales: Application to nanofilms, nanocomposites and thin nanocomposite films. *Applied Mathematical Modelling*, 40:2143–2160.)

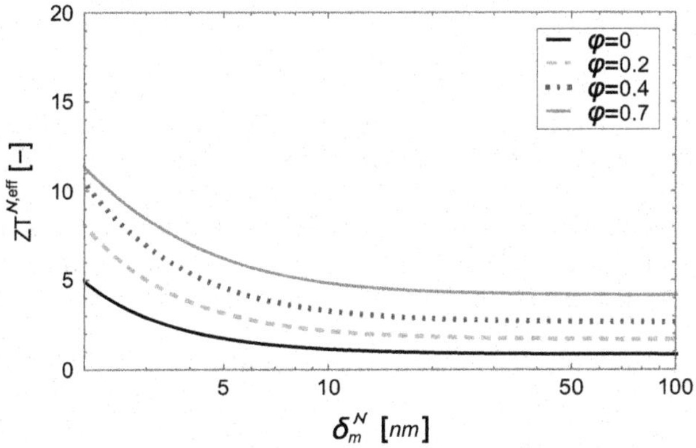

FIGURE 5.5
The dimensionless figure of merit $ZT^{\mathcal{N},\text{eff}}$ of thin films of 0.5-nm Bi nanoparticles embedded in a Bi_2Te_3 host matrix as a function of the nanocomposite film thickness $\delta_m^{\mathcal{N}}$, for volume fractions $\varphi = 0$, 0.2, 0.4 and 0.7. (From: Machrafi, H. 2016. An extended thermodynamic model for size-dependent thermoelectric properties at nanometric scales: Application to nanofilms, nanocomposites and thin nanocomposite films. *Applied Mathematical Modelling* 40:2143–2160.)

a Bi nanoparticle of radius 0.5 nm (smaller than both the phonon and electron mean free paths) within a host matrix Bi_2Te_3 of minimum thickness $\delta_{m,min}^{\mathcal{N}} = 2$ nm (smaller than the phonon mean free paths). We perform with these new settings the same calculations as shown in Figure 5.4a and present the results in Figure 5.5.

First of all, we can see high values of the figure of merit as high as $ZT^{\mathcal{N},\text{eff}} = 10$. As was mentioned in relation to Figures 5.1(IIf) and 5.3a, this value is rather high. However, for mathematical purposes, it is interesting to make such an extrapolation, which allows appreciating the possibility of the configuration proposed in this section. Indeed, Figure 5.5 shows that a much higher figure of merit can be obtained when combining the principle nanofilms and nanocomposites. Generally, this work has presented that in many ways [36] (nanofilms, nanocomposites or a combination thereof), thermoelectric properties tend to improve considerably.

References

[1] Baird, J.R., Fletcher, D.F., Haynes, B.S. 2003. Local condensation heat transfer rates in fine passages. *International Journal of Heat and Mass Transfer* 46:4453–4466.

[2] Lebon, J., Jou, D., Casas-Vázquez, J. 2008. *Understanding Non-equilibrium Thermodynamics*, first ed. Heidelberg: Springer-Verlag.

[3] He, W., Zhang, G., Zhang, X., Ji, J., Li, G., Zhao, X. 2015. Recent development and application of thermoelectric generator and cooler. *Applied Energy* 143:1–25.

[4] Pattamatta, A., Madnia, C.K. 2009. Modeling heat transfer in Bi_2Te_3-Sb_2Te_3 nanostructures. *International Journal of Heat and Mass Transfer* 52:860–869.

[5] Ezzat, M.A. 2011. Theory of fractional order in generalized thermoelectric MHD. *Applied Mathematical Modelling* 35:4965–4978.

[6] Cornett, J.E., Rabin, O. 2011. Thermoelectric figure of merit calculations for semiconducting nanowires. *Applied Physics Letters* 98:82184.

[7] Sellitto, A., Cimmelli, V.A., Jou, D. 2013. Thermoelectric effects and size dependency of the figure-of-merit in cylindrical nanowires. *International Journal of Heat and Mass Transfer* 57:109–116.

[8] Moreira, D.C., Sphaier, L.A., Reis, J.M.L., Nunes, L.C.S. 2011. Experimental investigation of heat conduction in polyester–Al_2O_3 and Polyester–CuO nanocomposites. *Experimental Thermal Fluid Science* 35:1458–1462.

[9] Chen, J., Li, J.L., Tao, Z.L., Zhang, L.Z. 2002. Reversible hydrogen and lithium storage of MoS_2 nanotubes. *International Journal Nanoscience* 1:295–302.

[10] Sun, T., Samani, M.K., Khosravian, N., Ang, K.M., Yan, Q., Tay, B.K., Hng, H.H. 2014. Enhanced thermoelectric properties of n-type $Bi_2Te_{2.7}Se_{0.3}$ thin films through the introduction of Pt nano inclusions by pulsed laser deposition. *Nano Energy* 8:223–230.

[11] Liu, W., Yan, X., Chen, G., Ren, Z. 2012. Recent advances in thermoelectric nanocomposites. *Nano Energy* 1:42–56.

[12] Alama, H., Ramakrishna, S. 2013. A review on the enhancement of figure of merit from bulk to nano-thermoelectric materials. *Nano Energy* 2:190–212.

[13] Lynch, M.E., Ding, D., Harris, W.M., Lombardo, J.J., Nelson, G.J., Chiu, W.K.S., Liu, M. 2013. Flexible multiphysics simulation of porous electrodes: Conformal to 3D reconstructed microstructures. *Nano Energy* 2:105–115.

[14] Choudhary, K.K., Prasad, D., Jayakumar, K., Varshney, D. 2009. Effect of embedding nanoparticles on thermal conductivity of crystalline semiconductors: Phonon scattering mechanism. *International Journal Nanoscience* 8:551–556.

[15] Jou, D., Casas-Vazquez, J., Lebon, G. 2010. *Extended Irreversible Thermodynamics*, fourth ed. Berlin: Springer-Verlag.

[16] Lebon, G., Machrafi, H., Grmela, M., Dubois, C. 2011. An extended thermodynamic model of transient heat conduction at sub-continuum scales. *Proceedings of the Royal Society A* 467:3241–3256.

[17] Cattaneo, C. 1948. Sulla conduzione del calore. Atti del Seminario Matematico e Fisico delle Università di Modena 3:83–101.

[18] Satyala, N., Rad, A.T., Zamanipour, Z., Norouzzadeh, P., Krasinski, J.S., Tayebi, L., Vashaee, D. 2014. Reduction of thermal conductivity of bulk nanostructured bismuth telluride composites embedded with silicon nano-inclusions. *Journal of Applied Physics* 115:044304.

[19] Satyala, N., Rad, A.T., Zamanipour, Z., Norouzzadeh, P., Krasinski, J.S., Tayebi, L., Vashaee, D. 2014. Influence of germanium nano-inclusions on the thermoelectric power factor of bulk bismuth telluride alloy. *Journal of Applied Physics* 115:204308.

[20] Gothard, N.W. 2008. The effects of nanoparticle inclusions upon the microstructure and thermoelectric transport properties of Bismuth Telluride-based composites. *PhD dissertation*, Clemson University, pp. 118–119.

[21] Cimpoiasu, E., Stern, E., Cheng, G., Munden, R., Sanders, A., Reed, M.A. 2006. Electron mobility study of hot-wall CVD GaN and InN nanowires. *Brazilian Journal of Physics* 36(3b):824–827.

[22] Foell, H. 2014. Semiconductors I, Chapter 2, lecture notes, University of Kiel.

[23] Goldsmid, H.J. 2006. Bismuth – The Thermoelectric Material of the Future? *25th International Conference on Thermoelectrics*, pp. 5–10, Vienna, Austria.

[24] Goldsmid, H.J. 2014. Bismuth telluride and its alloys as materials for thermoelectric generation. *Materials* 7:2577–2592.

[25] Cho, S., Kim, Y., Freeman, A.J., Wong, G.K.L., Ketterson, J.B., Olafsen, L.J., Vurgaftman, I., Meyer, J.R., Hoffman, C.A. 2001. Large magnetoresistance in postannealed Bi thin films. *Applied Physics Letters* 79:3651.

[26] Jin, J.S. 2014. Prediction of phonon and electron contributions to thermal conduction in doped silicon film. *Journal of Mechanical Science and Technology* 28:2287–2292.

[27] Madelung, O., Rössler, U., Schulz, M. 1998. *Bismuth (Bi)* Effective M*asses – Non-Tetrahedrally Bonded Elements and Binary Compounds I.* Berlin: Springer, pp. 1–11.

[28] Gamaly, E.G., Rode, A.V. 2009. Is the ultra-fast transformation of bismuth non-thermal? arXiv:0910.2150.

[29] Gamaly, E.G., Rode, A.V. 2013. Ultrafast electronic relaxation in superheated bismuth. *New Journal of Physics* 15:013035.

[30] Pélissier, J.L., Wetta, N. 2001. A model-potential approach for bismuth (I). Densification and melting curve calculation. *Physica A* 289:459–478.

[31] Issi, J.P., Luyckx, A. 1966. Size effect in the thermopower of Bismuth. *Physics Letters* 23:13–14.

[32] Jou, D., Sellitto, A., Cimmelli, V.A. 2014. Multi-temperature mixture of phonons and electrons and nonlocal thermoelectric transport in thin layers. *International Journal of Heat and Mass Transfer* 71:459–468.

[33] Satterthwaete, C.B., Ure jr., R.W. 1957. Electrical and thermal properties of Bi_2Te_3. *Physical Review* 108:1164–1170.

[34] Nan, C.W., Birringer, R., Clarke, D.R., Gleiter, H. 1997. Effective thermal conductivity of particulate composites with interfacial thermal resistance. *Journal of Applied Physics* 81:6692–6699.

[35] Ordonez-Miranda, J., Yang, R., Alvarado-Gil, J.J. 2011. On the thermal conductivity of particulate nanocomposites. *Applied Physics Letters* 98:233111.

[36] Machrafi, H. 2016. An extended thermodynamic model for size-dependent thermoelectric properties at nanometric scales: Application to nanofilms, nanocomposites and thin nanocomposite films. *Applied Mathematical Modelling* 40: 2143–2160.

6

Enhancement of the Thermal Conductivity in Nanofluids and the Role of Viscosity

6.1 Context

It has been shown that thermophysical properties and, in particular, heat transfer coefficients of base fluids can be modified significantly by introducing nanoparticles [1,2]. It is well known that thermal conductivity of some nanofluids and nanocomposites is considerably enhanced by adding nanoparticles. It is also admitted that the presence of nanoparticles influences considerably the thermomechanical properties of the basic fluid-like, in particular, thermal conductivity and viscosity. A huge amount of theoretical (e.g. [3,4]) and experimental [5] works have been devoted to the subject. Recently, there has been much attention paid to some particular heat transfer mechanisms in nanofluids, such as formation of a nano-liquid layer around the nanoparticles (liquid layering) [6–9], particle clustering [10–13] and Brownian motion of particles [14–16]. Contradictory conclusions have been formulated about the relative importance of the aforementioned mechanisms [1]. In many applications that seek to enhance the thermal conductivity, the nanofluids are in movement in order to serve a purpose. Such systems have met an increasing interest in several industrial applications, such as in biotechnology, nanotechnology and electromechanical systems, they have proved to be relevant in the developments of new drugs, paints and lubricants among others. In these fields, very often, the displacement of the nanofluids needs knowledge about the viscosity. We can consider nanofluids as binary mixtures, consisting of nanoparticles dispersed in a host fluid. The focus is put on the role of nanoparticles on the shear viscosity of the system, an impressive lot of works (e.g. [17–25]) have been published on the subject. It has been observed that the viscosity of nanofluids is much larger that that of the host fluid. The viscosity depends essentially on the temperature, the nature of the particles and the fluid, the volume fraction of particles and their size. The dependence of viscosity on the size of nanoparticles has been a subject of debate. For some authors [26,27], viscosity increases with increasing dimensions of nanoparticles while others [29–30] assert that viscosity diminishes with the increasing size. For sufficiently large particles, the dependence with respect to size becomes negligible. It is however worth to mention that no valuable theoretical considerations, outside molecular dynamic simulations [31], are able to explain such behaviours.

6.2 Influence of Several Heat Transfer Mechanisms

6.2.1 Hypotheses

The objective is to derive a closed-form expression for the effective thermal conductivity of nanofluids in the presence of various transport mechanisms. Thermal conductivity depends on several factors such as the nanoparticles' volume fraction, the thermal boundary resistance between fluid and particles and their shape and size. Here, for simplicity, the analysis is restricted to spherical rigid nanoparticles of the same dimension dispersed in a homogeneous fluid. The other working hypotheses are the following:

- Particles are homogeneously dispersed
- Temperature is fixed equal to the room temperature
- There is no electric surface charge
- Surfactants are absent.

Let us briefly comment about these hypotheses. The distribution of the dispersed particles is assumed to be uniform, i.e. independent of the spatial coordinates. Although it is true that the concentration of particles is able to modify the thermal conductivity, other factors such as the ones investigated in this work play a more important role. At low volume fraction of particles and weak concentration gradients, the influence of the spatial distribution of particles on heat transfer is negligible due to the smallness of the coupling coefficients of the Soret and Dufour effects.

In the following sections, the influence of several heat transfer mechanisms, with the aim to determine their relative importance, is investigated. The effects considered in the forthcoming are successively those caused by the presence of a liquid layer around the nanoparticles (liquid layering), particle agglomeration and Brownian motion of particles.

6.2.2 Liquid Layering

Because of the strong interatomic forces at the particle–fluid interface, some of the fluid molecules will attach themselves to the surface of the nanoparticle, forming an interfacial layer of thickness d with a higher density than the fluid and having the property approaching that of the solid phase of the fluid [32]. As this "solid" surrounding layer has its own thermal conductivity, it is expected that it will influence the heat transfer mechanism. A possible way to take into account the liquid-layer effect is to modify relation (3.1), so that we are led to [6–9]

$$\lambda_{ll}^{eff} = \lambda_f \frac{2\lambda_f + (1+2\alpha_l)\,\lambda_{pl} + 2\varphi\,(1+b)^3\left[(1-\alpha_l)\,\lambda_{pl} - \lambda_f\right]}{2\lambda_f + (1+2\alpha_l)\,\lambda_{pl} - \varphi\,(1+b)^3\left[(1-\alpha_l)\,\lambda_{pl} - \lambda_f\right]}, \tag{6.1}$$

where index "ll" stands for "liquid layer" and λ_{pl} for the thermal conductivity of the nanoparticle with the surrounding layer. The presence of the liquid layer of thickness d will modify the radius of the particle from a_p to $a_p + d$ and the volume fraction from φ to $\varphi(1+b)^3$ with $b = \frac{d}{a_p}$, the ratio of the liquid layer thickness and the nanoparticle radius [6] ($b \to 0$ means that the liquid-layer effect is not taken into account). Finally, the equivalent thermal conductivity of the heterogeneous system formed by the nanoparticle and the liquid layer is given by:

$$\lambda_{pl} = \lambda_p \frac{\gamma\left(2\,(1-\gamma) + (1+b)^3\,(1+2\gamma)\right)}{(\gamma-1) + (1+b)^3\,(1+2\gamma)}, \tag{6.2}$$

where $\gamma = \frac{\lambda_l}{\lambda_p}$, is the ratio of the thermal conductivities of the liquid layer (λ_l) and the nanoparticle λ_p [given by Eq. (3.14)]:

$$\lambda_p = \frac{3\lambda_p^0}{4\pi^2 Kn^2}\left[\frac{2\pi Kn}{\arctan\,(2\pi Kn)} - 1\right], \tag{6.3}$$

It is expected that the solid-like liquid layer has a thermal conductivity intermediate between that of the bulk liquid and that of the nanoparticle. Xie et al. [9] propose

$$\lambda_l = \frac{\lambda_f M^2}{(M - \ell)\ln(1 + M) + \ell M}, \tag{6.4}$$

where $M = (\lambda_p/\lambda_f)(1 + \ell) - 1$. It remains to determine the liquid layer thickness d around the nanoparticle. Based on the electron density profile at the interface, Hashimoto et al. [33] derived the interfacial layer thickness at the surface of a spherical micro-domain. Later, Li et al. [34] introduced the same model for determining the interfacial layer thickness of a pseudo solid–liquid system. In both works, $d = \sqrt{2\pi}\varsigma$, where ς is a characteristic length related to the diffuseness of the interfacial boundary, with a typical value that falls in the range 0.4–0.6 nm, from which follows that d is of the order of 1–1.5 nm. In addition, experimental results [35] and molecular dynamics' simulations [36] showed that the typical interfacial layer thickness between the solid (nanoparticles) and liquid phase is of the order of a few atomic distances, namely, 1–2 nm. Murshed et al. [37] also reported that the variation of the interfacial layer thickness (they considered 1–3 nm) does not significantly affect the thermal conductivity enhancement. Accordingly, we expect that there is not much difference between 1 and 1.5 nm for the interfacial layer (this has been also confirmed by a posteriori calculations). For the sake of completeness, we have studied the impact of choosing a higher value, say $d = 3$ nm (the maximum value being generally considered [37]). As commented in Section 6.4.3, a modification of the value of d has little influence so that it is reasonable to use $d = 1$ nm in the calculations as widely accepted in the literature [33,34,37].

6.2.3 Agglomeration of Particles

We now allow some of the particles to form clusters supposed to be uniformly dispersed within the fluid with their distributions independent of the space coordinates. To account for the agglomeration of nanoparticles [38–41], we will change the radius a_p of the particles into $a_{p,a}$ (radius of the agglomerate) in the expressions of α_a and ℓ_a, which reads, respectively, $\alpha_a = \frac{Rk_f}{a_{p,a}+d}$ and $\ell_a = \frac{d}{a_{p,a}}$. In these expressions, we have introduced the thickness d of the liquid layer because agglomeration and liquid layering are linked to one another, indeed liquid layers may also be found around agglomerates. The quantity $\varphi_a = \varphi\left(\frac{a_{p,a}}{a_p}\right)^{3-D}$ is the volume fraction of the agglomerates, with D the fractal index, whose typical value lies between 1.6 and 2.5 for aggregates of spherical nanoparticles [40]. The value for D is often taken as 1.8, and since the thermal conductivity appears to be rather insensitive to the value of D [39–41], we will here work with $D = 1.8$. This leads finally to the following expression for the effective thermal conductivity $\lambda_{ll,a}^{eff}$ taking into account the agglomeration and the liquid-layer effects:

$$\lambda_{ll,a}^{eff} = \lambda_f \frac{2\lambda_f + (1 + 2\alpha_a)\lambda_{a,l} + 2\varphi_a(1 + \ell_a)^3[(1 - \alpha_a)\lambda_{a,l} - \lambda_f]}{2\lambda_f + (1 + 2\alpha_a)\lambda_{a,l} - \varphi_a(1 + \ell_a)^3[(1 - \alpha_a)\lambda_{a,l} - \lambda_f]}. \tag{6.5}$$

In Eq. (6.5), we have introduced the quantity $\lambda_{a,l}$, which is the thermal conductivity of an agglomerate formed by the cluster of particles surrounded by a liquid layer. We use $\lambda_{a,l}$ for the expression established by Hui et al. [42] and adapt it by introducing the liquid-layer effect in the same way as done in the previous subsection, resulting into

$$\lambda_{a,l} = \frac{1}{4}\left[3\varphi_s(\lambda_p - \lambda_f) + (2\lambda_f - \lambda_p) + \sqrt{8\lambda_f\lambda_p + (3\varphi_s(\lambda_f - \lambda_p) + (\lambda_p - 2\lambda_f))^2}\right], \tag{6.6}$$

where $\varphi_s = \frac{\varphi}{\varphi_a}$ is the ratio of the volumes occupied by the particles and the aggregates. By letting d tend to zero, one obtains the effective thermal conductivity accounting for agglomeration only, as $\lambda^{eff}_{ll,a}|_{d\,\to\,0} \equiv \lambda^{eff}_{a}$. With $\lambda^{eff}_{ll,a}|_{r_{p,a}\,\to\,r_p} \equiv \lambda^{eff}_{ll}$, one finds back the expression of the effective thermal conductivity accounting only for the liquid layer effect. Finally, we recover Eq. (3.1) by noticing that $\lambda^{eff}_{ll,a}|_{r_{p,a}\,\to\,r_p\ \&\ d\,\to\,0} \equiv \lambda^{eff}$.

6.2.4 Brownian Motion

In this subsection, we consider another mechanism, namely the Brownian motion of nanoparticles; this effect finds its origin in the stochastic bombardment of the liquid host molecules and results in micro convections at the nanoscale and hence, enhancement of thermal interactions between the nanoparticles and the ambient fluid [14–16,43]. The contribution to the effective thermal conductivity may be expressed as an additional term λ_{Br} given by Jang and Choi [16]:

$$\lambda_{Br} = \frac{h d_H \varphi}{Pr_f}, \tag{6.7}$$

where Pr_f stands for the Prandtl number of the base fluid

$$Pr_f = \frac{\mu_f c_f}{k_f} \tag{6.8}$$

with μ_f and c_f being the kinematic viscosity and the specific heat capacity of the fluid, respectively, while d_H is the so-called hydrodynamic boundary layer thickness. Several options have been proposed for d_H, for instance, Jang and Choi [16] argued that d_H is of the order of three times the diameter of the fluid molecules. In a later publication [44], the same authors assume that it is comparable to the size of the fluid molecules. Nonetheless, Yu et al. [35] and Prasher et al. [45] take d_H equal to three times the molecule diameters as well. We adopt the same position here, meaning that $d_H \equiv 0.9$ nm for water, $d_H \equiv 1.2$ nm for ethylene glycol and $d_H \equiv 1$ nm obtained by taking the molar average value for a 50/50 w% water/ethylene glycol mixture.

Furthermore, the factor h in relation (6.7) is an overall heat transfer coefficient at the surface of the particles given by Jang and Choi [16]

$$h = \frac{Nu_f\,\lambda_f}{2\,(a_p + d)}, \tag{6.9}$$

with Nu_f designating the Nusselt number, which for a flow past a sphere is [16,44]

$$Nu_f = 2 + \frac{1}{2} Re_f Pr_f, \tag{6.10}$$

Re_f is the Reynolds number, given by

$$Re_f = \frac{k_B T \rho_f}{3\pi \mu_f^2 d_H}, \tag{6.11}$$

where ρ_f is the base fluid density, k_B the Boltzmann constant and T the temperature.

When liquid layering and agglomeration are negligible, the effective thermal conductivity λ^{eff}_{Br} of the nanofluid may be written as

$$\lambda^{eff}_{Br} = \lambda^{eff} + \lambda_{Br}, \tag{6.12}$$

with λ^{eff} expressed by relation (3.1).

When all the mechanisms considered in this work are taken into account, namely, liquid layering, agglomeration and Brownian motion, the total effective thermal conductivity (λ^{eff}_{tot}) is written as

$$\lambda^{eff}_{tot} = \lambda^{eff}_{ll,a} + \lambda_{Br}, \tag{6.13}$$

with $\lambda_{ll,a}^{eff}$ given by Eq. (6.5). Note that λ_{tot}^{eff} is not just simply the sum of the various mechanisms, taken individually, but rather includes all the interacting couplings. The addition of relation (6.3) to the equations in Sections 6.2.2 to 6.2.4 gives closure to the problem of determining the effective thermal conductivity of the nanofluid.

6.3 Viscosity of Nanofluids

6.3.1 Viscous Pressure Flux

Several theoretical and ad hoc expressions for the viscosity μ in terms of the particle's volume fractions φ have been proposed, among which the well-known Einstein formula [46]

$$\mu = \mu_f \left(1 + 2.5\varphi\right) \tag{6.14}$$

with μ_f denoting the viscosity of the host fluid, this expression is valid for dilute mixtures ($\varphi < 0.05$) with spherical particles. Other more sophisticated ad hoc relations have been formulated such as a quadratic dependence in the viscosity [47]

$$\mu = \mu_f \left(1 + a_1\varphi + a_2\varphi^2\right) \tag{6.15}$$

where a_1 and a_2 are ad hoc parameters taking different values according to the nature of the nanofluid. Other models like that of Krieger and Dougherty [24] have also been exploited.

The description of systems at subscales, such as nanoparticles, requires going beyond the classical theory of irreversible processes [48]. We follow the same procedure as is done for the thermal conductivity and the electrical conductivity. Since some important nuances exist for the viscosity, it is worth performing that procedure over again for the viscosity. Let us consider the flow of a viscous incompressible fluid at uniform temperature. The generalization to more complicated systems as fluid mixtures [49], polymer solutions [50], suspensions [51], porous media and others have been dealt with in detail in numerous publications and books. In the case of an incompressible fluid flow, the only relevant conserved variable is the specific internal energy e (per unit mass) whereas the corresponding flux variable is the viscous pressure tensor $\boldsymbol{P}$. It is a second-order symmetric traceless tensor and, in contrast to e, it is not a conserved quantity. As is the corner stone of the theory presented in the preceding chapters, we assume the existence of a specific non-equilibrium entropy function $\mathcal{S}$ depending both on e and $\boldsymbol{P}$ so that $\mathcal{S} = \mathcal{S}(e, \boldsymbol{P})$ or, in terms of time derivatives,

$$d_t\mathcal{S} = \frac{\partial \mathcal{S}}{\partial e} d_t e + \frac{\partial \mathcal{S}}{\partial \boldsymbol{P}} \otimes d_t \boldsymbol{P}, \tag{6.16}$$

where $\otimes$ stands for the inner product of the corresponding tensors, the symbol d_t designates the time derivative which is indifferently the material or the partial time derivative as the system is, respectively, in motion or at rest. It is assumed that η is a concave function of the variables and that it obeys a general time-evolution equation, which can be written in the form

$$\eta^s = \rho d_t\mathcal{S} + \nabla \cdot \boldsymbol{J}^s, \tag{6.17}$$

with η^s its rate of production per unit volume (in short, the entropy production) to be positive definite in order to satisfy the second principle of thermodynamics, ρ is the mass density of the nanofluid and the vector $\boldsymbol{J}^s$ is the entropy flux, the dot between ∇ and $\boldsymbol{J}^s$ denotes the scalar product. Let us define the temperature by $T^{-1} = \frac{\partial \mathcal{S}}{\partial e}$, which is assumed to be independent of the dissipative flux $\boldsymbol{P}$. Next, we select the constitutive equation for

$\frac{\partial \mathcal{S}}{\partial \boldsymbol{P}}$ as assumed to be given by the linear relation $\frac{\partial \mathcal{S}}{\partial \boldsymbol{P}} = -\frac{\gamma_1}{\rho}\boldsymbol{P}$, where γ_1 is a material coefficient depending generally on ρ and T. Moreover, γ_1 is positive definite in order to meet the property that $\mathcal{S}$ is maximum at local equilibrium, whereas the minus sign in front of $\gamma_1(T)\,\boldsymbol{P}$ has been introduced for convenience. Under these conditions, expression (6.16) can be written as

$$\rho d_t \mathcal{S} = -T^{-1}\boldsymbol{P} \otimes \boldsymbol{D} - \gamma_1 \boldsymbol{P} \otimes d_t \boldsymbol{P}, \tag{6.18}$$

after use has been made of the energy conservation law

$$\rho d_t e = -\boldsymbol{P} \otimes \boldsymbol{D}, \tag{6.19}$$

where $\boldsymbol{D}$ is the symmetric traceless velocity gradient tensor. At the actual order of approximation and in the absence of heat flux, the entropy flux is zero and the entropy production takes the form

$$\eta^s = -\boldsymbol{P} \otimes \left(T^{-1}\boldsymbol{D} + \gamma_1 d_t \boldsymbol{P}\right) \geq 0. \tag{6.20}$$

It is a bilinear relation in the flux $\boldsymbol{P}$ and the quantity represented by the two terms between the parentheses that is usually called the thermodynamic force $\boldsymbol{X}$. The simplest way to guarantee the positiveness of the entropy production η^s is to assume a linear flux–force relation of the form $\boldsymbol{X} = -\nu_1 \boldsymbol{P}$ with ν_1 a positive phenomenological coefficient, this procedure leads to the well-known Maxwell model

$$\tau_1 d_t \boldsymbol{P} = -\boldsymbol{P} - 2\mu \boldsymbol{D}, \tag{6.21}$$

after one has put $\frac{\gamma_1}{\nu_1} = \tau_1$ (relaxation time) and $\frac{1}{T\nu_1} = 2\mu$ (shear viscosity) and wherein τ_1 and μ are positive quantities as ν_1 and γ_1 have been proven to be positive coefficients. Letting τ_1 vanish, one finds back Newton's law $\boldsymbol{P} = -2\mu\boldsymbol{D}$. Although Maxwell's relation is useful at short time scales (high frequencies), it is not satisfactory with the purpose to describe short length scales wherein non-localities play a preponderant role, for instance, fluids in the presence of nanoparticles.

In more complex materials, such as in nanofluids, fluxes of higher order should be introduced as extra state variables. Non-local effects, which are important in the presence of nanoparticles, are elegantly accounted for, by appealing to a hierarchy of fluxes $\boldsymbol{P}^{(2)}$ $(\equiv \boldsymbol{P})$, $\boldsymbol{P}^{(3)}, \ldots\ \boldsymbol{P}^{(N)}$. Herein the second-order tensor $\boldsymbol{P}^{(2)}$ is identified with the usual viscous pressure tensor $\boldsymbol{P}$, and $\boldsymbol{P}^{(3)}$ (a tensor of rank three) is the flux of the pressure tensor, ...etc. Here for simplicity, we limit our developments to the use of $\boldsymbol{P}$ and $\boldsymbol{P}^{(3)}$ as flux state variables, but there will be no difficulty to include higher order tensors. In the next section, we consider up to the third-order approximation. In Section 6.3.3, we extend the development to an infinite order. Furthermore, the dependence of the viscosity on the nanoparticle volume fraction can be seen in two different ways. It can be seen to stem from an influence on the mean free path of the momentum carrier, which is proposed in Section 6.3.2, or it can be seen as a geometrical parameter, which is considered in Section 6.3.3.

6.3.2 Third-Order Approximation

Up to the third-order moment approximation, which is sufficient for the present purpose, the Gibbs equation generalizing expression (6.16) takes the form

$$d_t \mathcal{S}\left(e, \boldsymbol{P}, \boldsymbol{P}^{(3)}, \ldots\right) = T^{-1} d_t e - \frac{\gamma_1}{\rho} \boldsymbol{P} \otimes d_t \boldsymbol{P} - \frac{\gamma_2}{\rho} \boldsymbol{P}^{(3)} \otimes d_t \boldsymbol{P}^{(3)}, \tag{6.22}$$

while the entropy flux is given by

$$\boldsymbol{J}^s = \beta \boldsymbol{P}^{(3)} \otimes \boldsymbol{P}, \tag{6.23}$$

where β, a phenomenological coefficient is allowed to depend on e and the volume fraction of the particles but not on the flux variables. The entropy production (6.17) is obtained by substitution of $d_t\mathcal{S}$ and $\boldsymbol{J}^s$ from Eqs. (6.22) and (6.23), respectively, and elimination of d_te via the energy balance Eq. (6.19). The result is

$$\eta^s = -\boldsymbol{P} \otimes \left(T^{-1}\boldsymbol{D} + \gamma_1 d_t\boldsymbol{P} - \beta\nabla \cdot \boldsymbol{P}^{(3)}\right) - \boldsymbol{P}^{(3)} \otimes \left(\gamma_2 d_t\boldsymbol{P}^{(3)} - \beta\nabla\boldsymbol{P}\right) \geq 0. \quad (6.24)$$

The above bilinear expression in fluxes and forces (the quantities between parentheses) suggests the following linear flux–force equation

$$-T^{-1}\boldsymbol{D} - \gamma_1 d_t\boldsymbol{P} + \beta\nabla \cdot \boldsymbol{P}^{(3)} = \nu_1\boldsymbol{P}, \quad (6.25)$$

$$-\gamma_2 d_t\boldsymbol{P}^{(3)} + \beta\nabla\boldsymbol{P} = \nu_2\boldsymbol{P}^{(3)}, \quad (6.26)$$

where γ_n, β and ν_n (n = 1, 2) are phenomenological coefficients allowed to depend, in particular, on the temperature and/or the relative volume fraction of the constituents. Relations (6.25) and (6.26) can also be viewed as time-evolution equations for the fluxes $\boldsymbol{P}$ and $\boldsymbol{P}^{(3)}$. Making use of Eqs. (6.25) and (6.26), expression (6.24) of the entropy production reads as

$$\eta^s = \nu_1\boldsymbol{P} \otimes \boldsymbol{P} + \nu_2\boldsymbol{P}^{(3)} \otimes \boldsymbol{P}^{(3)} \geq 0, \quad (6.27)$$

from which it follows that $\nu_1 \geq 0$ and $\nu_2 \geq 0$ to satisfy the positiveness of the entropy production. Identifying the quantities $\frac{\gamma_1}{\nu_1} = \tau_1$ and $\frac{\gamma_2}{\nu_2} = \tau_2$ with τ_1 and τ_2, the positive relaxation times of the pressure tensors $\boldsymbol{P}$ and $\boldsymbol{P}^{(3)}$, respectively, and, assuming that $\tau_2 \ll \tau_1$, as confirmed by the kinetic theory of gases, Eq. (6.26) leads to

$$\boldsymbol{P}^{(3)} = \frac{\beta}{\nu_2}\nabla\boldsymbol{P}. \quad (6.28)$$

Substitution of this result in Eq. (6.25) and making use of the identification $2\mu = \frac{1}{\nu_1 T}$ yields

$$\tau_1 d_t\boldsymbol{P} + \boldsymbol{P} = -2\mu\boldsymbol{D} + \frac{\beta^2}{\nu_1\nu_2}\nabla^2\boldsymbol{P}. \quad (6.29)$$

The last term accounts for the non-local effects related to the presence of nanoparticles. In previous works [12,52,53], general relations between the various phenomenological coefficients were obtained. Through dimensional analysis, it was inferred that $\beta = -\tau_1\nu_1$ and $\nu_2 = \left(\frac{\tau_1}{L}\right)^2\nu_1$, with L denoting a characteristic reference length, related to the mean free path ℓ of the momentum carriers. Moreover, in a nanofluid (a two-component system), the reference length L will generally be dependent on the volume fraction of the nanoparticles, i.e. $L = L(\ell, \varphi)$. At the simplest first-order approximation in both ℓ and φ, one has $L = \ell\varphi$. In virtue of the above results, we may write

$$\frac{\beta^2}{\nu_1\nu_2}\,(\geq 0) = \ell^2\varphi^2. \quad (6.30)$$

This coefficient fulfils the following three requirements: it is positive definite, it has the dimension of a length to the square and it vanishes in the absence of nanoparticles, i.e for $\varphi = 0$. Substituting expression (6.30) in Eq. (6.29), one obtains the time-evolution equation of the pressure tensor $\boldsymbol{P}$,

$$\tau_1 d_t\boldsymbol{P} + \boldsymbol{P} = -2\mu\boldsymbol{D} + \ell^2\varphi^2\nabla^2\boldsymbol{P}. \quad (6.31)$$

Introduction of the factor φ^2 in Eq. (6.31) is justified because, as recalled above, a priori all the phenomenological coefficients are φ-dependent as we are faced with a two-component system formed by a fluid and nanoparticles. A dependence in φ^2 is not new, and it was also shown in the constitutive equation of the pressure tensor in polymer blends (e.g. [50,54]). It is the introduction in Eq. (6.31) of the factor φ which accounts explicitly for the dependence of the constitutive equation on the nanoparticles. For $\varphi = 0$, and at steady states, expression (6.31) reduces to the classical Newton law $\boldsymbol{P} = -2\eta\boldsymbol{D}$ for a Newtonian fluid. Relation (6.31) is the key one of the formalism in this subsection and is the equivalent of the Guyer–Krumhansl [55] equation, which is used to describe non-local effects in heat transport [56–58]. After application of the spatial Fourier transforms

$$\hat{\boldsymbol{P}}(\boldsymbol{k},t) = \int_{-\infty}^{\infty} \boldsymbol{P}(\boldsymbol{r},t)\, e^{-i\boldsymbol{k}\cdot\boldsymbol{r}} d\boldsymbol{r} \tag{6.32}$$

and

$$\hat{\boldsymbol{v}}(\boldsymbol{k},t) = \int_{-\infty}^{\infty} \boldsymbol{v}(\boldsymbol{r},t)\, e^{-i\boldsymbol{k}\cdot\boldsymbol{r}} d\boldsymbol{r} \tag{6.33}$$

to relation (6.31), with $\boldsymbol{k}$ designating the wave-number vector, $\boldsymbol{r}$ the position vector and $\mathrm{v}(\boldsymbol{r}, t)$ the vectorial velocity field, one is led to the following time-evolution equation of the Fourier transformed fluxes

$$\tau_1 d_t \hat{\boldsymbol{P}} + \hat{\boldsymbol{P}} = -i\boldsymbol{k}\mu\hat{\boldsymbol{v}} - \ell^2 k^2 \varphi^2 \boldsymbol{P}. \tag{6.34}$$

where $k^2 = \boldsymbol{k} \cdot \boldsymbol{k}$. Restricting the analysis to steady states, which is sufficient for our purpose to determine the effective viscosity in terms of the volume fraction and size of the nanoparticles, expression (6.34) writes as

$$\hat{\boldsymbol{P}}\left(1 + \ell^2 k^2 \varphi^2\right) = -i\boldsymbol{k}\mu\hat{\boldsymbol{v}}. \tag{6.35}$$

suggesting that the effective viscosity of the nanofluid [using Eq. (6.14)] is given by

$$\mu^{eff} = \frac{\mu_f\,(1 + 2.5\varphi)}{1 + \ell^2 k^2 \varphi^2}, \tag{6.36}$$

An important feature that distinguishes the viscosity of nanofluids from classical particle dispersion in fluids is the effect caused by the presence of an amorphous overlayer encapsulating the nanoparticles [6–9,48,59–62]. As a consequence of liquid layering, the radius a_p of the particles will be modified from a_p to $a_p + \mathscr{d}$ [see discussion under Eq. (6.4)] and the volume fraction of the particles from φ to an effective volume fraction $\varphi\,(1 + \mathscr{d}/a_p)^3$ with $\mathscr{d}$ standing for the thickness of the surrounding layer.

Selecting the wave number k in expression (22) is a delicate task. Indeed, k will be related to a characteristic length that takes into account the presence of nanoparticles. Here, we find it rather natural to take $a_p + \mathscr{d}$ as a reference length. A similar attitude is followed in works about the effect of nanoparticles' dispersion on the thermal conductivity coefficient of nanofluids and nanocomposites [62–65]. Accordingly, the wave number k will therefore be given by

$$k = \frac{2\pi}{a_p + \mathscr{d}}. \tag{6.37}$$

To summarize, the final expression of the effective viscosity of the nanofluid will take the following form obtained by combining Eqs. (6.36) and (6.37),

$$\mu^{eff} = \mu_f \frac{1 + 2.5\varphi\,(1 + \mathscr{d}/a_p)^3}{1 + 4\pi^2\varphi^2\,(1 + \mathscr{d}/a_p)^4\,(\ell^2/a_p^2)}. \tag{6.38}$$

For $\varphi = 0$, μ^{eff} reduces to μ_f as expected.

6.3.3 Complete Expression

Some simplifications were made in Section 6.3.2, which implied using an approximate characteristic length and a third-order approximation for the viscous pressure flux. In this section, we take over from Eq. (6.22) to Eq. (6.23) and extend it to the nth order of fluxes, which gives [interestingly, also in analogy with Eqs. (2.30) and (2.31)]

$$d_t\mathcal{S}\left(e, \boldsymbol{P}, \ldots, \boldsymbol{P}^{(n)}\right) = T^{-1}d_te - \frac{\gamma_1}{\rho}\boldsymbol{P}\otimes d_t\boldsymbol{P} - \frac{\gamma_2}{\rho}\boldsymbol{P}^{(3)}\otimes d_t\boldsymbol{P}^{(3)} - \cdots - \frac{\gamma_{n-1}}{\rho}\boldsymbol{P}^{(n)}\otimes d_t\boldsymbol{P}^{(n)}, \tag{6.39}$$

$$\boldsymbol{J}^s = \beta_1\boldsymbol{P}^{(3)}\otimes\boldsymbol{P} + \cdots + \beta_{n-2}\boldsymbol{P}^{(n)}\otimes\boldsymbol{P}^{(n-1)}, \tag{6.40}$$

where $n = 3, 4, \ldots, N$. The entropy production (6.17) is obtained by substitution of $d_t\mathcal{S}$ and $\boldsymbol{J}^s$ from Eqs. (6.39) and (6.40), respectively, and elimination of d_te via the energy balance (6.19). The result is

$$\eta^s = -\boldsymbol{P}\otimes\left(T^{-1}\boldsymbol{D} + \gamma_1 d_t\boldsymbol{P} - \beta_1\nabla\cdot\boldsymbol{P}^{(3)}\right) - \cdots - \sum_{n=3}^{N}\boldsymbol{P}^{(n)}\otimes\left(\gamma_{n-1}d_t\boldsymbol{P}^{(n)} - \beta_{n-1}\nabla\cdot\boldsymbol{P}^{(n+1)} - \beta_{n-2}\nabla\boldsymbol{P}^{(n-1)}\right) \geq 0. \tag{6.41}$$

The above bilinear expression in fluxes and forces (the quantities between parentheses) suggests the following linear flux–force equation

$$-T^{-1}\boldsymbol{D} - \gamma_1 d_t\boldsymbol{P} + \beta_1\nabla\cdot\boldsymbol{P}^{(3)} = \nu_1\boldsymbol{P}, \tag{6.42}$$

$$-\gamma_{n-1}d_t\boldsymbol{P}^{(n)} + \beta_{n-1}\nabla\cdot\boldsymbol{P}^{(n+1)} + \beta_{n-2}\nabla\boldsymbol{P}^{(n-1)} = \nu_{n-1}\boldsymbol{P}^{(n)}, \tag{6.43}$$

where γ_n, β_n and ν_n $(n = 1, 2, \ldots)$ are phenomenological coefficients and are determined in the same manner as in Section 6.3.2. We extend Eq. (6.43) to an infinite order and apply Eqs. (6.32) and (6.33) to them, which gives finally

$$\tau d_t\hat{\boldsymbol{P}} + \hat{\boldsymbol{P}} = -i\boldsymbol{k}\mu^{eff}(\boldsymbol{k})\,\hat{\boldsymbol{v}}. \tag{6.44}$$

where $\mu^{eff}(\boldsymbol{k})$ is a wavelength-dependent viscosity, taking the form of a continued fraction expansion in analogy with Eq. (2.37) for the thermal conductivity:

$$\mu^{eff}(\boldsymbol{k}) = \cfrac{\mu}{1 + \cfrac{\boldsymbol{k}^2 l_1^2}{1 + \cfrac{\boldsymbol{k}^2 l_2^2}{1 + \cfrac{\boldsymbol{k}^2 l_3^2}{1 + \cdots}}}}, \tag{6.45}$$

where l_n is the correlation length of order n defined by $l_n^2 = \beta_n^2/(\mu_n\mu_{n+1})$. Here, it is assumed that the relaxation times τ_n $(n > 1)$ corresponding to higher order fluxes are negligible with respect to $\tau_1 \equiv \tau$, which is a hypothesis generally admitted in kinetic theories. In the present problem, there is only one dimension. In the presence of nanoparticles, it can be suggested to choose the hydrodynamic radius, $\boldsymbol{a}_H$. It can be easily deduced that

$$\boldsymbol{a}_H \equiv \frac{2a_p}{3}\frac{(1-\varphi)}{\varphi}. \tag{6.46}$$

It is then natural to define $\boldsymbol{k} = k \equiv 2\pi/\boldsymbol{a}_H$. The correlation lengths are selected, just as in Chapter 2 for the thermal conductivity, as $l_n^2 = a_{n+1}l^2$, with $a_n = n^2/(4n^2-1)$ and l identified as the mean free path of the momentum carriers independently of the order of approximation, i.e. ℓ. Finally, using Eqs. (6.46) and (6.14), Eq. (6.45) reduces to an asymptotic limit, giving

$$\mu^{eff} = \mu_f \frac{3\left(1+2.5\varphi\left(1+d/a_p\right)^3\right)}{9\pi^2\left(\frac{\varphi(1+d/a_p)^3}{1-\varphi(1+d/a_p)^3}\right)^2 Kn^2}\left[\frac{3\pi\frac{\varphi(1+d/a_p)^3}{1-\varphi(1+d/a_p)^3}Kn}{\arctan\left(3\pi\frac{\varphi(1+d/a_p)^3}{1-\varphi(1+d/a_p)^3}Kn\right)}-1\right]. \tag{6.47}$$

where $Kn = \ell/a_p$. For small volume fractions, it can be easily seen that

$$\mu^{eff} = \mu_f \frac{3\left(1+2.5\varphi\left(1+d/a_p\right)^3\right)}{9\pi^2\varphi^2\left(1+d/a_p\right)^6 Kn^2}\left[\frac{3\pi\varphi\left(1+d/a_p\right)^3 Kn}{\arctan\left(3\pi\varphi\left(1+d/a_p\right)^3 Kn\right)}-1\right]. \tag{6.48}$$

6.4 Discussion and Case Studies for the Thermal Conductivity

6.4.1 Thermal Conductivities of Alumina–Water, Alumina–Ethylene Glycol and Alumina–50/50 w% Water/Ethylene Glycol Mixture Nanofluids

In this section, the various contributions to the effective thermal conductivity λ_{tot}^{eff} as a function of the volume fraction of nanoparticles and their size in the following situations are discussed, respectively:

- λ^{eff}: without the considered heat transfer mechanisms [Eq. (3.1)]
- λ_{ll}^{eff}: liquid-layering alone [Eq. (6.1)]
- λ_{a}^{eff}: agglomeration alone [Eq. (6.5) wherein $d = 0$]
- λ_{Br}^{eff}: Brownian motion alone [Eq. (6.12)]
- λ_{tot}^{eff}: incorporating all the considered heat transfer mechanisms [Eq. (6.13)]

Note that the maximum volume fraction value may not be extended to $\varphi = 1$ but is limited by the maximum packing of spheres, namely $\varphi_{max} = \pi/\sqrt{18}$. The calculations are performed for suspensions of alumina spherical nanoparticles in water, ethylene glycol and a 50/50 w% water/ethylene glycol mixture, respectively. The theory is of course by no means limited to these fluids which have been selected because they allow a direct comparison with experimental data [12,66–71]. The material properties of alumina used for the calculations are given in Table 6.1 and refer to the room temperature (taken here to be 300 K).

The thermal conductivities λ_f of the base fluids are 0.613, 0.252 and 0.4 W m^{-1} K^{-1} for water, ethylene glycol and the 50/50 w% water/ethylene glycol mixture, respectively [44,73].

TABLE 6.1
Room Temperature Material Properties for Alumina (Al_2O_3) Particles [72].

Material	Heat Capacity [MJ (m^{-3} K^{-1})]	Group Velocity [m s^{-1}]	Mean Free Path [nm]
Al_2O_3	3.04	7009	5.08

Water and ethylene glycol are common heat transfer fluids, but they present the characteristic to have limited heat transfer capabilities [66]. This explains the interest in enhancing their heat transfer properties by adding nanoparticles. The results are presented in Figures 6.1–6.3, with the thickness of the liquid boundary layer given by $d = 1$ nm. The experimental values are taken from [12,66–71].

The general trend is that the thermal conductivity is enhanced by more than 20% for an increase of 10% of the volume fraction. Experimental data are not available for any particle size and volume fraction and, therefore, the theoretical predictions are compared with several different experimental sources. A satisfactory agreement with the available experimental data is observed. The full model λ_{tot}^{eff} fits reasonably well with the experiments

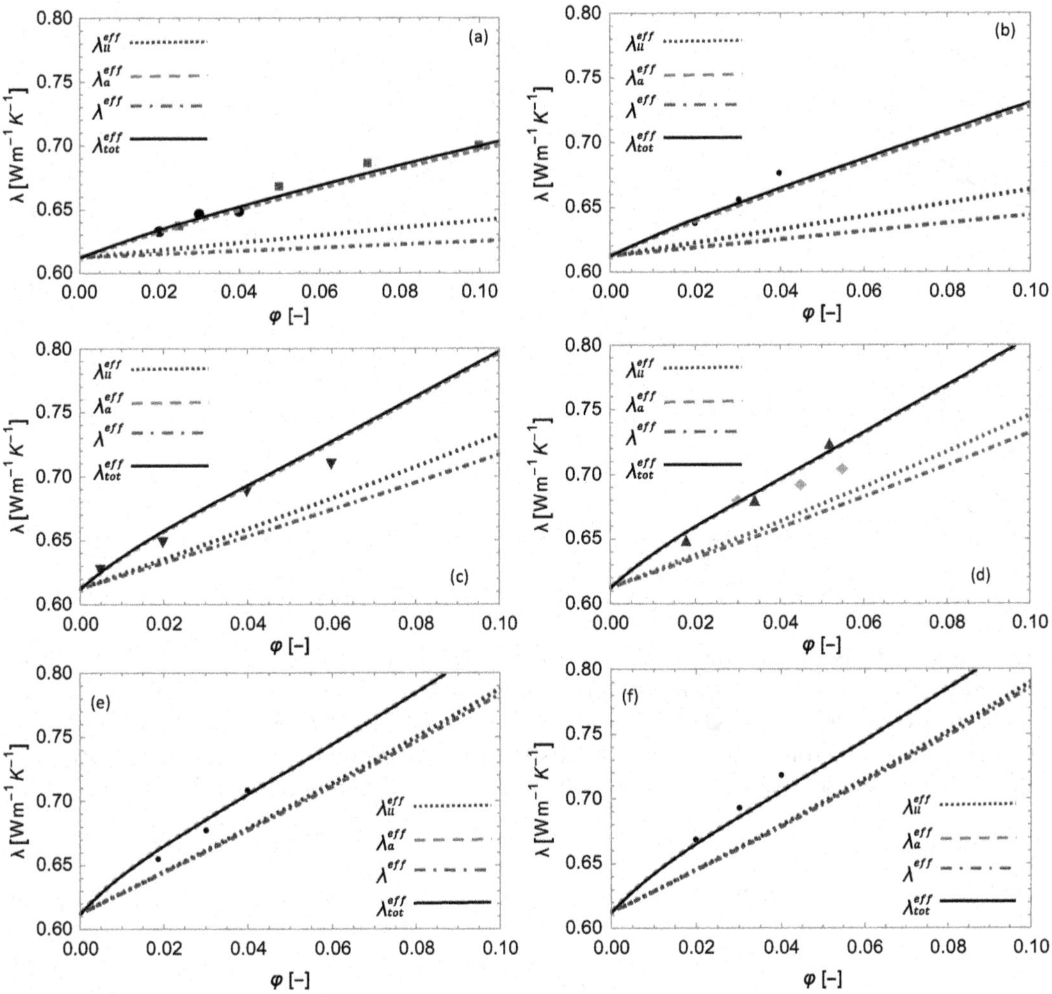

FIGURE 6.1
Effective thermal conductivities (λ_{ll}^{eff}, λ_{a}^{eff}, λ^{eff} and λ_{tot}^{eff}) of alumina–water nanofluids as a function of the volume fraction for different nanoparticle radii (a_p): (a) 6 nm, (b) 8 nm, (c) 23.5 nm, (d) 30 nm, (e) 122.5 nm and (f) 141 nm. Experimental values are drawn from [12] ■, [66] ♦, [67] ▲, [69] • and [70] ▼. (Modified from: Machrafi, H., Lebon, G. 2016. The role of several heat transfer mechanisms on the enhancement of thermal conductivity in nanofluids. *Continuum Mechanics and Thermodynamics* 28:1461–1475.)

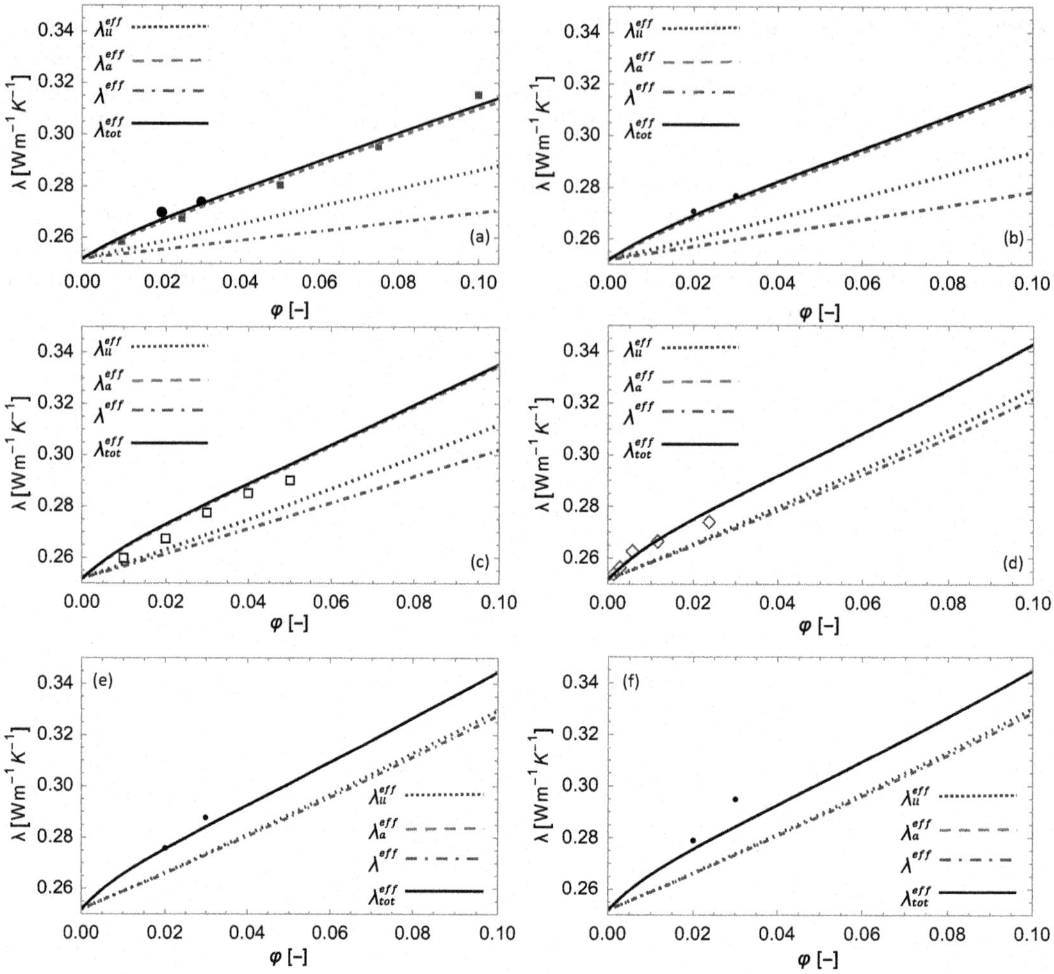

FIGURE 6.2
The same as in Figure 6.1 for alumina–ethylene glycol nanofluids with nanoparticle radii (a_p): (a) 6 nm, (b) 8 nm, (c) 19 nm, (d) 60 nm, (e) 122.5 nm and (f) 141 nm. Experimental values are drawn from [12] ■, [68] ◇, [69] • and [71] □. (Modified from: Machrafi, H., Lebon, G. 2016. The role of several heat transfer mechanisms on the enhancement of thermal conductivity in nanofluids. *Continuum Mechanics and Thermodynamics* 28:1461–1475.)

except at one point for ethylene glycol, namely $a_p = 141$ nm, $\varphi = 0.035$ (Figure 6.2f). This may be due to the uncertainty about the degree of agglomeration whose value is not experimentally univocally determined [11]. Nevertheless, it can be said that λ_{tot}^{eff} is of the same order of magnitude as the experimental data. It is interesting to see to which extent the results, predicted by considering only one single heat transfer mechanism, namely either λ_{ll}^{eff}, λ_a^{eff} or λ_{Br}^{eff}, may be different from λ_{tot}^{eff}. Our calculations have shown that the difference between λ_{Br}^{eff} and λ^{eff}, expressed by Eq. (3.1), is too small to be reproducible on the graphs, so that we feel legitimate to take $\lambda_{Br}^{eff} \approx \lambda^{eff}$ and to represent only λ^{eff}. The validity of this assumption is discussed in the next subsection. We observe that λ_a^{eff} is closer to λ_{tot}^{eff} than λ_{ll}^{eff}, whatever the sizes of the nanoparticle. As the nanoparticle radius increases, the difference $\lambda_a^{eff} - \lambda^{eff}$ increases, while $\lambda_{ll}^{eff} - \lambda^{eff}$ decreases. One important feature is that

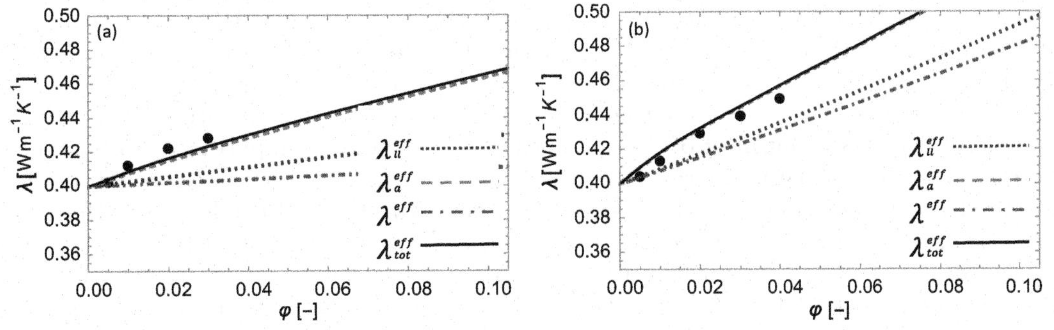

FIGURE 6.3
The same as in Figure 6.1 for alumina–50/50 w% water/ethylene glycol nanofluids with nanoparticle radii (a_p): (a) 5 nm and (b) 25 nm. Experimental values • are drawn from [73]. (Modified from: Machrafi, H., Lebon, G. 2016. The role of several heat transfer mechanisms on the enhancement of thermal conductivity in nanofluids. *Continuum Mechanics and Thermodynamics* 28:1461–1475.)

different particle dimensions lead to different outcomes. The crucial role played by the particle size justifies that it be further investigated. Beforehand, let us comment about the controversial role of Brownian motion on heat transfer in nanofluids.

6.4.2 Note on the Brownian Motion

The behaviour of the effective thermal conductivity of Al_2O_3–ethylene glycol and Al_2O_3–water nanofluids versus the particle size is analysed. Two different volume fractions (1% and 5%) are considered.

In Figure 6.4, the values of the effective thermal conductivity (in the absence of agglomeration and liquid-layering effects) are compared with and without Brownian motion for alumina particles dispersed in water and ethylene glycol, respectively.

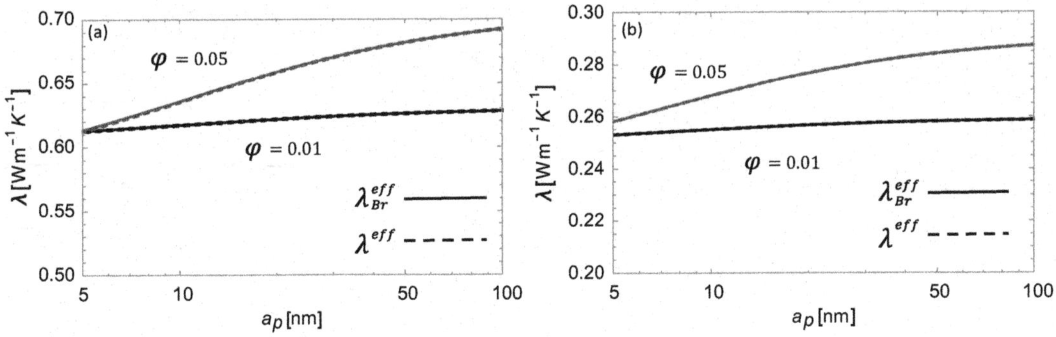

FIGURE 6.4
Effective thermal conductivity λ^{eff}_{Br}, compared to λ^{eff} versus the nanoparticle radius at volume fractions of 1% (lower curves) and 5% (upper curves), for alumina nanoparticles in (a) water (solid lines) and (b) ethylene glycol (dashed lines). (Modified from: Machrafi, H., Lebon, G. 2016. The role of several heat transfer mechanisms on the enhancement of thermal conductivity in nanofluids. *Continuum Mechanics and Thermodynamics* 28:1461–1475.)

It appears that Brownian motion (λ_{Br}^{eff}) does not significantly modify (less than 1%) the basic thermal conductivity (λ^{eff}). It may therefore be concluded from our analysis that the Brownian motion is of no great influence, at least for the considered nanofluids. The influence of Brownian motion on heat transfer in nanofluids was investigated in several papers [14–16,74–81], but most of these studies disagree about its importance. For instance, on the basis of the kinetic theory, Jang and Choi [16] claimed that the Brownian motion is one of the key mechanisms in the enhancement of the thermal conductivity. However, for others, like Gupta and Kumar [76], Nie et al. [78] and Evans et al. [81], the Brownian motion does not play a significant role in agreement with the results of the present work. Nonetheless, it is true that for nanofluids with a low viscous base fluid or aerosols, the Brownian effect may become relevant.

6.4.3 Thermal Conductivity as a Function of the Particle Size

The abovementioned considerations allow us to omit the Brownian motion and to focus on the agglomeration and liquid-layering effects. In Figure 6.5, as a function of the particle radius, the separate influences of agglomeration (λ_a^{eff}), nano-layering (λ_{ll}^{eff}) and the agglomeration–nano-layering coupling ($\lambda_{ll,a}^{eff}$) for water and ethylene glycol are shown; two values of the particles' volume fraction and two values of the liquid boundary layer thickness are considered.

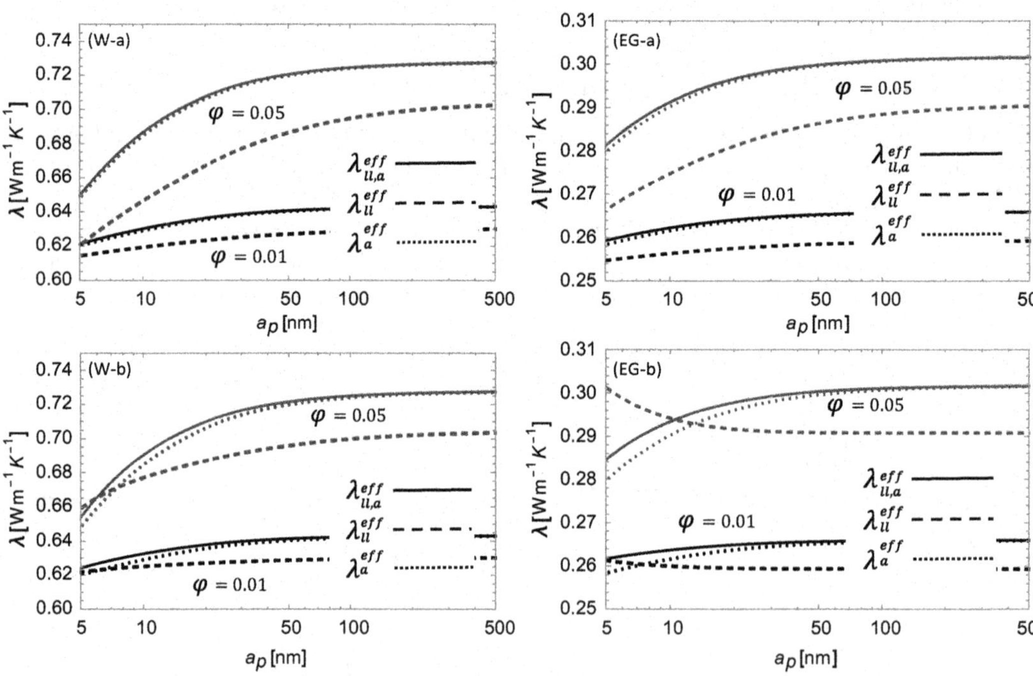

FIGURE 6.5
Effective thermal conductivity $\lambda_{ll,a}^{eff}$ (solid lines) compared to λ_{ll}^{eff} (dashed lines) and λ_a^{eff} (dotted lines) versus the nanoparticle radius at volume fractions of 1% (lower curves) and 5% (upper curves), for (a) $d = 1$ nm and (b) $d = 3$ nm. The nanofluids consist of alumina nanoparticles in water (W) and ethylene glycol (EG). (Modified from: Machrafi, H., Lebon, G. 2016. The role of several heat transfer mechanisms on the enhancement of thermal conductivity in nanofluids. *Continuum Mechanics and Thermodynamics* 28:1461–1475.)

The general tendency is the increase of the thermal conductivity with, from one side, the volume fraction and, from the other side, with the particle size. The presence of an asymptotic value is confirmed for $a_p > 100$ nm. This is understandable because, for relatively large particles with respect to the mean free paths, the size will no longer influence the value of the thermal conductivity. Indeed, at large sizes, the Knudsen number becomes very small [$\lambda_p \to \lambda_p^0$ in Eq. (6.3)], indicating that heat transport is no longer of ballistic but rather of diffusive nature, governed by Fourier's law. The results in Figure 6.5 show that the behaviour of the thermal conductivity, especially at $a_p \approx O(1\text{–}10)$ nm, is strongly dependent on the thickness of the interfacial layer and the rate of agglomeration. For particles in the range of molecular scale, it is observed experimentally [69] and computationally [82] that the thermal conductivity increases with increasing size, as seen in Figures 6.5 (W-a) and 6.5 (EG-a). This behaviour was confirmed by several experimental works [12,71,83]. However, it was mentioned by Keblinski et al. [2] that the thermal conductivity may rather decrease for increasing particle sizes in the presence of highly conducting liquid layers around the nanoparticles, such a conclusion is supported by our developments (in absence of agglomeration) as shown by the dashed curves of Figures 6.5 (EG-b). At larger particle dimensions ($a_p \geq 10$ nm), the thickness $\mathscr{d}$ of the interfacial liquid hardly influences the value of the effective thermal conductivity, from which it can be deduced that, at relatively large sizes, agglomeration appears to be the dominant effect, as confirmed by the experiments [2].

6.4.4 Complementary Comments

To shed a different light on the previously mentioned developments, we have calculated the so-called enhancement factor *EF* (in %), defined as the ratio of the thermal conductivity enhancement, $\lambda_i^{eff} - \lambda^{eff}$ (with the subscript i designating the mechanism under study: either Brownian motion, liquid layering or particles agglomeration) to the overall enhancement defined as $\left(\lambda_{ll}^{eff} - \lambda^{eff}\right) + \left(\lambda_a^{eff} - \lambda^{eff}\right) + \left(\lambda_{Br}^{eff} - \lambda^{eff}\right)$.

The results of the previous figures are confirmed for alumina-in-water (Figure 6.6) and ethylene glycol (Figure 6.7). It is observed that the role of Brownian motion is minute and that both liquid layering and agglomeration are the leading heat transfer mechanisms,

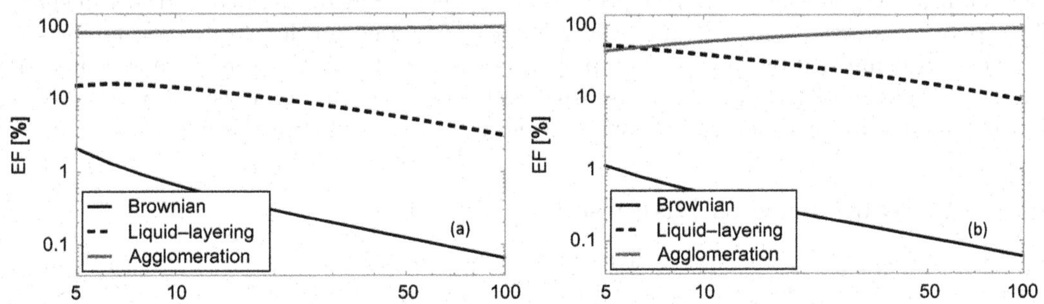

FIGURE 6.6
Enhancement factor *EF* for alumina nanoparticles in water calculated for the different heat transfer mechanisms as a function of the nanoparticle radius for (a) $\mathscr{d} = 1$ nm and (b) $\mathscr{d} = 3$ nm. The volume fraction is 1%. (Modified from: Machrafi, H., Lebon, G. 2016. The role of several heat transfer mechanisms on the enhancement of thermal conductivity in nanofluids. *Continuum Mechanics and Thermodynamics* 28:1461–1475.)

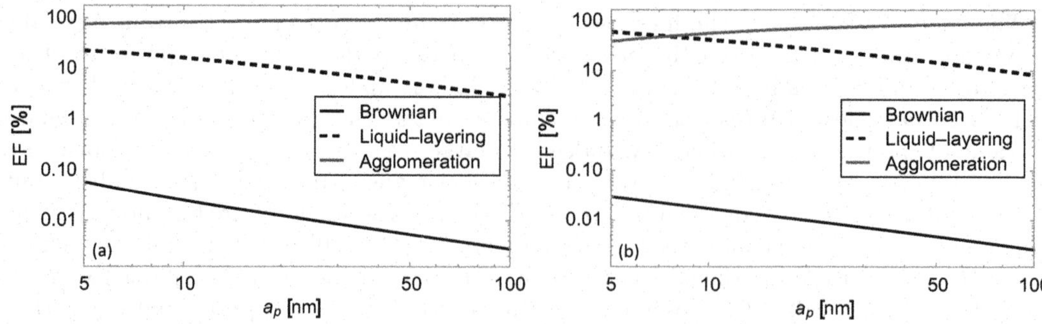

FIGURE 6.7
The same as in Figure 6.6 for alumina nanoparticles in ethylene glycol. (Modified from: Machrafi, H., Lebon, G. 2016. The role of several heat transfer mechanisms on the enhancement of thermal conductivity in nanofluids. *Continuum Mechanics and Thermodynamics* 28:1461–1475.)

depending on specific conditions: the thickness of the liquid layering and the nanoparticle size. Similar conclusions can be drawn for the 50/50 w% water/ethylene glycol mixture.

6.5 Discussion and Case Studies for the Viscosity

6.5.1 Alumina Al_2O_3 Particles in Water

For alumina nanoparticles dispersed in water, it is found [38] that the thickness of the boundary layer around Al_2O_3 particles is roughly $d = 4$ nm. The mean free path ℓ is taken to be that of the average distance between water molecules [84], i.e. $\ell = 0.31$ nm. In Figure 6.8, the theoretical effective viscosity of the nanofluid is compared to experimental values [85] for two different radii of Al_2O_3 particles: $a_p = 4$ and 21.5 nm. For the sake of completeness, we have also drawn the results provided by Einstein's theory. It is still shown that viscosity is increasing with increasing volume fraction and decreases with larger size. The latter is confirmed by the results reported in Figure 6.9, where we have represented the dependence of the relative effective viscosity on the particle radius. It is also worth noting that our model fits satisfactorily the experimental data available in the literature.

6.5.2 Li Nanoparticles Dispersed in Liquid Ar

Next to often used nanofluids, such as alumina-in-water ones, it is interesting to apply the theory from Sections 6.2 and 6.3 to a different situation: lithium nanoparticles (Li) in liquid argon (Ar) (under pressure at ambient temperature). The results are compared with a recent analysis by Rudyak and Krasnotlutskii [86]. The nanoparticles are assumed to be rigid spheres uniformly dispersed in the fluid carrier, the temperature is fixed at $T =$ 300 K and the volume fraction of particles is small, ranging from 0.01 to 0.6. The radius a_p of the nanoparticles Li is ranging from 1 to 4 nm. As for the mean free path ℓ, we start from Matthiessen's rule stating that $1/\ell = 1/\ell_{Ar-Ar} + 1/\ell_{Li-Li} + 1/\ell_{Ar-Li}$, with ℓ, ℓ_{Li-Li} and ℓ_{Ar-Li} designating the mean free path associated to collisions between Ar–Ar, Li–Li and Ar–Li molecules, respectively. Referring to the kinetic theory [87], the mean free path

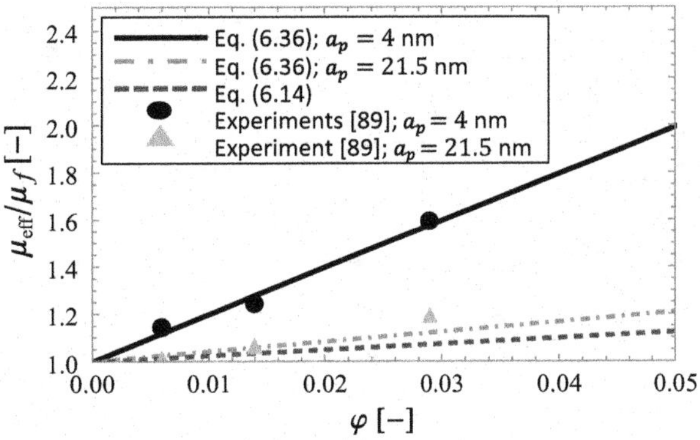

FIGURE 6.8
Relative viscosity as a function of the volume fraction of Al_2O_3 nanoparticles in water at ambient conditions, comparing our model [Eq. (6.36)] with Eq. (6.14) and experimental values [84] for two nanoparticles radii: $a_p = 4$ and 21.5 nm. (Modified from: Lebon, G., Machrafi, H. 2018. A thermodynamic model of nanofluid viscosity based on a generalized Maxwell-type constitutive equation. *Journal of Non-Newtonian Fluid Mechanics* 253:1–6).

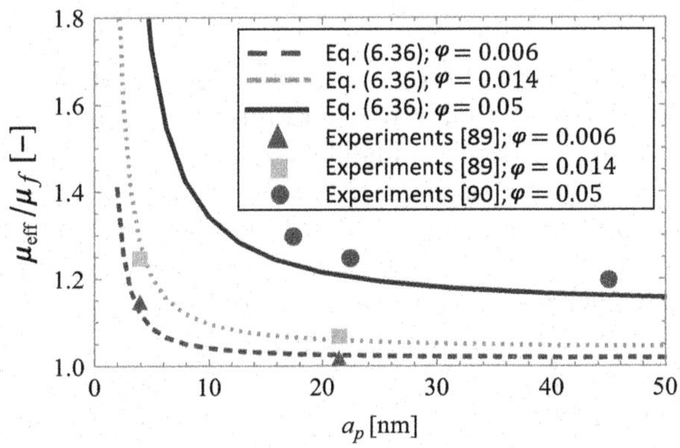

FIGURE 6.9
Relative effective viscosity as a function of the radius of Al_2O_3 nanoparticles in water at ambient conditions, comparing our model [Eq. (6.36)] with experimental values [84,85] for three different volume fractions of $\varphi = 0.006$, 0.014 and 0.05. (Modified from: Lebon, G., Machrafi, H. 2018. A thermodynamic model of nanofluid viscosity based on a generalized Maxwell-type constitutive equation. *Journal of Non-Newtonian Fluid Mechanics* 253:1–6.)

is inversely proportional to the number of particles per unit volume, it follows that the main contribution will arise from the term $1/\ell_{Ar-Ar}$. It can be checked numerically that by including the other contributions of Matthiessen's law, the modifications are minute without an influence on the final results; we, therefore, select as a first approximation $\ell = \ell_{Ar-Ar}$. The mean free path is found to be given by $\ell_{Ar-Ar} = \left(\sqrt{3/2} - 1\right) r_{Ar} \approx 0.043$ nm, after taking for r_{Ar} the Van der Waals radius of Ar, $r_{Ar} = 0.19$ nm. The value of ℓ_{Ar-Ar} takes into account the radius of an interstitial sphere that would fit between closely packed spheres of Ar without distorting the structure explicitly; this particular value has been selected because of the strong packing of the Ar molecules considered in [86] to which our results will be compared. Since no value for the thickness d of the boundary layer surrounding the

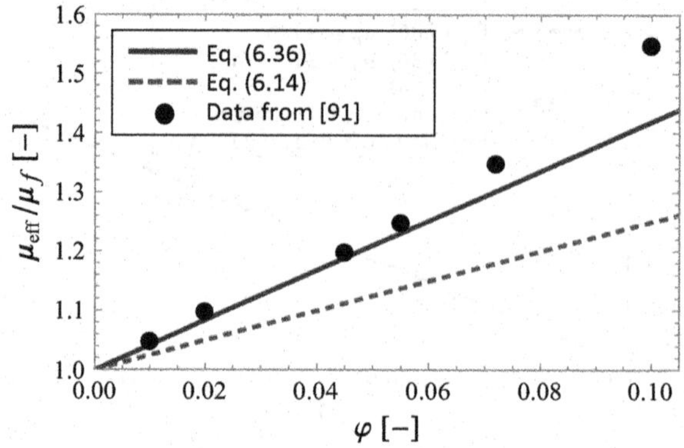

FIGURE 6.10
Relative effective viscosity as a function of the volume fraction of Li nanoparticles with $a_p = 1$ nm in 300 K liquid argon, comparing our model [Eq. (6.36)] with Einstein's relation [Eq. (6.14), dotted line] and data from [86] (full circles). (Modified from: Lebon, G., Machrafi, H. 2018. A thermodynamic model of nanofluid viscosity based on a generalized Maxwell-type constitutive equation. *Journal of Non-Newtonian Fluid Mechanics* 253:1–6.)

Li particles in argon is found in the literature, we have assumed that $\mathscr{d} = 0.2$ nm, which is of the order of magnitude of the Van der Waals radius of Ar, which, by the way, lies within the range of values proposed for other fluids [33–36,88,89].

The dependence of the relative viscosity μ^{eff}/μ_f on the volume fraction of Li particles is shown in Figure 6.10, assuming that the radius of Li nanoparticles is $a_p =$ nm, the thickness of the bounding layer $\mathscr{d} = 0.2$ nm and the mean free path in liquid Ar is $\ell = 0.043$ nm. The results are compared with experimental data and with Einstein's model [46], which accounts only for the volume fraction of dispersed particles.

Numerical simulations indicate a drastic increase of effective viscosity with the volume fraction of nanoparticles in full agreement with the theoretical models developed earlier. This result may be interpreted by the fact that by increasing the volume fraction, one increases the interfacial surface between the particles and the fluid giving rise to a larger resistance to deformation, whence a higher viscosity. Our results are also shown to be in good accord with those obtained by Rudyak and Krasnolutskii [86] using a molecular dynamics method. It is seen that for volume fractions higher than 0.06, the agreement between the model and the data from [86] is less satisfactory, which can be understood by recalling that Einstein's model is only valid for small volume fractions.

6.6 Closing Notions on the Use of Nanofluids

Having elaborated on the thermal conductivity and viscosity of nanofluids as well its behaviour that is quite well understood [60,90], we should mention that a combination of the two is what mostly affects industrial application. As such, we can define a dimensionless ratio

$$f_{\lambda/\mu} = \frac{\lambda^{\mathit{eff}}}{\mu^{\mathit{eff}} C_p} \tag{6.49}$$

TABLE 6.2
The Dimensionless Ratio $f_{\lambda/\mu}$ as a Function of φ for Aqueous Al_2O_3 Nanofluids with $a_p = 6$ nm, $d = d_H = 1$ nm and $s = 0$.

φ	λ^{eff} [W K^{-1} m^{-1}]	μ^{eff} [mPa s]	$\langle C_p \rangle$ [kJ kg^{-1} K^{-1}]	$f_{\lambda/\mu}$
0	0.613	1.00	4.17	0.147
0.01	0.632	1.04	4.14	0.147
0.02	0.648	1.09	4.10	0.145
0.03	0.663	1.13	4.07	0.144
0.04	0.678	1.17	4.04	0.143
0.05	0.692	1.22	4.00	0.142
0.06	0.706	1.26	3.97	0.141
0.07	0.720	1.30	3.93	0.141
0.08	0.734	1.34	3.90	0.140
0.09	0.748	1.39	3.87	0.139
0.1	0.763	1.43	3.83	0.139

where C_p is the mean heat capacity per unity mass. This ratio is a qualitative indication of what volume fraction could be the optimum—highest possible thermal conductivity with a viscosity that is not too high in order to limit the difficulty related to nanofluid flow (e.g. viscous losses). It is interesting to assess this ratio as a function of the volume fraction. As the alumina-in-water nanofluid has been considered in Sections 6.4.1 and 6.5.1, the ratio (6.49) is evaluated for some volume fractions in Table 6.2.

As can be seen, too high-volume fractions could, depending on the application of nanofluids, have implications on the flow becoming too viscous. Nonetheless, as can be seen from Table 6.2, one still may use higher volume fractions, such that the benefits of a nanofluid (e.g. thermal conductivity if it concerns a cooling application) outweigh the increase of viscosity. This is only valid if the nanoparticle size is not too small (around 1 or 2 nm), which, as Figure 6.9 shows, results into a strong increase of the viscosity.

References

[1] Marín, E., Bedoya, A., Alvarado, S., Calderón, A., Ivanov, R., Gordillo-Delgado, F. 2014. An explanation for anomalous thermal conductivity behaviour in nanofluids as measured using the hot-wire technique. *Journal of Physics D: Applied Physics* 47:085501.

[2] Keblinski, P., Phillpot, S.R., Choi, S.U.S., Eastman, J.A. 2002. Mechanisms of heat flow in suspensions of nano-sized particles (nanofluids). *International Journal of Heat and Mass Transfer* 45:855–863.

[3] Keblinski, P., Eastman, J.A., Cahill, D.G. 2005. Nanofluids for thermal transport. *Materials Today* 8:36–44.

[4] Wang, X.Q., Mujumdar, A.S. 2008. A review of nanofluids – Part I: Theoretical and numerical investigations. *Brazilian Journal of Chemical Engineering* 25:613–630.

[5] Iranidokht, V., Hamian, S., Mohammadi, N., Shafii, M.B. 2013. Thermal conductivity of mixed nanofluids under controlled pH conditions. *International Journal of Thermal Science* 74:63–71.

[6] Xie, H., Fujii, M., Zhang, X. 2005. Effect of interfacial nanolayer on the effective thermal conductivity of nanoparticle-fluid mixture, *International Journal of Heat and Mass Transfer* 48:2926–2932.

[7] Yu, W., Choi, S.U.S. 2003. The role of interfacial layers in the enhanced thermal conductivity of nanofluids: A renovated Maxwell model. *Journal of Nanoparticle Research* 5:167–171.

[8] Pasrija, R., Srivastava, S. 2014. The interfacial layer effect on thermal conductivity of nano-colloidal dispersions. *International Journal of Applied Physics and Mathematics* 4:1–4.

[9] Kole, M., Dey, T.K. 2011. Role of interfacial layer and clustering on the effective thermal conductivity of CuO–gear oil nanofluids. *Experimental Thermal and Fluid Science* 35:1490–1495.

[10] Chen, H., Ding, Y., Tan, C. 2007. Rheological behaviour of nanofluids. *New Journal of Physics* 9:367–390.

[11] Kochetov, R., Korobko, A.V., Andritsch, T., Morshuis, P.H.F., Picken, S.J., Smit, J.J. 2011. Modelling of the thermal conductivity in polymer nanocomposites and the impact of the interface between filler and matrix. *Journal of Physics D: Applied Physics* 44:395401.

[12] Timofeeva, E.V., Gavrilov, A.N., McCloskey, J.M., Tolmachev, Y.V. 2007. Thermal conductivity and particle agglomeration in alumina nano-fluids: Experiment and theory. *Physical Review E* 76:061203.

[13] Anoop, K.B., Kabelac, S., Sundararajan, T., Das, S.K. 2009. Rheological and flow characteristics of nanofluids: Influence of electroviscous effects and particle agglomeration. *Journal of Applied Physics* 106:034909.

[14] Xiao, B., Yang, Y., Chen, L. 2013. Developing a novel form of thermal conductivity of nanofluids with Brownian motion effect by means of fractal geometry. *Powder Technology* 239:409–414.

[15] Azizian, R., Doroodchi, E., Moghtaderi, B. 2012. Effect of nanoconvection caused by Brownian motion on the enhancement of thermal conductivity in nanofluids. *Industrial and Engineering Chemistry Research* 51:1782–1789.

[16] Jang, S.P., Choi, S.U.S. 2004. Role of Brownian motion in the enhanced thermal of nanofluids. *Applied Physics Letters* 84:4316–4318.

[17] Rudyak, V.Y. 2013. Viscosity of nanofluids—Why it is not described by the classical theories. *Advances in Nanoparticles* 2:266–279.

[18] Shanker, N.V., Shekar Reddy, M.C., Basavo Rao, V.V. 2012. On prediction of viscosity of nanofluids for low volume fractions of nanoparticles. *International Journal of Engineering Research and Technology* 1:1–10.

[19] Masomi, M., Sohrabi, N., Behzadmehr, A. 2009. A new model for calculating the effective viscosity of nanofluids. *Journal of Applied Physics D* 42:055501.

[20] Mishra, P.C., Mukherjee, S., Nayak, S.K., Panda, A. 2014. A brief review on viscosity of nanofluids. *International Nano Letters* 4:109–120.

[21] Prasher, P., Song, D., Wang, J. 2006. Measurements of nanofluid viscosity and its implications for thermal applications. *Applied Physics Letters* 89:133108.

[22] Graham, A. 1981. On the viscosity of suspensions of solid spheres. *Applied Scientific Research* 37:275–286.

[23] Mahbubul, I.M., Saidur, R., Amalina, M.A. 2012. Latest developments on the viscosity of nanofluids. *International Journal of Heat and Mass Transfer* 55:874–885.

[24] Krieger, J.M., Dougherty, T.J. 1957. A mechanism for non-Newtonian flow in suspensions of rigid spheres. *Transactions of the Society of Rheology* 3:137–152.

[25] Szymański, J., Wilk, A., Hołyst, R., Roberts, G., Sinclair, K., Kowalski, A. 2008. Micro- and macro-shear viscosity in dispersed lamellar phases. *Journal of Non-Newtonian Fluid Mechanics* 148:134–140.

[26] He, Y., Jin, Y., Chen, H., Ding, Y., Cang, D., Lu, H. 2007. Heat transfer and flow behaviour of aqueous suspensions of TiO_2 nanoparticles (nanofluids) flowing upward through a vertical pipe. *International Journal of Heat and Mass Transfer* 50:2272–2281.

[27] Nguyen, C.T., Desgranges, F., Roy, G., Galanis, N., Mare, T., Boucher, S., Mintsa, H. 2007. Temperature and particle-size dependent viscosity data for water-based nanofluids—hysteresis phenomenon. *International Journal of Heat and Fluid Flow* 28:1492–1506.

[28] Namburu, P.K., Kulkarni, D.P., Dandekar, A., Das, D.K. 2007. Experimental investigation of viscosity and specific heat and Silicon Dioxide nanofluids. *Micro & Nano Letters* 2:67–71.

[29] Chevalier, J., Tillement, O., Ayela, F. 2008. Rheological properties of nanofluids flowing through microchannels. *Applied Physics Letters* 91:233103.

[30] Timofeeva, E.V., Smith, D.S., Yu, W., France, D.M., Singh, D., Routbo, J.L. 2010. Particle size and interfacial effects on thermo-physical and heat transfer characteristics of water-based α-SiC nanofluids. *Nanotechnology* 21:215703.

[31] Rudyak, V.Y., Belkin, A.A., Egorov, V.V. 2009. On the effective viscosity of nanosuspensions. *Technical Physics* 54:1102–1109.

[32] Michaelides, E.E. 2013. Transport properties of nanofluids. A critical review. *Journal of Non-Equilibrium Thermodynamics* 38:1–79.

[33] Hashimoto, T., Fujimura, M., Kawai, H. 1980. Domain-boundary structure of styrene-isoprene block co-polymer films cast from solutions. *Macromolecules* 13:660–669.

[34] Li, Z.H., Gong, Y.J., Pu, M., Wu, D., Sun, Y.H., Wang, J., Liu, Y., Dong, B.Z. 2001. Determination of interfacial layer thickness of a pseudo two-phase system by extension of the Debye equation. *Journal of Physics D: Applied Physics* 34:2085–2088.

[35] Yu, C.J., Richter, A.G., Datta, A., Durbin, M.K., Dutta, P. 2000. Molecular layering in a liquid on a solid substrate: An X-ray reflectivity study. *Physica B* 283:27–31.

[36] Xue, L., Keblinski, P., Phillpot, S.R., Choi, S.U.S., Eastman, J.A. 2004. Effect of liquid layering at the liquid-solid interface on thermal transport. *International Journal of Heat and Mass Transfer* 47:4277–4284.

[37] Murshed, S.M.S., Leong, K.C., Yang, C. 2009. A combined model for the effective thermal conductivity of nanofluids. *Applied Thermal Engineering* 29:2477–2483.

[38] Barnes, H.A., Hutton, J.F., Walters, K. 1989. *An Introduction to Rheology.* Amsterdam: Elsevier.

[39] Goodwin, J.W., Hughes, R.W. 2000. *Rheology for Chemists—An Introduction.* Cambridge: The Royal Society of Chemistry.

[40] Wang, B.X., Zhou, L.P., Peng, X.F. 2003. A fractal model for predicting the effective thermal conductivity of liquid with suspension of nanoparticles. *International Journal of Heat and Mass Transfer* 46:2665–2672.

[41] Prasher, R., Evans, W., Meakin, P., Fish, J., Phelan, P., Keblinski, P. 2006. Effect of aggregation on thermal conduction in colloidal nanofluids. *Applied Physics Letters* 89:143119.

[42] Hui, P.M., Zhang, X., Markworth, A.J., Stroud, D. 1999. Thermal conductivity of graded composites: Numerical simulations and an effective medium approximation. *Journal of Materials Science* 34:5497–5503.

[43] Prasher, R., Bhattacharya, P., Phelan, P.E. 2006. Brownian-motion-based convective-conductive model for the effective thermal conductivity of nanofluids. *Journal of Heat Transfer* 128:588–595.

[44] Jang, S.P., Choi, S.U.S. 2007. Effects of various parameters on nanofluid thermal conductivity. *Journal of Heat Transfer* 129:617–623.

[45] Prasher, R., Bhattacharya, P., Phelan, P.E. 2005. Thermal conductivity of nanoscale colloidal solutions (nanofluids). *Physical Review Letters* 94:025901.

[46] Einstein, A. 1906. Eine neue bestimmung der molekul-dimension. *Annalen der Physik* 34:289–306.

[47] Batchelor, G.K. 1977. Effect of Brownian-motion on the bulk stress in a suspension of spherical particles. *Journal of Fluid Mechanics* 83:97–117.

[48] Prigogine, I. 1961. *Introduction to Thermodynamics of Irreversible Processes.* New York: Interscience.

[49] Boukary, M., Lebon, G. 1986. A comparative analysis of binary fluid mixtures by extended thermodynamics and the kinetic theory. *Physica* 137A:546–572.

[50] Jou, D., Casas-Vazquez, J., Criado-Sancho, M. 2011. *Thermodynamics of Fluids under Flow*, Second ed. Heidelberg: Springer.

[51] Depireux, N., Lebon, G. 2001. An extended thermodynamic modelling of non-Fickian diffusion. *Journal of Non-Newtonian Fluid Mechanics* 96:105–117.

[52] Lebon, G., Jou, D. 2014. Early history of extended irreversible thermodynamics (1953–1983): An exploration beyond local equilibrium and classical transport theory. *European Physics Journal H* 40:205–240.

[53] Cimmelli, V.A., Jou, D., Ruggeri, T., Van, P. 2014. Entropy principle and recent results in non-equilibrium theories. *Entropy* 16:1756–1807.

[54] Clarke, N., McLeish, T.C.B. 1998. Shear flow on phase separation of entangled polymer blends. *Physical Review* 57:R3731–R3734.

[55] Guyer, R.A., Krumhansl, J.A. 1966. Solution of the linearized Boltzmann phonon equation. *Physical Review* 148:766–778.

[56] Jou, D., Casas-Vàzquez, J., Lebon, G. 2010. *Extended Irreversible Thermodynamics*, fourth ed. New York: Springer.

[57] Lebon, G., Jou, D., Casas-Vazquez, J. 2008. *Understanding Non-Equilibrium Thermodynamics.* Berlin: Springer.

[58] Machrafi, H. 2016. Heat transfer at nanometric scales described by extended irreversible thermodynamics. *Communication Applied Industrial Mathematics* 7:177–195.

[59] Choi, S.U.S. 1995. Enhancing thermal conductivity of fluids with nanoparticles, In: Siginer, D.A., Wang, H.P. (Eds.), *Developments Applications of Non-Newtonian Flows* 231:99–105. New-York: ASME.

[60] Machrafi, H., Lebon, G. 2016. The role of several heat transfer mechanisms on the enhancement of thermal conductivity in nanofluids. *Continuum Mechanics and Thermodynamics* 28:1461–1475.

[61] Michaelidis, E.E. 2013. Transport properties of nanofluids. A critical review. *Journal of Non-Equilibrium Thermodynamics* 38:1–31.

[62] Alvarez, F.X., Jou, D. 2007. Memory and the non-local effects in heat transport from diffusive and ballistic regimes. *Applied Physics Letters* 90:083109.

[63] Lebon, G., Machrafi, H., Grmela, M. 2015. An extended irreversible thermodynamic modelling of size-dependent thermal conductivity of spherical nanoparticles dispersed in homogeneous media. *Proceedings of Royal Society A* 471:20150144.

[64] Sellitto, A., Cimmelli, V.A., Jou, D. 2016. *Mesoscopic Theories of Heat Transport in Nanosystems.* Berlin: Springer.

[65] Lebon, G., Machrafi, H. 2019. Effective thermal conductivity of nanostuctures: A review, *Atti Accademia Peloritana dei Perilocanti* 96:A14.

[66] Jamal-Abadi, M.T., Zamzamian, A.H. 2013. Optimization of thermal conductivity of Al_2O_3 nanofluid by using ANN and GRG methods. *International Journal Nanoscience Nanotechnology* 9:177–184.

[67] Xie, H., Wang, J., Xi, T., Liu, Y., Ai, F., Wu, Q. 2002. Thermal conductivity enhancement of suspensions containing nanosized alumina particles. *Journal Applied Physics* 91:4568.

[68] Witharana, S., Weliwata, J.A. 2012. Suspended nanoparticles as a way to improve thermal energy transfer efficiency, *ICIAFS conference*, Beijing, China.

[69] Beck, M.P., Yuan, Y., Warrier, P., Teja, A.S. 2009. The effect of particle size on the thermal conductivity of alumina nanofluids. *Journal Nanoparticle Research* 11:1129–1136.

[70] Li, C.H., Peterson, G.P. 2007. The effect of particle size on the effective thermal conductivity. *Journal Applied Physics* 101:044312.

[71] Kaviany, M. 2011. *Essentials of Heat Transfer*. Cambridge: Cambridge University Press.

[72] Lee, S., Choi, S.U.S., Li, S., Eastman, J.A. 1999. Measuring thermal conductivity of fluids containing oxide nanoparticles. *Journal Heat Transfer* 121:280–289.

[73] Beck, M.P., Yuan, Y., Warrier, P., Teja, A.S. 2010. The thermal conductivity of alumina nanofluids in water, ethylene glycol, and ethylene glycol + water mixtures. *Journal Nanoparticle Research* 12:1469–1477.

[74] Wang, X.W., Xu, X.F., Choi, S.U.S. 1999. Thermal conductivity of nanoparticle-fluid. *Journal Thermophysics Heat Transfer* 13:474–480.

[75] Koo, J., Kleinstreuer, C. 2004. A new thermal conductivity model for nanofluids. *Journal Nanoparticle Research* 6:577–588.

[76] Gupta, A., Kumar, R. 2007. Role of Brownian motion on the thermal conductivity enhancement of nanofluids. *Applied Physics Letters* 91:223102.

[77] Das, S.K., Choi, S.U.S., Yu, W., Pradeep, T. 2008. *Nanofluids: Science and Technology*. Hoboken, NJ: Wiley.

[78] Nie, C., Marlow, W., Hassan, Y.A. 2008. Discussion of proposed mechanism of thermal conductivity enhancement in nanofluids. *International Journal of Heat and Mass Transfer* 51:1342–1348.

[79] Shima, P.D., Philip, J., Raj, B. 2009. Role of microconvection induced by Brownian motion of nanoparticles in the enhanced thermal conductivity of stable nanofluids. *Applied Physics Letters* 94:223101.

[80] Kleinstreuer, C., Feng, Y. 2011. Experimental and theoretical studies of nanofluid thermal conductivity enhancement: A review. *Nanoscale Letters* 6:229.

[81] Evans, W., Fish, J., Keblinski, P. 2006. Role of Brownian motion hydrodynamics on nanofluid thermal conductivity. *Applied Physics Letters* 88:093116.

[82] Fang, K.C., Weng, C.I., Ju, S.P. 2006. An investigation into the structural features and thermal conductivity of silicon nanoparticles using molecular dynamics simulations. *Nanotechnology* 17:3909–3914.

[83] Sridhara, V., Satapathy, L.N. 2011. Al_2O_3-based nanofluids: A review. *Nanoscale Letters* 6:456.

[84] Pastoriza-Gallego, M.J., Casanova, C., Páramo, R., Barbés, B., Legido, J.L., Piñeiro, M.M. 2009. A study on stability and thermophysical properties (density and viscosity) of Al_2O_3 in water nanofluid. *Journal of Applied Physics* 106:064301.

[85] Lu, W.Q., Fan, Q.M. 2008. Study for the particle's scale effect on some thermophysical properties of nanofluids by a simplified molecular dynamics method. *Engineering Analysis with Boundary Elements* 32:282–289.

[86] Rudyak, V.Y., Krasnolutskii, S.L. 2014. Dependence of the viscosity of nanofluids on nanoparticle size and material. *Physics Letters A* 378:1845–1849.

[87] Chapman, S., Cowling, T.C. 1970. *The Mathematical Theory of Non-Uniform Gases*. Cambridge: Cambridge University Press.

[88] Firlar, E., Cinar, S., Kashyap, S., Akinc, M., Prozorov, T. 2015. Direct visualization of the hydration layer on alumina nanoparticles with the fluid cell STEM in situ. *Scientific Reports* 5:9830.

[89] Perkins, S.J. 1986. Protein volumes and hydration effects. The calculations of partial specific volumes, neutron scattering match points and 280-nm absorption coefficients for proteins and glycoproteins from amino acid sequences. *European Journal of Biochemistry* 157:169–180.

[90] Lebon, G., Machrafi, H. 2018. A thermodynamic model of nanofluid viscosity based on a generalized Maxwell-type constitutive equation. *Journal of Non-Newtonian Fluid Mechanics* 253:1–6.

7

Nanoporous Flow and Permeability

7.1 Porous Flow

In nanoporous materials, the permeability of the pores plays a fundamental role in the derivation of the constitutive relations and the fluid flow characteristics. In the present approach, we will focus on the permeability of nanopores and derive its expression in terms of some relevant material parameters, such as the mean free path of the fluid particles, the hydrophobicity of the wall (slip-factor) and the pore size. It should also be stressed that at small characteristic lengths, the no-slip boundary condition between the fluid and the solid is no longer valid because in nanomaterials, the boundary layer, also called the Knudsen layer, has a characteristic length comparable to the dimensions of the systems, and therefore, its influence will be felt everywhere inside the whole system. The situation is comparable to that observed in rarefied gases and microfluidics, where it is currently admitted that slipping is important at the boundaries. Here, we will adhere to this point of view, and the dependence of permeability with respect to this parameter coupled to the pore size will be the main subject of this work.

The porous medium will be modelled as a two-phase system, constituting a (incompressible) fluid flowing through nano (rigid) elements. Following the lines of thought effective properties, one is led to a momentum equation for the fluid flow generalizing Darcy's constitutive law written as

$$\boldsymbol{u} = -\frac{K_{\mathit{eff}}}{\mu}\nabla p, \tag{7.1}$$

where the vector $\boldsymbol{u}$ is a characteristic velocity (the so-called seepage or Darcy velocity identified as the mean volumetric flow rate per unit area), μ is the dynamic viscosity of the fluid, K_{eff} the effective permeability and ∇p the pressure gradient along the fluid flow. It is important to note that the use of Darcy's law is limited to macroscopic pores where the influence of the wall is negligible and low Reynolds flows.

Our objective is to determine the effective permeability coefficient of nanopores in terms of relevant characteristics of the system as the mean free path of the fluid particles and the slippage length at the walls. Two particular nanopore's configurations, namely cylindrical and parallelepiped pores will be investigated.

The methodology developed in the forthcoming consists in calculating the average velocity flow, or flow rate, in the abovementioned geometrical pore configurations, as a function of the constant pressure difference along the symmetry axis. By comparison with Darcy's equation (7.1), one is then able to propose an expression of the effective permeability in terms of the parameters introduced in the model.

7.2 Nanoporous Flow

Fluid motion through a porous medium of porosity ε (the ratio of the volume occupied by the pores and the total volume) is modelled as a binary system constituted by a viscous incompressible fluid of mass density ρ_f moving with a velocity $\boldsymbol{v}_f$ in a non-deformable solid of mass density ρ_s moving at velocity $\boldsymbol{v}_s$. The diffusion flux of the fluid is defined as

$$\boldsymbol{J} = \varepsilon \rho_f \left(\boldsymbol{v}_f - \boldsymbol{v}\right), \tag{7.2}$$

with $\boldsymbol{v}$ the barycentric velocity, given by

$$\rho \boldsymbol{v} = \varepsilon \rho_f \boldsymbol{v}_f + (1 - \varepsilon)\, \rho_s \boldsymbol{v}_s, \tag{7.3}$$

where the total mass density ρ is $\rho = \varepsilon \rho_f + (1 - \varepsilon)\, \rho_s$.

7.2.1 Extended Constitutive Equation of the Mass Flux

For pedagogical purpose, let us study the simple problem of matter diffusion in a two-component mixture of mass fractions c_1 and c_2, the temperature T is assumed to be uniform. The main idea is to elevate the diffusion fluxes $\boldsymbol{J}_1$ and $\boldsymbol{J}_2$ to the status of independent variables, at the same level as the classical concentration variables. According to the definition of the barycentric velocity, it is directly seen that $\boldsymbol{J}_1 + \boldsymbol{J}_2 = 0$. Moreover, since $c_1 + c_2 = 1$, it follows that the set of independent variables is given by c_1 and $\boldsymbol{J}_1$. Assuming that the entropy $\mathcal{S}$ per unit mass of the system depends on both kinds of variables, one has $\mathcal{S} = \mathcal{S}(c_1, \boldsymbol{J}_1)$ or, in terms of the material time derivative

$$\frac{d\mathcal{S}}{dt} = \frac{\partial \mathcal{S}}{\partial c_1}\frac{dc_1}{dt} + \frac{\partial \mathcal{S}}{\partial \boldsymbol{J}_1} \cdot \frac{d\boldsymbol{J}_1}{dt} = -\frac{\chi}{T}\frac{dc_1}{dt} - \gamma_1 \boldsymbol{J}_1 \cdot \frac{d\boldsymbol{J}_1}{dt}, \tag{7.4}$$

where the use has been made of the classical definition $\frac{\chi}{T} = -\frac{\partial \mathcal{S}}{\partial c_1}$, with χ designating the difference $\chi_1 - \chi_2$ between the chemical potentials of both the constituents and where it has been assumed that $\frac{\partial s}{\partial \boldsymbol{J}_1}$ is a linear function of $\boldsymbol{J}_1$ with γ a phenomenological coefficient to be positive to guarantee that $\mathcal{S}$ is maximum at equilibrium [1]. Entropy is also assumed to obey a time evolution equation of the general form

$$\rho\frac{d\mathcal{S}}{dt} = -\nabla \cdot \boldsymbol{J}^s + \eta^s, \text{ with } \eta^s \geq 0. \tag{7.5}$$

η^s is the rate of entropy production imposed to be positive definite in virtue of the second law of thermodynamics and $\boldsymbol{J}^s$ is the entropy flux classically given by

$$\boldsymbol{J}^s = -\frac{\chi}{T}\boldsymbol{J}_1. \tag{7.6}$$

This result (Eq. 7.5) is easily obtained by setting $\gamma_1 = 0$ in Eq. (7.4) and substituting $\frac{dc_1}{dt}$ by the mass conservation law

$$\rho\frac{dc_1}{dt} = -\nabla \cdot \boldsymbol{J}_1. \tag{7.7}$$

By comparison with the time evolution (7.5) of $\mathcal{S}$, it is then directly checked that the expression of $\boldsymbol{J}^s$ is given by Eq. (7.6), whereas the entropy production is [1,2]

$$\eta^s = -\boldsymbol{J}_1 . \frac{\nabla \chi}{T} \geq 0. \tag{7.8}$$

However, in the presence of non-localities, which are especially relevant in micro and nanosystems, it is rather natural to admit that $\boldsymbol{J}^s$ depends, in addition, on the gradients of the diffusion flux $\boldsymbol{J}_1$, for example,

$$\boldsymbol{J}^s = -\frac{\chi}{T}\boldsymbol{J}_1 + \beta_1 \boldsymbol{J}_1 \cdot \nabla \boldsymbol{J}_1, \tag{7.9}$$

where β_1 is a coefficient to be determined later on. The final task consists in deriving the time evolution equation of the state variables. The one corresponding to the classical mass fraction variable is given by Eq. (7.7), whereas the time evolution equation of the diffusion flux is obtained by substituting Eqs. (7.4) and (7.9) in Eq. (7.5). The corresponding entropy production is now given by

$$\sigma^s = \boldsymbol{J}_1 \cdot \left(-\frac{\nabla \chi}{T} - \gamma_1 \frac{d\boldsymbol{J}_1}{dt} + \beta_1 \nabla^2 \boldsymbol{J}_1 \right) + \beta_1 \nabla \boldsymbol{J}_1 \otimes \nabla \boldsymbol{J}_1 \geq 0, \tag{7.10}$$

with $\otimes$ standing for the tensorial product. The simplest way guaranteeing the positiveness of the relation (7.10) is to assume that there exists a linear relation between the flux $\boldsymbol{J}_1$ and its conjugated force represented by the terms between parenthesis and that β_1 is a positive factor. To summarize, one is led to

$$\boldsymbol{J}_1 = \frac{1}{\nu_1} \left(-\frac{\nabla \chi}{T} - \gamma_1 \frac{d\boldsymbol{J}_1}{dt} + \beta_1 \nabla^2 \boldsymbol{J}_1 \right), \quad \gamma \geq 0 \tag{7.11}$$

where ν_1 is a positive phenomenological coefficient in order to meet the condition $\eta^s \geq 0$. Expressing the chemical potential χ in terms of c_1 leading to $\nabla \chi = (\partial \chi / \partial c_1) \nabla c_1$ and introducing the notations

$$\frac{\gamma_1}{\nu_1} = \tau, \; \frac{\beta_1}{\nu_1} = \ell^2 \frac{1}{T\nu_1} \frac{\partial \chi}{\partial c_1} = \rho D, \tag{7.12}$$

Expression (7.11) takes the more familiar form

$$\tau \frac{\partial \boldsymbol{J}_1}{\partial t} + \boldsymbol{J}_1 = -\rho D \nabla c_1 + \ell^2 \nabla^2 \boldsymbol{J}_1, \tag{7.13}$$

where τ has the dimension of time and can be interpreted as the time relaxation of the diffusion flux, ℓ has the dimension of length and can be seen as the mean free path of the component. Finally, letting τ and ℓ tending to zero, relation (7.13) reduces to Fick's law

$$\boldsymbol{J}_1 = -\rho D \nabla c_1, \tag{7.14}$$

with D standing for the classical diffusion coefficient.

In the forthcoming, we will identify component 1 with the fluid and restrict the analysis to small velocities so that the material time derivative $\frac{d}{dt} = \frac{\partial}{\partial t} + \boldsymbol{v} \cdot \nabla$ can be substituted by the partial time derivative ∂_t and all the non-linear terms in the velocity (as $\boldsymbol{v} \cdot \nabla \boldsymbol{v}$) can be neglected. Designating the kinematic viscosity, accordingly, the basic general time evolution of the diffusion flux of the fluid will read as

$$\tau \partial_t \boldsymbol{J} + \boldsymbol{J} = -\rho D \nabla c + \ell^2 \nabla^2 \boldsymbol{J}, \tag{7.15}$$

where $\boldsymbol{J}$ is given by expression (7.2). This kind of constitutive equations can also be obtained for several other physical phenomena, and the method is proven to be validated [3].

7.2.2 The Basic Momentum Equation

Equation (7.15) is the keystone of the future developments. Substituting the definition (7.2) of the mass flux in Eq. (7.15) and taking into account the incompressibility of the fluid, one is led to

$$\tau\varepsilon\rho_f\partial_t\left(\boldsymbol{v}_f-\boldsymbol{v}\right)=-\rho D\nabla c-\rho_f\varepsilon\left(\boldsymbol{v}_f-\boldsymbol{v}\right)+\ell^2\nabla^2\rho_f\varepsilon\left(\boldsymbol{v}_f-\boldsymbol{v}\right). \tag{7.16}$$

To eliminate the term in $\partial_t\boldsymbol{v}$, we make use of the momentum equation

$$\rho\partial_t\boldsymbol{v}=-\nabla p+\nabla\cdot\boldsymbol{\sigma}, \tag{7.17}$$

where p is the hydrostatic pressure, σ the stress tensor given by Newton's constitutive law $\boldsymbol{\sigma}=\mu\nabla\boldsymbol{v}$ as the fluid is assumed to be Newtonian and μ designating the kinematic viscosity. In Eq. (7.17), external body forces are omitted. The system under study consisting in a binary mixture of fluid and solid, it is justified to formulate the momentum equation in terms of the barycentric velocity rather than the fluid velocity. In the particular case that the solid is at rest ($\boldsymbol{v}_s=0$), $\boldsymbol{v}$ is directly related to $\boldsymbol{v}_f$ with $\boldsymbol{v}=\varepsilon\frac{\rho_f}{\rho}\boldsymbol{v}_f$ in virtue of Eq. (7.3). Using this result and the momentum equation to eliminate $\boldsymbol{v}$ and $\partial_t\boldsymbol{v}$ in Eq. (7.16), one obtains the following time evolution equation of the fluid flow through the pores,

$$\rho\partial_t\boldsymbol{v}_f=-\nabla p+\mu\varepsilon\frac{\rho_f}{\rho}\nabla^2\boldsymbol{v}_f-\frac{(1-\varepsilon)\,\rho_s}{\tau}\boldsymbol{v}_f+\ell^2\frac{(1-\varepsilon)\,\rho_s}{\tau}\nabla^2\boldsymbol{v}_f, \tag{7.18}$$

In Eq. (7.18), the term involving diffusion has been omitted as it is generally negligible. In view of future developments, let us introduce the so-called absolute permeability, K_0, of the porous medium, defined through

$$\frac{K_0}{\mu}\equiv\frac{\varepsilon}{1-\varepsilon}\frac{\tau}{\rho_s}. \tag{7.19}$$

This result stems from comparison of a steady-state ($\partial_t\boldsymbol{v}_f=0$) and local $\left(\nabla^2\boldsymbol{v}_f=0\right)$ version of Eq. (7.18) with Darcy's law [Eq. (7.1)] after that $\boldsymbol{u}$ has been identified as $\boldsymbol{u}=\boldsymbol{v}_f\varepsilon$. Note that the ratio of the absolute permeability and the viscosity depends essentially on the relative volume and mass fraction of both the constituents. Under steady conditions and in terms of K_0, expression (7.18) reads as

$$-\nabla p-\frac{\mu\varepsilon}{K_0}\boldsymbol{v}_f+\mu\varepsilon\left(\frac{\rho_f}{\rho}+\frac{\ell^2}{K_0}\right)\nabla^2\boldsymbol{v}_f=0. \tag{7.20}$$

Relation (7.20) is the basic equation of our work expressing the velocity field through nanopores under steady conditions. The first term at the left-hand side is the classical pressure gradient term, the second represents an extra contribution to the momentum balance due to porosity, the third term is associated to the fluid viscosity and the last one is a consequence of the nano-properties of the pores. By omitting this last term ($\ell^2/L^2_{ref}\ll 1$, with L_{ref} a reference length), we obtain a Brinkman-like equation. On the other hand, by letting in (7.20) K_0 tend to infinity, one finds back Navier–Stokes relation. Finally, omitting non-local contribution $\left(\ell^2/L^2_{ref}\ll 1\right)$ and assuming that $K_0/L^2_{ref}\ll 1$, it is found that

$$\varepsilon\boldsymbol{v}_f=-\frac{K_0}{\mu}\nabla p, \tag{7.21}$$

and, after substituting in Eq. (7.21) $\boldsymbol{v}_f$ by its mean value $\langle\boldsymbol{v}_f\rangle$ as defined below, one finds back Darcy's law (7.1).

The particular cases discussed previously show the flexibility of the formalism, which is valid for both the porous and the non-porous systems and at both the nanoscale and macroscales. It is worth stressing that in this work. Darcy's relation is introduced as a particular case of the momentum equation rather than a phenomenological relation, such as Fourier's, Fick's or Ohm's laws.

At nano-length scales, the boundary layer between the fluid and the solid's wall has a characteristic length comparable to the pore dimensions, and therefore, its influence will be felt in the whole material. The usual no-slip boundary condition is no longer valid, and the slip flows may become relevant; the situation is similar to that observed in microfluidics and in rarefied gas dynamics wherein slippery conditions at the walls are important.

Our objective in the forthcoming is to determine the effective permeability coefficient in terms of relevant characteristics of the system as the mean free path of the fluid particles, the slippage length at the walls and the size of the nanopores. This is achieved by calculating the fluid velocity field as a function of the imposed pressure gradient and by identifying the effective permeability K_{eff} by strict comparison with Darcy's expression (7.21), i.e. being equivalent to the velocity divided by $-\frac{\nabla p}{\mu\varepsilon}$. Two particular configurations will be considered, namely fluid flow through porous nanoducts of circular and rectangular cross sections.

7.2.3 Absolute Permeability

The expression for the absolute permeability can be found by taking the asymptotic limit $\varepsilon \to 1$ of a porous medium and stating that this should be equal to Poiseuille flow through a large cylinder (with, of course, absence of porous material, i.e. $\rho_s \to 0$) with an equivalent overall flow rate or mean velocity. For a large (well above nanoscopic dimensions) cylinder, this means that Eq. (7.9) becomes

$$\frac{\partial p}{\partial x} - \mu \frac{1}{r} \frac{\partial}{\partial r}\left(r \frac{\partial v_f}{\partial r} \right) = 0 \tag{7.22}$$

which represents a simple steady-state Poiseuille flow. The solution of Eq. (7.22) for a non-slip boundary (at $r = R$) and a maximum velocity in the middle of the cylinder ($r = 0$) is

$$v_f = \frac{\left(r^2 - R^2\right)}{4\mu} \frac{\partial p}{\partial x} \tag{7.23}$$

The average velocities are defined by

$$\langle v_f \rangle = \frac{1}{\pi R^2} \int_0^R 2\pi r v_f\left(r\right) dr, \tag{7.24}$$

for the circular pores, and

$$\langle v_f \rangle = \frac{1}{WH} \int_0^W \int_0^H v_f\left(z\right) dy dz = \frac{1}{H} \int_0^H v_f\left(z\right) dz, \tag{7.25}$$

for the parallelepiped pores. The mean velocity for Eq. (7.23) [using Eq. (7.24) and taking $\frac{\partial p}{\partial x} \equiv -\frac{\Delta p}{L}$] is then

$$\langle v_f \rangle \left(R\right) = \frac{R^2}{8\mu} \frac{\Delta p}{L} \tag{7.26}$$

For the porous medium, Darcy's law predicts that the mean velocity from Eq. (7.21), with the asymptotic limit $\varepsilon \to 1$, is

$$\langle v_f \rangle = \frac{K_0}{\mu} \frac{\Delta p}{L} \tag{7.27}$$

Equalling Eqs. (7.26) and (7.27) leads finally to the absolute permeability K_0 of the cylindrical pore

$$K_0 = R^2/8 \tag{7.28}$$

It should be noted that K_0 is implicitly dependent on the porosity because the pore size represents actually a mean hydraulic radius, which in practice is a function of the characteristics of the porous material. There is therefore no contradiction between relations (7.28) and (7.19). The same procedure can be repeated for the parallelepiped configuration, resulting into

$$K_0 = H^2/12. \tag{7.29}$$

7.3 Effective Permeability

In the following, the fluid velocity is supposed to remain constant in the flow direction (say the x-direction) due to the assumption that the pore length L is much larger than the lateral dimensions normal to the fluid flow (represented by the aforementioned reference length L_{ref}). For a circular nanoduct of radius R, this means that $L \gg R$. For the parallelepiped pore, it is assumed that perpendicularly to the flow direction, a selected dimension, say the width W (defined in the y-direction), is much larger than the other one, say the height H (defined in the z-direction), so that $L \gg W \gg H$. This amounts to consider a flow between two parallel plates separated by a distance H. Introducing relevant boundary conditions, we will in the next subsections determine the one-dimensional fluid velocity profile as a function of the perpendicular coordinates ($\boldsymbol{v}_f \Rightarrow v_f(r)$ or $v_f(z)$), respectively, r designating the radial coordinate for the cylindrical pores and z the distance measured along the height of the parallelepiped pores. The average velocities are then calculated using Eqs. (7.24) and (7.25), for the cylindrical and parallelepiped pores, respectively. The corresponding flow rates are $Q_R = \pi R^2 \langle v_f \rangle$ and $Q_H = WH \langle v_f \rangle$, respectively. In the two next subsections, we will determine the effective permeability of the nanopores with circular and rectangular cross sections, respectively. Since the purpose of the present work is to obtain an analytic expression of the effective permeability, we have restricted our approach to a one-dimensional configuration.

7.3.1 Nanopores with Circular Cross Sections

Assuming that the velocity profile only changes in the radial direction and remains uniform in the axial direction, Eq. (7.20) becomes

$$\frac{\varepsilon\mu}{K_0} v_f - \mu\varepsilon \left(\frac{\rho_f}{\rho} + \frac{\ell^2}{K_0} \right) \frac{1}{r} \frac{\partial}{\partial r} \left(r \frac{\partial v_f}{\partial r} \right) = -\frac{\partial p}{\partial x}, \tag{7.30}$$

At $r = 0$, the centre of the pore, the velocity is assumed to be maximum, meaning that

$$\frac{\partial v_f}{\partial r}|_{r=0} = 0. \tag{7.31}$$

It is well recognized that in nanopores, the no-slip condition is no longer valid. Here, we will substitute it by the following second-order slip boundary condition at $r = R$ (e.g. [4–8]). Similarly, slip boundary conditions were also introduced by [9,10] in their study of heat transport at nanoscales. The boundary condition is

$$v_f(R) = -C_1 \ell_s \frac{\partial v_f}{\partial r}|_{r=R} - \beta_S C_2 \ell_s^2 \frac{1}{R} \frac{\partial}{\partial r} \left(r \frac{\partial v_f}{\partial r} \right)|_{r=R}, \tag{7.32}$$

where ℓ_s is the slipping length whereas C_1 and C_2 are coefficients, which are often taken to be constant. In reality, they are dependent on the system's geometry and fluid/porous matrix properties, as some studies reveal [11,12]. These coefficients, called "slip correction factors" (SCF), are a function of the material properties and the system's geometry. The quantity β_S is introduced in Eq. (7.32) in order to consider simultaneously both first-order ($\beta_S \equiv 0$) and second-order slip boundary conditions ($\beta_S \equiv 1$).

Let a constant pressure difference $\Delta p\,(= p\,(x = L) - p\,(x = 0))$ act along the symmetry axis, i.e. $\frac{\partial p}{\partial x} \equiv -\frac{\Delta p}{L}$ and introduce two non-dimensional numbers $B_s \equiv \frac{\ell_s}{R}$ and $Kn \equiv \frac{\ell}{R}$. The first number, B_s, stands for the non-dimensional slippage friction factor and the second one, Kn, is the Knudsen number associated to the molecular mean free path. We are now able to solve Eq. (7.30) for $v_f\,(r)$, the result being

$$v_f(r) = \frac{K_0}{\varepsilon\mu}\frac{\Delta p}{L}$$
$$\left(1 - \frac{2\left(\rho_f \frac{K_0}{R^2} + \rho Kn^2\right)\ {}_0\mathcal{F}_1\left[1, \frac{\rho}{4\rho_f\frac{K_0}{R^2}+4\rho Kn^2}\frac{r^2}{R^2}\right]}{C_1\rho B_s\ {}_0\mathcal{F}_1\left[2, \frac{\rho}{4\rho_f\frac{K_0}{R^2}+4\rho Kn^2}\right] + 2\left(\rho_f\frac{K_0}{R^2} + \rho\left(\beta_S C_2 B_s^2 + Kn^2\right)\right)\ {}_0\mathcal{F}_1\left[1, \frac{\rho}{4\rho_f\frac{K_0}{R^2}+4\rho Kn^2}\right]}\right) \tag{7.33}$$

where ${}_0\mathcal{F}_1$ is the confluent regularized hypergeometric function. The values of the slip correction factors C_1 and C_2, introduced in the boundary condition (7.32), will be determined by imposing that the velocity is minimum, truly zero, at slip length $\ell_s = -R$. Zero velocity, through Eq. (7.1), is equivalent to zero effective permeability. A second condition is necessary to assure mathematically that zero permeability is the strict minimum. In summary, C_1 and C_2 should be thus that

$$\lim_{\ell_s \to -R} K_{eff} = 0\ ; \qquad \lim_{\ell_s \to -R} \frac{\partial K_{eff}}{\partial \ell_s} = 0 \tag{7.34}$$

Using Eq. (7.34), one finds the expressions for the SCF, given in Table 7.1.

Taking the mean value of v_f and multiplying by ε, we obtain the expression of the seepage velocity u. We are now in a position to compare our result with Darcy's law (7.1) and to derive the corresponding expression of the effective permeability, which is given by

TABLE 7.1
SCF Corresponding to Eq. (7.33).

Related to Equation	C_1	C_2
(7.33)	$\frac{{}_0\mathcal{F}_1\left[3, \frac{2\rho}{\rho_f+8\rho Kn^2}\right]}{{}_0\mathcal{F}_1\left[2, \frac{2\rho}{\rho_f+8\rho Kn^2}\right]}$	$\frac{{}_0\mathcal{F}_1\left[3, \frac{2\rho}{\rho_f+8\rho Kn^2}\right]}{8\ {}_0\mathcal{F}_1\left[1, \frac{2\rho}{\rho_f+8\rho Kn^2}\right]}$
(7.33a)	$\frac{2Kn\ \mathfrak{B}\left[2, \frac{1}{Kn}\right]}{\mathfrak{B}\left[1, \frac{1}{Kn}\right]}$	$\frac{Kn^2\mathfrak{B}\left[2, \frac{1}{Kn}\right]}{\mathfrak{B}\left[0, \frac{1}{Kn}\right]}$
(7.33c)	$\frac{{}_0\mathcal{F}_1\left[3, \frac{2\rho}{\rho_f}\right]}{{}_0\mathcal{F}_1\left[2, \frac{2\rho}{\rho_f}\right]}$	$\frac{{}_0\mathcal{F}_1\left[3, \frac{2\rho}{\rho_f}\right]}{8\ {}_0\mathcal{F}_1\left[1, \frac{2\rho}{\rho_f}\right]}$

$$K_{eff} = \frac{R^2}{8}$$

$$\left(1 - \frac{2\left(\rho_f \frac{K_0}{R^2} + \rho Kn^2\right) {}_0\mathcal{F}_1\left[2, \frac{\rho}{4\rho_f \frac{K_0}{R^2} + 4\rho Kn^2}\right]}{C_1 \rho B_s \, {}_0\mathcal{F}_1\left[2, \frac{\rho}{4\rho_f \frac{K_0}{R^2} + 4\rho Kn^2}\right] + 2\left(\rho_f \frac{K_0}{R^2} + \rho\left(\beta_S C_2 B_s^2 + Kn^2\right)\right) {}_0\mathcal{F}_1\left[1, \frac{\rho}{4\rho_f \frac{K_0}{R^2} + 4\rho Kn^2}\right]}\right) \tag{7.35}$$

where K_0 has been substituted by the result (7.28) and C_1 and C_2 are given in Table 7.1. The porosity dependence of the effective permeability results from the presence of the terms K_0 (here $R^2/8$) and ρ [see under Eq. (7.3)]. It follows from relation (7.35) that the effective permeability is a function of the fluid and solid densities, the porosity, the pore radius, the fluid molecules interactions via the Knudsen number Kn and the slip of molecules through the non-dimensional number B_s.

It may be of interest for application purposes to derive particular expressions of the effective permeability in some asymptotic cases say: (a) important ($Kn \gg 1$) and negligible non-local effects ($Kn \ll 1$) respectively, (b) first-, second-order or no-slip conditions or (c) high or low absolute permeability $\left(K_0/L_{ref}^2 \gg 1 \text{ or } \ll 1\right)$ and their combinations thereof. Three particular situations are examined in Table 7.1. The particular situations corresponding to the absence of slipping ($B_s \to 0$) or second-order contribution ($\beta_S \to 0$) can easily be derived from Table 7.2 and has therefore not been explicitly considered. For the sake of clarity, the various types of models associated to the above particular asymptotic cases are explicitly mentioned.

In Table 7.2, $\mathfrak{B}$ represents the Bessel-I function.

7.3.2 Nanopores with Parallelepiped Cross Sections

Let us consider a one-dimensional flow along the x-axis in a parallelepiped duct of lateral dimension $W \gg H$, with H designating the thickness, and assume a steady-state situation. Under steady conditions and absence of external forces, the momentum Eq. (7.20) reads as

$$\frac{\varepsilon\mu}{K_0} v_f - \mu\varepsilon\left(\frac{\rho_f}{\rho} + \frac{\ell^2}{K_0}\right)\frac{\partial^2 v_f}{\partial z^2} = -\frac{\partial p}{\partial x}, \tag{7.36}$$

We follow the same procedure as in the previous sub-section with the maximum fluid velocity at half the height of the channel, i.e.

$$\frac{\partial v_f}{\partial z}|_{z=H/2} = 0. \tag{7.37}$$

The boundary conditions are

$$\text{at } z = H/2\colon \frac{\partial v_f}{\partial z} = 0 \,(\text{maximum velocity}), \tag{7.38}$$

$$\text{at } z = 0\colon v_f = C_1 \ell_s \frac{\partial v_f}{\partial z} - \beta_S C_2 \ell_s^2 \frac{\partial^2 v_f}{\partial z^2}, \tag{7.39}$$

Calculating the average fluid velocity via Eq. (7.25) and using Eq. (7.1), we are able to identify the effective permeability as

TABLE 7.2
Asymptotic Expressions of Eq. (7.35) with Second-Order Slip Conditions.

Limits	Equation	Model	K_{eff} with $B_s \equiv \ell_s/R$ and $Kn \equiv \ell/R$
$K_0/L_{ref}^2 \ll 1, Kn \gg 1$	(7.35a)	Darcy model for nanopores	$\frac{R^2}{8}\left(\frac{Kn^2\mathfrak{B}\left[2,\ \frac{1}{Kn}\right]+C_1 B_s Kn\mathfrak{B}\left[1,\ \frac{1}{Kn}\right]+\beta_S C_2 B_s^2\mathfrak{B}\left[0,\ \frac{1}{Kn}\right]}{C_1 B_s Kn\mathfrak{B}\left[1,\ \frac{1}{Kn}\right]+(\beta_S C_2 B_s^2+Kn^2)\mathfrak{B}\left[0, \frac{1}{Kn}\right]}\right)$
$K_0/L_{ref}^2 \ll 1, Kn \ll 1$	(7.35b)	Darcy model for macropores	$\frac{R^2}{8}$
$K_0/L_{ref}^2 = O\left(1\right),\ Kn \ll 1$	(7.35c)	Brinkman-like model for porous systems	$\frac{R^2}{8}\left(1-\frac{\rho_f\ {}_0\mathcal{F}_1\left[2, \frac{2\rho}{\rho_f}\right]}{4C_1\rho B_s\ {}_0\mathcal{F}_1\left[2, \frac{2\rho}{\rho_f}\right]+(\rho_f+8\rho\beta_S C_2 B_s^2)\ {}_0\mathcal{F}_1\left[1, \frac{2\rho}{\rho_f}\right]}\right)$

$$K_{eff} = \frac{H^2}{12} \left(1 - \frac{\frac{2}{\sqrt{\rho}}\left(\frac{K_0}{H^2}\rho_f + \rho Kn^2\right)^2 \sinh\left(\frac{1}{2}\sqrt{\frac{\rho}{\frac{K_0}{H^2}\rho_f + \rho Kn^2}}\right)}{\sqrt{\frac{K_0}{H^2}\rho_f + \rho Kn^2}\left(\frac{K_0}{H^2}\rho_f + \rho\left(Kn^2 + \beta_S C_2 B_s^2\right)\right)\cosh\left(\frac{1}{2}\sqrt{\frac{\rho}{\frac{K_0}{H^2}\rho_f + \rho Kn^2}}\right) + \sqrt{\rho}C_1 B_s\left(\frac{K_0}{H^2}\rho_f + \rho Kn^2\right)\sinh\left(\frac{1}{2}\sqrt{\frac{\rho}{\frac{K_0}{H^2}\rho_f + \rho Kn^2}}\right)} \right) \tag{7.40}$$

The values of the slip correction factors C_1 and C_2, introduced in the boundary condition (7.39) are, this time, obtained by imposing that the velocity is minimum, truly zero, at slip length $\ell_s = -H/2$. In the same manner as in Eq. (7.34), the use is made of

$$\lim_{\ell_s \to -H/2} K_{eff} = 0; \qquad \lim_{\ell_s \to -H/2} \frac{\partial K_{eff}}{\partial \ell_s} = 0 \tag{7.41}$$

which leads to the factors C_1 and C_2 given in Table 7.3

The corresponding expressions in the same asymptotic cases as for Eq. (7.35) are shown in Table 7.4. Here, it is assumed that $W \gg H$ (the case $W \ll H$ leads of course to the same result, replacing H by W). For W of the same order of magnitude as H, $W = O(H)$, Eq. (7.40) and the asymptotic derivations in Table 7.4 remain applicable by replacing H by $\hat{H} \equiv \frac{HW}{H+W}$.

Moreover, in view of application to experiments, we need not only the expression of the effective permeability, but also that of the effective viscosity, since non-local effects influence the viscosity as well (e.g. [13,14]). Therefore, instead of Eq. (7.1), we will rather use

$$\boldsymbol{u} = -\frac{K_{eff}}{\mu_{eff}} \nabla p, \tag{7.42}$$

where μ_{eff} is the effective viscosity taking into account the size effects that nano-confinement has on a pure liquid.

7.3.3 Effective Viscosity

It is well recognized that the viscosity of a fluid flowing in nanopores will be influenced by the pore characteristics. In Chapter 6, the effect of the presence of nanostructures dispersed in a fluid was studied within the framework of Extended Thermodynamics. The principles and methods used there can be implemented in the present chapter as well. The main steps of the analysis may be summarized as follows.

1. The space of the state variables is extended by including the pressure tensor denoted $\boldsymbol{P}^{(1)}$ and his higher moments $\boldsymbol{P}^{(2)}, \ldots, \boldsymbol{P}^{(n)}$ with n tending to infinity.

2. These extra variables obey a hierarchy of linearized time evolution equations of the general form

$$\beta_{n-1}\nabla \boldsymbol{P}^{(n-1)} - \gamma_n \partial_t \boldsymbol{P}^{(n)} + \beta_n \nabla \cdot \boldsymbol{P}^{(n+1)} = \mathrm{v}_n \boldsymbol{P}^{(n)} \; n = 1, 2, \; \ldots \tag{7.43}$$

where γ_n, β_n and v_n are phenomenological coefficients related to relaxation times of the variables, correlation length and transport coefficients.

TABLE 7.3
SCF Corresponding to Eq. (7.40).

Related to Equation	C_1	C_2
(7.40)	$2\sqrt{\frac{\rho_f+12\rho Kn^2}{3\rho}}\coth\left(\sqrt{\frac{3\rho}{\rho_f+12\rho Kn^2}}\right)-\frac{2\rho_f}{3\rho}-12Kn^2$	$\left(\frac{\rho_f+12\rho Kn^2}{3\rho}\right)\left(1-\sqrt{\frac{\rho_f+12\rho Kn^2}{3\rho}}\tanh\left(\sqrt{\frac{3\rho}{\rho_f+12\rho Kn^2}}\right)\right)$
(7.40a)	$2\sqrt{\frac{\rho_f}{3\rho}}\coth\left(\sqrt{\frac{3\rho}{\rho_f}}\right)-\frac{2\rho_f}{3\rho}$	$\frac{\rho_f}{3\rho}\left(1-\sqrt{\frac{\rho_f}{3\rho}}\tanh\left(\sqrt{\frac{3\rho}{\rho_f}}\right)\right)$
(7.40c)	$4Kn\ \coth\left(\frac{1}{2Kn}\right)-8Kn^2$	$4Kn^2-8Kn^3\tanh\left(\frac{1}{2Kn}\right)$

TABLE 7.4
Asymptotic Evaluations for Eq. (7.40) with Second-Order Slip Conditions.

Limits	Equation	Model	K_{eff} with $B_s\equiv\ell_s/H$ and $Kn\equiv\ell/H$
$K_0/L_{ref}^2\ll 1, Kn\gg 1$	(7.40a)	Darcy model for nanopores	$\frac{H^2}{12}\left(1-\frac{2Kn^3\ \tanh\left(\frac{1}{2Kn}\right)}{Kn^2+\beta_S C_2 B_s^2+C_1 B_s Kn\ \tanh\left(\frac{1}{2Kn}\right)}\right)$
$K_0/L_{ref}^2\ll 1, Kn\ll 1$	(7.40b)	Darcy model for macropores	$\frac{H^2}{12}$
$K_0/L_{ref}^2=O(1), Kn\ll 1$	(7.40c)	Brinkman-like model for porous system	$\frac{H^2}{12}\left(1-\frac{\rho_f\sqrt{\rho_f}}{6\rho\sqrt{\rho_f}C_1B_s+\left(12\rho\sqrt{3\rho}\beta_S C_2 B_s^2+\rho_f\sqrt{3\rho}\right)\coth\left(\sqrt{\frac{3\rho}{\rho_f}}\right)}\right)$

3. Applying a Fourier transform to the set (7.43), one obtains a generalized Newton's constitutive law which, for steady situations, reads as

$$\widehat{\boldsymbol{P}}(\boldsymbol{k}) = -i\boldsymbol{k}\mu_{eff}(\boldsymbol{k})\widehat{\boldsymbol{v}}(\boldsymbol{k}), \tag{7.44}$$

with an upper hat designating Fourier's transforms, $\boldsymbol{v}$ the velocity field and $\boldsymbol{k}$ the wave number whereas the viscosity $\mu_{eff}(\boldsymbol{k})$ is expressed by the following $\boldsymbol{k}$-dependent continued fraction

$$\mu_{eff}(\boldsymbol{k}) = \frac{\mu}{1+\frac{k^2 l_1^2}{1+\frac{k^2 l_2^2}{1+\ldots}}}, \tag{7.45}$$

where μ is the viscosity of the bulk fluid in the absence of nanostructures and l_n $(n = 1, 2, \ldots)$ are correlation lengths defined by $l_n^2 = \beta_n^2/\gamma_n\gamma_{n+1}$ (e.g. [1]).

4. In the applications considered previously, there is only one reference length so that it is natural to select a one-dimensional wave number k as given by $k = 2\pi/R$ and $k = 2\pi/H$ for cylindrical and parallelepiped pores, respectively. Moreover, introducing a reference length l, identified as the mean free path of the fluid particles through $l_n^2 = l^2 n^2/(4n^2 - 1)$ (a well-known kinetic relation) and letting $n \to \infty$, one obtains the final expression of the effective viscosity in terms of Kn, namely

$$\mu_{eff} = \frac{3\mu}{4\pi^2 Kn^2}\left(\frac{2\pi\, Kn}{\text{Arctan}\,(2\pi\, Kn)} - 1\right). \tag{7.46}$$

7.4 Asymptotic Limits

Before comparing our model to experimental data in the next section, it is interesting to have a general insight on the behaviour of the permeability in terms of the size of the system and the parameters ℓ and ℓ_s. Our objective is to search for mathematical coherency and asymptotic behaviours. The influence of R (or H), ℓ and ℓ_s on the effective permeability will be investigated via the dimensionless numbers Kn and B_s. Since the influence of ε has already been studied intensively in the past, it will be disregarded here and will be imposed a priori equal to $\varepsilon = 0.5$. For the sake of concision, we introduce the notion of the relative effective permeability obtained by dividing the effective permeability by the absolute one (K_0):

$$K_{eff}^{rel} = \frac{K_{eff}}{K_0}. \tag{7.47}$$

This parameter is typically a measure of non-locality and fluid slip at the wall. In Tables 7.5 and 7.6 are given the asymptotic values of K_{eff}^{rel} as a function of Kn and B_s for the cylindrical and rectangular pores, respectively. We examine successively the asymptotic cases $Kn \to 0$, (macroscopic scale) and $Kn \to \infty$ (nanoscale), coupled with $B_s \to 0$ (no slip) and $B_s \to \infty$ (full slip), respectively. The value $Kn \to 0$, disregarding slip ($B_s \to 0$), corresponds mathematically to $R \to \infty$ or $H \to \infty$, whereas $Kn \to \infty$ and $B_s \to \infty$ amounts at taking $R \to 0$ or $H \to 0$. Only second-order slip is considered here ($\beta_S \to 1$). The limiting values $B_s = -1$ (cylinder) or $B_s = -\frac{1}{2}$ (parallelepiped) describe fluids asymptotically at rest.

In order to apprehend the behaviour of the permeability, the lengths ℓ and ℓ_s are selected as: $\ell = 1$ nm and $\ell_s = 4$ nm (so that $B_s \neq 0$) together with $\frac{\rho_f}{\rho} = \frac{1}{2}$ (which amounts to $\rho_s = 3\rho_f$ and $\varepsilon = 0.5$). In Figures 7.1 and 7.2 are plotted the permeability K_{eff}^{rel} as a function of the dimensions R (cylinder) and H (parallelepiped).

TABLE 7.5
Asymptotic Values of $K_{eff}^{rel}(R)$.

Equation	$Kn \to 0$	$Kn \to \infty$	$Kn \to 0$	$Kn \to \infty$
	$B_s \to 0$	$B_s \to 0$	$B_s \to \infty$	$B_s \to \infty$
Equivalent limits	$R \to \infty$	$\ell_s \ll R \ll \ell$	$\ell_s \gg R \gg \ell$	$R \to 0$
(7.35)	$1 - \frac{{}_0\mathcal{F}_1[2,2\rho/\rho_f]}{{}_0\mathcal{F}_1[1,2\rho/\rho_f]}$	0	1	$\frac{C_2 B_s^2}{Kn^2+C_2 B_s^2}$
(7.35a)	1	0	1	$\frac{C_2 B_s^2}{Kn^2+C_2 B_s^2}$
(7.35b)	1	1	1	1
(7.35c)	$1 - \frac{{}_0\mathcal{F}_1[2,2\rho/\rho_f]}{{}_0\mathcal{F}_1[1,2\rho/\rho_f]}$	$1 - \frac{{}_0\mathcal{F}_1[2,2\rho/\rho_f]}{{}_0\mathcal{F}_1[1,2\rho/\rho_f]}$	1	1

TABLE 7.6
Asymptotic Values of $K_{eff}^{rel}(H)$.

Equation	$Kn \to 0$	$Kn \to \infty$	$Kn \to 0$	$Kn \to \infty$
	$B_s \to 0$	$B_s \to 0$	$B_s \to \infty$	$B_s \to \infty$
Equivalent limits	$H \to \infty$	$\ell_s \ll H \ll \ell$	$\ell_s \gg H \gg \ell$	$H \to 0$
(7.40)	$1 - \sqrt{\frac{\rho_f}{3\rho}}\tanh\left(\sqrt{\frac{3\rho}{\rho_f}}\right)$	0	1	$\frac{C_2 B_s^2}{Kn^2+C_2 B_s^2}$
(7.40a)	1	0	1	$\frac{C_2 B_s^2}{Kn^2+C_2 B_s^2}$
(7.40b)	1	1	1	1
(7.40c)	$1 - \sqrt{\frac{\rho_f}{3\rho}}\tanh\left(\sqrt{\frac{3\rho}{\rho_f}}\right)$	$1 - \sqrt{\frac{\rho_f}{3\rho}}\tanh\left(\sqrt{\frac{3\rho}{\rho_f}}\right)$	1	1

Some comments are in form. First, we observe a similar behaviour for both the configurations. Moreover, the values of K_{eff}^{rel} remain unchanged, when R, or H, take the values smaller or larger than those given in Figures 7.1 and 7.2. Let us focus on the limiting values $R(H) \to 0$ and $R(H) \to \infty$. At $R \to \infty$, one has $K_{eff}^{rel} = 1 - \frac{{}_0\mathcal{F}_1[2,4]}{{}_0\mathcal{F}_1[1,4]} \approx 0.57$ and $K_{eff}^{rel} = 1 - \frac{\tanh(\sqrt{6})}{\sqrt{6}} \approx 0.60$ for the cylindrical and parallelepiped pores, respectively. For $R \to 0$, it is found that $K_{eff}^{rel} = \frac{C_2\ell_s^2}{\ell^2+C_2\ell_s^2} \approx 0.67$ with $C_2 = \frac{1}{8}$ for the cylindrical pores and $\frac{C_2\ell_s^2}{\ell^2+C_2\ell_s^2} \approx 0.84$ with $C_2 = \frac{1}{3}$ for the parallelepiped pores.

Second, note that for $R \to \infty$ and $H \to \infty$, the standard and non-local Darcy models (black dot dashed and green dashed curves) tend to the asymptotic value of 1, while our model (black solid curves) and Brinkman's one (red dotted curves) lead to the aforementioned asymptotic values. For $R \to 0$ and $H \to 0$, one finds back the same asymptotic values for the present and the non-local Darcy models, whereas the Brinkman and standard Darcy's models tend asymptotically to 1. Figures 7.1 and 7.2 are only drawn for values of R and H ranging from 10^{-12} to 10^{-6}, because outside this domain, the variations of K_{eff}^{rel} are minute.

Third, our model presents typically a maximum value at a characteristic length which is neither the case for Brinkman's nor non-local Darcy's models.

The reason why our model does not tend to unity at small sizes is due to non-local effects, while slipping effects prevent K_{eff}^{rel} to go to zero (see columns 4 and 5 of Tables 7.5

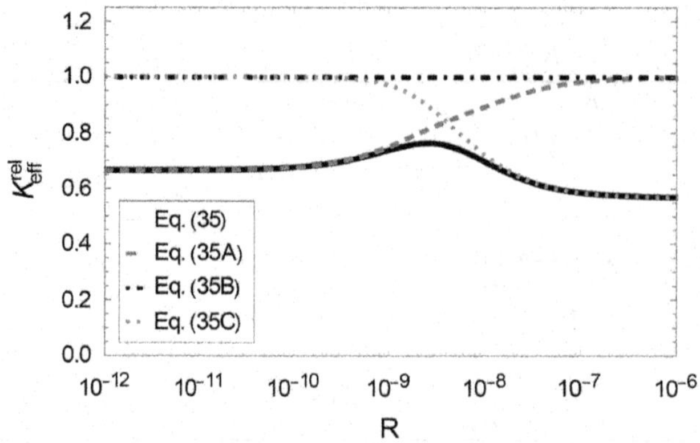

FIGURE 7.1
$K_{eff}^{rel}(R)$ for Eqs. (7.35)–(7.35c), for $\ell = 1\,\text{nm}$, $\ell_s = 4\,\text{nm}$, $\rho_s = 3\rho_f$ and $\varepsilon = 0.5$. ——— Our model, -·-·-·- Darcy standard, - - - - Darcy non-local and ·········· Brinkmann. (Modified from: Machrafi, H., Lebon, G. 2018. Fluid flow through porous and nanoporous media within the prisme of extended thermodynamics: emphasis on the notion of permeability. *Microfluidics and Nanofluidics* 22:65.)

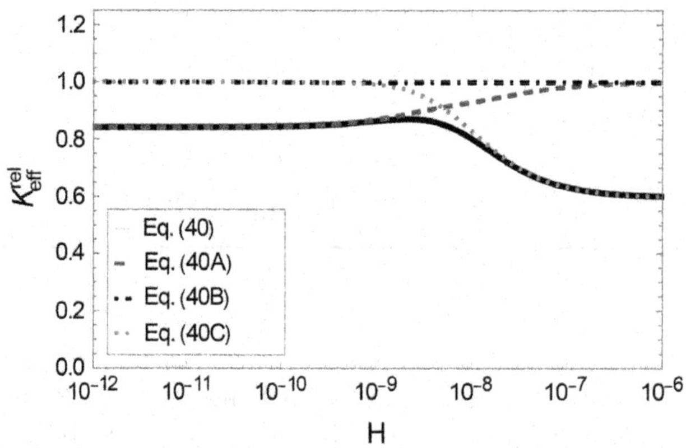

FIGURE 7.2
$K_{eff}^{rel}(H)$ for Eqs. (7.40)–(7.40c), for $\ell = 1\,\text{nm}$, $\ell_s = 4\,\text{nm}$, $\rho_s = 3\rho_f$ and $\varepsilon = 0.5$. ——— Our model, -·-·-·- Darcy standard, - - - - Darcy non-local and ·········· Brinkmann. (Modified from: Machrafi, H., Lebon, G. 2018. Fluid flow through porous and nanoporous media within the prisme of extended thermodynamics: emphasis on the notion of permeability. *Microfluidics and Nanofluidics* 22:65.)

and 7.6). These results are confirmed by the asymptotic value of $\frac{C_2 B_s^2}{Kn^2 + C_2 B_s^2}$, which is non-zero in the presence of slipping and equal to 1 in the absence of non-local effects. Note, however, that, in contrast with the results predicted by Darcy's model, K_{eff}^{rel} does not tend to unity at large sizes. This reduction of permeability can easily be understood. Indeed, Darcy's approach ignores viscous effects, which, at large sizes, are governed by parabolic-like velocity profiles, whose value in the mean is lower than that of plug-like flows in Darcy's models. By increasing the dimension, either R or H, the viscous drag becomes dominant with respect to the slipping effects, resulting in a lower permeability.

7.5 Case Study: Flow in Nanoporous Glass

In this subsection, our model will be validated against experimental results. In that respect, it is convenient to define an overall flow rate (not the flow rate through a single pore) through the porous medium, i.e. $Q_{tot} = \frac{A\varepsilon u}{T_p}$, with u designating the seepage velocity, T_P the porous medium tortuosity and $A = \pi\mathcal{R}^2$, with $\mathcal{R}$ the radius of the porous material as a whole (for the cylindrical configuration). Making use of Eq. (7.1) for u, one is led to

$$Q_{tot} = \pi\mathcal{R}^2 \frac{\varepsilon}{T_P} \frac{K_{eff}}{\mu_{eff}} \frac{\Delta p}{L}, \tag{7.48}$$

where K_{eff} is given by Eq. (7.35) and μ_{eff} by Eq. (7.46). The values of Q_{tot} derived from our model are compared with experimental data [11,12] for water and n-hexane flowing through nanoporous Vycor glass. Table 7.7 gives the material properties for this case study.

In the literature (e.g. [15–17]), the mean free path in fluids is often identified with the intermolecular distance of the molecules. It is worth to stress that for liquids, the intermolecular distance is often of the same order of magnitude as the molecule size, suggesting that the molecule size is a pertinent approximation. This motivated our choice for water and n-hexane in Table 7.7. The slip lengths are obtained from experimental measurements in stagnant fluid layers in nanoscale conduits for water [18] and n-hexane [19] in contact with silica channel walls. The values found here are of the same order of magnitude as the ones calculated a posteriori by [12]. The other material properties for the nanoporous system are taken from the literature [11,12].

The volumetric flows Q_{tot} of water and n-hexane, respectively, as a function of the externa applied pressure drop, are represented in Figures 7.3 and 7.4. For the sake of comparison, we have also plotted the values obtained from Darcy's model.

A comparison between theoretical and experimental results shows a satisfactory agreement. It is worth to stress that this chapter has shown that the model expressed by Eq. (7.35) [suggesting also the use of Eq. (7.40)] predicts much better results than Darcy's law [20].

TABLE 7.7
Material Properties of Water, n-Hexane and Vycor Glass.

Material Properties	Water	n-Hexane
ℓ [nm]	0.3	0.6
ℓ_s [nm]	−0.7	−0.5
ρ_f [kg m^{-3}]	1,000	655
μ_f at 25°C [Pa s]	$8.9*10^{-4}$	$3.0*10^{-4}$
	Vycor glass	
ρ_s [kg m^{-3}]	2,650	
ε [−]	0.32	
$\mathcal{R}$ [m]	$3*10^{-3}$	
L [m]	$4*10^{-3}$	
T_p [−]	3.6	
R [nm]	3.4 and 5	

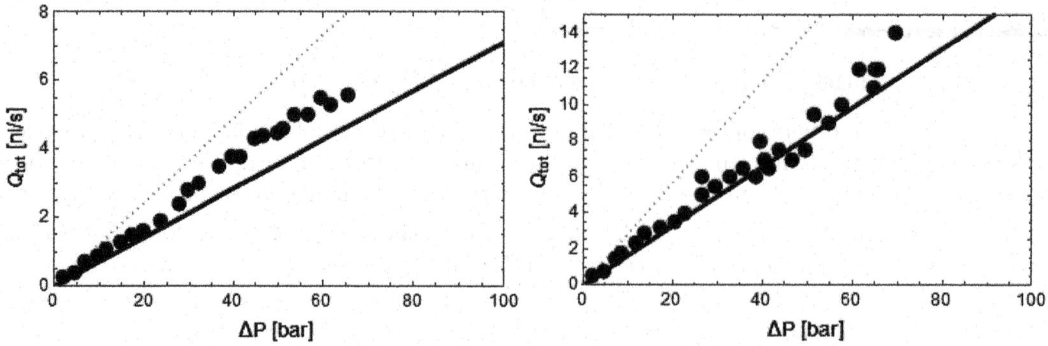

FIGURE 7.3
Flow rate of water (in nL/s) through nanoporous Vycor glass with nanopores of dimensions (a) 3.4 nm and (b) 5 nm and porosity of 0.32. Solid circles represent experimental data, the solid line corresponds to our model Eq. (7.35), whereas the red dotted line refers to Darcy's law. (From: Machrafi, H., Lebon, G. 2018. Fluid flow through porous and nanoporous media within the prisme of extended thermodynamics: emphasis on the notion of permeability. *Microfluidics and Nanofluidics* 22:65.)

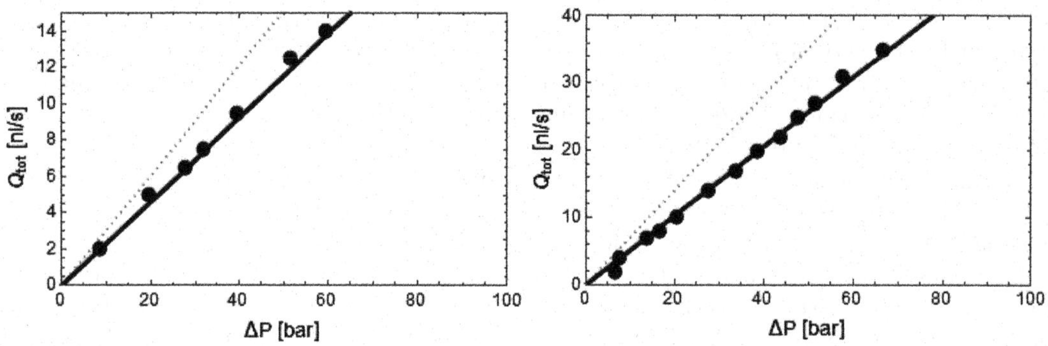

FIGURE 7.4
Flow rate of n-hexane through nanoporous Vycor glass with nanopores of dimensions (a) 3.4 nm and (b) 5 nm and porosity of 0.32. Solid circles represent experimental data, the solid line corresponds to our model Eq. (7.35) and the red dotted line to Darcy's law. (From: Machrafi, H., Lebon, G. 2018. Fluid flow through porous and nanoporous media within the prisme of extended thermodynamics: emphasis on the notion of permeability. *Microfluidics and Nanofluidics* 22:65.)

References

[1] Jou, D., Casas-Vàzquez, J., Lebon, G. 2010. *Extended Irreversible Thermodynamics*, fourth ed. Berlin: Springer.

[2] DeGroot, S.R., Mazur, P. 1962. *Non-equilibrium Thermodynamics.* Amsterdam: North-Holland Publishing.

[3] Machrafi, H., Lebon, G. 2016. General constitutive equations of heat transport at small length and high frequencies with extension to mass and electrical scales transport. *Applied Mathematics Letters* 22:30–37.

[4] Priezjev, N.K., Troian, S.M. 2006. Influence of wall rougness on the slip behavior at liquid/solid interfaces: Molecular-scale simulations versus continuum predictions. *Journal of Fluid Mechanics* 554:25–46.

[5] Priezjev, N.K. 2013. Molecular dynamics simulations of Couette flows with slip boundary conditions. *Microfluidics and Nanofluidics* 14:225–233.

[6] Yong, X, Zhang, L.T. 2013. Slip in nanoscale shear flow mechanisms of interfacial friction. *Microfluidics and Nanofluidics* 14:229–308.

[7] Manjare, M., Ting, W.Y., Yang, B., Zhao, Y.D. 2014. Hydrophobic catalytic Janus motors: Slip boundary condition and enhanced catalytic reaction rate. *Applied Physics Letters* 104:054102.

[8] Cherevko, V., Kizilova, N. 2017. Complex flows of immiscible microfluids and nanofluids with velocity slip boundary conditions, in: Nanophysics, Nanomaterials, Interface Studies, and Applications, *International Conference Nanotechnology Nanomaterials*, pp. 207–228, Lviv, Ukraine.

[9] Lebon, G. 2014. Heat conduction at micro and macro scales: A review through the prism of Extended Irreversible Thermodynamics. *Journal of Nonequilibrium Thermodynamics* 39:35–59.

[10] Sellitto, A., Cimmelli, V.A., Jou, D. 2016. *Mesoscopic Theories of Heat Transport in Nanosystems*. Berlin: Springer.

[11] Gruener, S., Huber, P. 2011. Imbibition in mesoporous silica: Rheological concepts and experiments on water and a liquid crystal. *Journal of Physics: Condensed Matter* 23:184109.

[12] Gruener, S., Wallacher, D., Greulich, S., Busch, M., Huber, P. 2016. Hydraulic transport across hydrophilic and hydrophobic nanopores: Flow experiments with water and n-hexane. *Physical Review E* 93:013102.

[13] Machrafi, H., Lebon, G. 2016. The role of several heat transfer mechanisms on the enhancement of thermal conductivity in nanofluids. *Continuum Mechanics and Thermodynamics* 28:1461–1475.

[14] Lebon, G., Machrafi, H. 2018. A thermodynamic model of nanofluid viscosity based on a generalized Maxwell-type constitutive equation. *Journal of Non-Newtonian Fluid Mechanics* 253:1–6.

[15] Arlemark, E.J., Dadzie, S.K., Reese, J.M. 2010. An extension to the Navier-Stokes equations to incorporate gas molecular collisions with boundaries. *Journal of Heat Transfer* 132:041006.

[16] Saeki, A., Koizumi, Y., Aida, T., Seki, S. 2012. Comprehensive approach to intrinsic charge carrier mobility in conjugated organic molecules, macromolecules, and supramolecular architectures. *Accounts of Chemical Research* 45:1193–1202.

[17] Hus, M., Urbic, T. 2012. Strength of hydrogen bonds of water depends on local environment. *The Journal of Chemical Physics* 136:144305.

[18] Alibakhshi, M.A., Xie, Q., Li, Y., Duan, C. 2016. Accurate measurement of liquid transport through nanoscale conduits. *Scientific Reports* 6:24936.

[19] Qiao, S.Z, Bhatia, S.K., Nicholson, D. 2004. Study of hexane adsorption in nanoporous MCM-41 Silica. *Langmuir* 20:389–395.

[20] Machrafi, H., Lebon, G. 2018. Fluid flow through porous and nanoporous media within the prisme of extended thermodynamics: Emphasis on the notion of permeability. *Microfluidics and Nanofluidics* 22:65.

8

Opto-Thermoelectric Coupling for Photovoltaic Energy

8.1 State of the Art

Developments in renewable energy seek to alleviate the global energy crisis and reduce its impact on the environment. One way is to use solar energy. By means of photovoltaic (PV) solar cells, photonic energy is mainly converted into electricity and waste heat. PV cells have relatively low conversion efficiency because they can only utilize part of the incident solar energy due to its given band gap and often require hybrid configurations [1]. One way to increase the efficiency of PV cells is using thermal management by means of heat sinks [2]. Otherwise, the waste heat can be used in order to be converted to more electricity via thermoelectric (TE) devices [3–6]. As a common PV cell converts a large amount of solar irradiant energy into heat, a hybrid PV cell and TE device (PVTE) may be a prospective way to improve the overall efficiency of solar energy [7]. One form of PVTE systems uses the so-called spectrum splitting concentrating system, where the photons with an energy out of the PV working wave band are incident to the TE devices, thereby generating electricity via the TE effect [8–10]. This system is complex, and the heat produced from the PV is still not used. Connecting the TE device directly at the dark side of the PV cell is simpler, and theoretically, all thermal energy can be used by the TE device [11], of which, the efficiency can be even increased by cooling the TE device [7,12]. It is the latter hybrid system that we consider in this work. As the efficiency in TE devices are proportional (though not necessarily linearly) on mainly the temperature difference across the device and the figure of merit, both are to be increased. The temperature difference stems from the operating system, while the figure of merit depends on the material properties. For this purpose, we propose to use a nanocomposite TE device, so that the figure of merit can be considerably increased, even doubled [5,6]. The optimization of the overall efficiency of the cooled PVTE systems is quite complicated, and many works have investigated this mostly by experiments [13–16] or by numerical methods [7–9]. Hereby, it appeared that working at nanoscale lengths, the efficiency could be increased [14]. Analytical models have also been developed, focussing sometimes on optical and electrical modelling [15], but at the cost of simplified physical phenomena [13,15,16]. We intend to develop an analytical model that is easy to use but still capture complex coupled optic, thermal and electric phenomena that are present in cooled PVTE systems with nanocomposite materials. Such a model can also be used as a support for understanding the mechanisms responsible for the performance of PV and TE devices.

As for the PV cell, the model takes into account the thickness dependence of the generation rate, the surface recombination velocity of electron and hole carriers (and indirectly passivation of dangling bonds at the cell's surface) and several recombination mechanisms (Shockley–Read–Hall [SRH], Auger and surface). As for the TE device, the model takes into account, the nanoparticle size, both phonon and electron scattering in both the nanoparticles and the bulk matrix, the nanoparticle volume fraction. In order to maximize the TE efficiency, not only should the figure of merit be increased by the introduction of

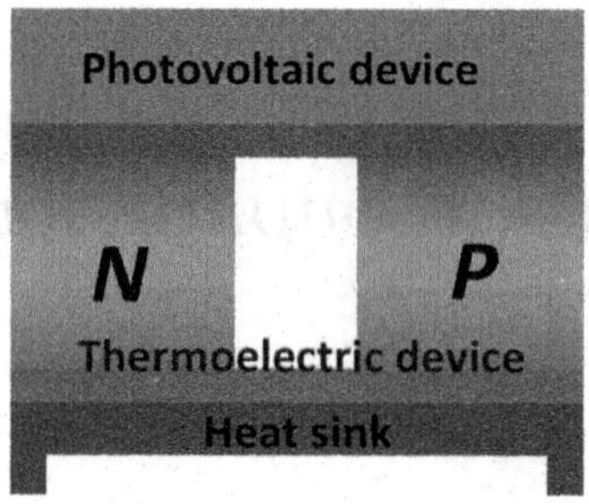

FIGURE 8.1
Schematic representation of a cell unit of the cooled PV TE hybrid system. (From: Machrafi, H. 2017. Enhancement of a photovoltaic cell performance by a coupled cooled nanocomposite thermoelectric hybrid system, using extended thermodynamics. *Current Applied Physics* 17:890–911.)

nanoparticles [6], but the temperature difference across it should also be optimized by means of a cooling (CL) device right under the TE device. By means of overall thermal modelling, it is the purpose to find optimal conditions for the PVTECL hybrid system. The PV power, the TE efficiency and the overall efficiency will be studied as a function of several parameters. Figure 8.1 shows a schematic representation of one cell unit of the hybrid system. The PV device is at the illuminated part on top of the system. Right under it, there is the TE device, consisting of an n- and p-leg. The hybrid system is cooled by means of a sun energy powered heat sink.

8.2 Nanostructured TE Model

8.2.1 TE Efficiency

Nanostructured materials can be used in different forms. Here, we are interested in nanocomposites. Nanocomposites are generally made out of a homogeneous matrix in which nanoparticles are dispersed. For nanoparticles with characteristic lengths of the same order of magnitude or smaller than the phonon and electron mean free paths, the Fourier theory, based on the classical approach of thermodynamics, is not able to predict the thermal interactions in the nanoparticles as well as between the nanoparticle and the bulk material. Therefore, we propose to investigate the problem of heat conduction in nanostructured TE materials following the principles set out in Chapters 2, 3 and 5 and also treated in the literature [5,17,18]. In this approach, the heat flux is elevated to the status of independent variable at the same footing as the temperature. The same goes for the electrical current.

We start with the phonon/electron model, presented by Eqs. (5.1)–(5.6). Equation (5.6) is rewritten, using our theory, following the dimensional form of (1.26) (with $N = 1$), which leads to

$$\tau_{ph}\frac{\partial \boldsymbol{q_{ph}}}{\partial t} + \boldsymbol{q_{ph}} = -\lambda_{ph}\nabla T + \ell_{ph}^2\nabla^2\boldsymbol{q_{ph}}, \tag{8.1a}$$

$$\tau_e\frac{\partial \boldsymbol{q_e}}{\partial t} + \boldsymbol{q_e} = -\left(\lambda_e + S\Pi\sigma_e\right)\nabla T + \ell_e^2\nabla^2\boldsymbol{q_e} + \Pi\sigma_e\boldsymbol{E}, \tag{8.1b}$$

$$\tau_I\frac{\partial \boldsymbol{I}}{\partial t} + \boldsymbol{I} = \sigma_e\left(\boldsymbol{E} - S\nabla T\right) + \ell_I^2\nabla^2\boldsymbol{I}, \tag{8.1c}$$

where τ, ℓ, Π, S, λ and σ are the relaxation time, mean free path, the Peltier coefficient, the Seebeck coefficient, the thermal conductivity and the electric conductivity, respectively, T the temperature, and the subscripts *ph* and *e* standing for phonon and electron, respectively. It should be noted that in Eq. (8.1a), the phonon transport is assumed to be dominated by normal phonon scattering, where Umklapp scattering is neglected [19]. The second term at the right-hand side stands for the non-locality. We suppose that the system is quasi-stationary, so that the energy balances are given by

$$\nabla \cdot \boldsymbol{q_{ph}} = 0, \tag{8.2a}$$

$$\nabla \cdot \boldsymbol{q_e} = \boldsymbol{E} \cdot \boldsymbol{I}, \tag{8.2b}$$

$$\nabla \cdot \boldsymbol{I} = 0 \tag{8.2c}$$

Using Eqs. (8.2), Eqs. (8.1) become with quasi-stationarity

$$\boldsymbol{q_{ph}} = -\lambda_{ph} \nabla T \tag{8.3a}$$

$$\boldsymbol{q_e} = -\lambda_e \nabla T + (\boldsymbol{E} - S\nabla T)\, \Pi \sigma_e \tag{8.3b}$$

$$\boldsymbol{I} = \boldsymbol{E}\sigma_e - S\sigma_e \nabla T \tag{8.3c}$$

From the abovementioned equations, we can deduce that

$$\boldsymbol{E} = \frac{\boldsymbol{I}}{\sigma_e} + S\nabla T \tag{8.4}$$

Inserting Eq. (8.4) in Eq. (8.3b), we find that for a one-dimensional system (having only scalar values) and boundary conditions $T(z=0) = T_c$ and $T(z=L) = T_h$ (with $\Delta T = T_h - T_c$):

$$q_{ph} = \lambda_{ph} \frac{\Delta T}{L} \tag{8.5a}$$

$$q_e = \lambda_e \frac{\Delta T}{L} + \Pi I \tag{8.5b}$$

Note by convention that the temperature gradient and electric field have the same direction, which is opposite to the one of the temperature difference, heat and current density fluxes. The TE efficiency is given by

$$\eta_{te} = \frac{P_e}{\dot{Q}} \tag{8.6}$$

The electric power output is given by

$$P_e = \boldsymbol{I} \cdot \int_0^L \boldsymbol{E} dy = I\Delta T S - I^2 L \sigma_e^{-1} \tag{8.7}$$

The total heat supplied, averaged over the TE element's length is given by

$$\dot{Q} = \frac{1}{L} \int_0^L \boldsymbol{q} dy = \lambda_{ph} \frac{\Delta T}{L} + \lambda_e \frac{\Delta T}{L} + \Pi I \tag{8.8}$$

Defining $\lambda_{tot} \equiv \lambda_{ph} + \lambda_e$, the efficiency is finally given by

$$\eta_{te} = \frac{SI\Delta T - I^2 L \sigma_e^{-1}}{\lambda_{tot} \frac{\Delta T}{L} + \Pi I} \tag{8.9}$$

The TE efficiency depends on an unknown current density, created by a temperature gradient. Defining $\Pi \equiv TS$ (recalling that material properties are taken at $T = T_{ref}$), it is easy to see that an optimal current density, taking $\partial \eta_{te}/\partial I = 0$, can be found to be

$$I_{opt} = \frac{\lambda_{tot} \Delta T \left(\sqrt{1+ZT} - 1\right)}{STL}, \tag{8.10}$$

so that a maximum TE efficiency can be found

$$\eta_{te,max} = \frac{\Delta T}{T_{te,h}} \frac{ZT + 2\left(1 - \sqrt{1+ZT}\right)}{ZT}, \tag{8.11}$$

where

$$ZT = T\frac{S^2}{\lambda_{tot}/\sigma_e} \tag{8.12}$$

the TE figure of merit. Here, it should kept in mind that $T_{h,te}$ is the upper temperature of the TE device, which will be derived from the bottom temperature of the PV device $T_{pv,h}$ in Section 8.5.2. As such, we define $\Delta T \equiv T_{te,h} - T_{te,c}$, where $T_{te,c}$ is the bottom temperature of the TE device. It can be seen that $\eta_{te,max}$ can be increased by increasing ZT. In principle, these equations are developed and valid for both the n- and p-legs of the TE element. We can define a TE efficiency for both the legs altogether by defining in Eqs. (8.11) and (8.12)

$$S = S_p - S_n \tag{8.13}$$

$$\frac{\lambda_{tot}}{\sigma_e} = \left(\sqrt{\frac{\lambda_{ph,p} + \lambda_{e,p}}{\sigma_{e,p}}} + \sqrt{\frac{\lambda_{ph,n} + \lambda_{e,n}}{\sigma_{e,n}}}\right)^2 \tag{8.14}$$

where the subscripts p and n denote the p- and n-legs of the TE element, respectively. Since the TE legs are made out of nanocomposites, the aforementioned material properties are dependent on the matrix and nanoparticle ones.

8.2.2 TE Material Properties for the Nanocomposite Legs

Equations (5.25)–(5.44) present the theory for TE nanocomposite materials. We treat here a reduced version thereof, for increased clarity, with necessary adaptations. The TE device has nanocomposite legs i, with $i = p,\ n$. The thermal conductivity of one leg is given by

$$\lambda_{c,i} = \lambda^m_{c,i} \frac{2\lambda^m_{c,i} + (1 + 2\alpha_{c,i})\,\lambda^p_{c,i} + 2\varphi\left[(1 - \alpha_{c,i})\,\lambda^p_{c,i} - \lambda^m_{c,i}\right]}{2\lambda^m_{c,i} + (1 + 2\alpha_{c,i})\,\lambda^p_{c,i} - \varphi\left[(1 - \alpha_{c,i})\,\lambda^p_{c,i} - \lambda^m_{c,i}\right]} \tag{8.15}$$

This expression is valid for both the phonons ($c = ph$) and the electrons ($c = e$). The analogy of the electron contribution with that of the phonon one is generally proposed throughout this development. From a physical point of view, the phonons and the electrons are considered as gas-like constituents, which behave as such in that they "flow" through the material lattice. We assume thusly that they also follow the same thermodynamic principles. The superscripts m and p concern the matrix and nanoparticle properties, φ the nanoparticle volume fraction and α is a dimensionless parameter describing the nanoparticle-matrix interaction:

$$\alpha_{c,i} = R_{c,i}\lambda^m_{c,i}/r_p. \tag{8.16}$$

Here, a_p is the nanoparticle radius ($d_p = 2a_p$ is the nanoparticle diameter) and the quantity R is the thermal boundary resistance coefficient given by

$$R_{c,i} = 4/c^m_{c,i}v^m_{c,i} + 4/c^p_{c,i}v^p_{c,i}, \tag{8.17}$$

where c_c is the phonon-specific heat and v_c the phonon group velocity. The phonon thermal conductivity of the bulk matrix is given by the classical Boltzmann phonon expression:

$$\lambda_{c,i}^{m} = \frac{1}{3}\left(C_{c,i}^{m} v_{c,i}^{m} \ell_{c,i}^{m}\right)|_{T_{ref}}, \tag{8.18}$$

where T_{ref} is the reference temperature, say the room temperature. Within the matrix, the phonons experience phonon–phonon interactions and the mean free path is given by the Matthiessen's rule:

$$\frac{1}{\ell_{c,i}^{m}} = \frac{1}{\ell_{c,i,b}^{m}} + \frac{1}{\ell_{c,i,coll}^{m}}. \tag{8.19}$$

with $\ell_{c,i,b}^{m}$ designating the mean free path in the bulk matrix and $\ell_{c,i,coll}^{m}$ the supplementary contribution due to the interactions at the particle–matrix interface given by

$$\ell_{c,i,coll}^{m} = \frac{4a_p}{3\varphi}, \tag{8.20}$$

The electric conductivity can be found through the electron thermal conductivity

$$\sigma_{e,i} = \frac{\lambda_{e,i}}{L_e T} \tag{8.21}$$

where L_e is the Lorentz number and T the absolute temperature. We also note that

$$\sigma_{e,i}^{m} = \lambda_{e,i}^{m}/L_e T_{amb} \tag{8.22}$$

The Lorentz number is determined by

$$L_e = \frac{\pi^2}{3}\left(\frac{k_B}{e_c}\right)^2, \tag{8.23}$$

where k_B is Boltzmann's constant and e_c the elementary charge. The Seebeck coefficient relates a temperature gradient with an electric current, albeit not directly. Indeed, the systems (8.1) and (8.3) show well that the thermal and electric conductivities precede the temperature gradient, while the Seebeck coefficient only precedes the temperature gradient in the form of a product with the electric conductivity. This motivates writing

$$(S_i\sigma_{e,i}) = \left(S_i^m\sigma_{e,i}^m\right)\frac{2\left(S_i^m\sigma_{e,i}^m\right) + (1+2\alpha_e)\left(S_i^p\sigma_{e,i}^p\right) + 2\varphi\left[(1-\alpha_e)\left(S_i^p\sigma_{e,i}^p\right) - \left(S_i^m\sigma_{e,i}^m\right)\right]}{2\left(S_i^m\sigma_{e,i}^m\right) + (1+2\alpha_e)\left(S_i^p\sigma_{e,i}^p\right) - \varphi\left[(1-\alpha_e)\left(S_i^p\sigma_{e,i}^p\right) - \left(S_i^m\sigma_{e,i}^m\right)\right]} \tag{8.24}$$

where S_i^m is the bulk value of the Seebeck coefficient of the matrix. The overall Seebeck coefficient of one leg can then be easily obtained a posteriori, by defining

$$S_i = (S_i\sigma_{e,i})/\sigma_{e,i}, \tag{8.25}$$

The TE model is complete when we find an expression for the phonon and the electron TE conductivity and the Seebeck coefficient of the nanoparticles, i.e. $\lambda_{ph,i}^{p}$, $\lambda_{e,i}^{p}$ and S_i^p. In order to find these, we have to take into account the size effects at nanoscale. The non-local effects that are introduced in this chapter for the phonon contribution also apply for the electron contribution of the thermal conductivity. As the non-local effects also apply for the electron contribution, the electrical conductivity can be treated in the same way in the same framework as well as the product of the Seebeck coefficient and the electric conductivity. The phonon thermal conductivity at nanoscale is given by

$$\lambda_{c,i}^{p} = \lambda_{c,i}^{p,0}\ f\left(Kn_c\right), \tag{8.26}$$

where $\lambda_{c,i}^{p,0}$ is the value of the phonon/electron thermal conductivity for the bulk material from which the nanoparticle is made of, the 0 indicating a reference value:

$$\lambda_{c,i}^{p,0} = \frac{1}{3}\left(c_{c,i}^{p} v_{c,i}^{p} \ell_{c,i}^{p}\right)|_{T_{ref}} \tag{8.27}$$

The Knudsen number is given by

$$Kn_c = \ell_{c,i}^{p}/a_p \tag{8.28}$$

We note that

$$\sigma_{e,i}^{p} = \lambda_{c,i}^{p}/L_e T \tag{8.29}$$

We have shown in a previous work [6] that

$$S_i^p \sim \left(\sigma_{e,i}^{p}\right)^{-1} \tag{8.30}$$

The only quantity still to be found is $f(Kn_c)$ a correction factor, taking into account the dimension of the nanoparticles. We derive this correction factor in section 8.3.

8.3 Nanoscale Material Properties

Let us consider for the purposes of this work a rigid and an isotropic body (with constant density), which is crossed by a heat flux $\boldsymbol{q}$ and an electric flux $\boldsymbol{I}$. Then, the relevant conserved variables are the internal energy u (or the temperature T) and the electric charge density ϱ_e, whereas the energy flux (here the heat flux vector $\boldsymbol{q}$) and the electric flux (here the electric current density $\boldsymbol{I}$) are the non-conserved flux variables so that the space of state variables is $\boldsymbol{V} = (u, \varrho_e, \boldsymbol{q}, \boldsymbol{I})$. However, in more complex materials, such as in nanomaterials, it is necessary to introduce fluxes of higher order. Therefore, we assume the existence of an entropy function $\mathcal{S}(\boldsymbol{V})$ [13], depending on the whole set $\boldsymbol{V}$ of variables: here $\mathcal{S} = \mathcal{S}(u, \varrho_e, \boldsymbol{q}, \boldsymbol{I})$, or in terms of time derivatives,

$$d_t\mathcal{S}(u, \varrho_e, \boldsymbol{q}, \boldsymbol{I}) = \frac{\partial \mathcal{S}}{\partial u} d_t u + \frac{\partial \mathcal{S}}{\partial \boldsymbol{q}} \cdot d_t \boldsymbol{q} + \frac{\partial \mathcal{S}}{\partial \varrho_e} d_t \varrho_e + \frac{\partial \mathcal{S}}{\partial \boldsymbol{I}} \cdot d_t \boldsymbol{I} \tag{8.31}$$

where u and $\mathcal{S}$ are measured per unit volume and a dot stands for the scalar product. The symbol d_t designates the time derivative which is indifferently the material or the partial time derivative as the system is, respectively, in motion or at rest. Here, $\mathcal{S}$ is assumed to be a concave function of the variables in order to guarantee stability of the equilibrium state obeying at the same time a general time-evolution equation of the form

$$d_t\mathcal{S} = -\nabla \cdot \boldsymbol{J}^s + \eta^s \quad (\eta^s \geq 0), \tag{8.32}$$

whose rate of production per unit volume η^s (in short, the entropy production) is positive definite to satisfy the second principle of thermodynamics, while the quantity $\boldsymbol{J}^s$ is the entropy flux. Let us define the local non-equilibrium temperature by $T^{-1}(u) = \partial\mathcal{S}/\partial u$ (fundamental thermodynamic relation at constant volume) and define $\partial\mathcal{S}/\partial\varrho_e = -T^{-1}\chi_e$, where χ_e is the chemical potential of electrons. Let us also select the simplified constitutive equations for $\partial\mathcal{S}/\partial\boldsymbol{q}$ and $\partial\mathcal{S}/\partial\boldsymbol{I}$ as given by $\partial\mathcal{S}/\partial\boldsymbol{q} = -\left(\gamma_q^q \boldsymbol{q} + \gamma_q^I \boldsymbol{I}\right)$ and $\partial\mathcal{S}/\partial\boldsymbol{I} = -\left(\gamma_I^q \boldsymbol{q} + \gamma_I^I \boldsymbol{I}\right)$, respectively. There, $\gamma^q(T)$ and $\gamma^I(T)$ are the material coefficients depending generally on the temperature, where the subscripts indicate the correspondence to the variable changing the entropy. These material coefficients are positive definite so that the property that s is maximum at equilibrium is met. With these conditions, expression (8.31), referred to as the Gibbs equation, can be written as

$$\begin{aligned} d_t\mathcal{S}(u,\varrho_e,\boldsymbol{q},\boldsymbol{I}) =& T^{-1}d_tu - \left(\gamma^{\boldsymbol{q}}_{\boldsymbol{q},1}\boldsymbol{q} + \gamma^{\boldsymbol{I}}_{\boldsymbol{q},1}\boldsymbol{I}\right)\cdot d_t\boldsymbol{q} \\ & - \chi_e T^{-1} d_t\varrho_e - \left(\gamma^{\boldsymbol{q}}_{\boldsymbol{I},1}\boldsymbol{q} + \gamma^{\boldsymbol{I}}_{\boldsymbol{I},1}\boldsymbol{I}\right)\cdot d_t\boldsymbol{I}, \end{aligned} \tag{8.33}$$

where $\gamma^{i}_{j,1}(i = \boldsymbol{q}, \boldsymbol{I}$ and independently $j = \boldsymbol{q}, \boldsymbol{I})$ are phenomenological crossing coefficients identified later on. However, expression (8.33) does not account for non-local effects. These non-local effects are elegantly introduced by appealing to a hierarchy of fluxes $\boldsymbol{Q}^{(1)}, \boldsymbol{Q}^{(2)}, \ldots, \boldsymbol{Q}^{(n)}$ with $\boldsymbol{Q}^{(1)}$ identical to the heat flux vector $\boldsymbol{q}$, $\boldsymbol{Q}^{(2)}$ (a tensor of rank two) is the flux of $\boldsymbol{q}$, $\boldsymbol{Q}^{(3)}$ the flux of $\boldsymbol{Q}^{(2)}$ and so on. The same is done for the electric current density, introducing the fluxes $\boldsymbol{I}^{(1)}(\boldsymbol{I} \equiv \boldsymbol{I}^{(1)}), \boldsymbol{I}^{(2)}, \ldots, \boldsymbol{I}^{(n)}$. Up to the nth-order flux, the Gibbs equation generalizing relation (8.33) becomes

$$\begin{aligned} & d_t\mathcal{S}\left(u,\varrho_e,\boldsymbol{q},\boldsymbol{Q}^{(2)},\ldots,\boldsymbol{Q}^{(n)},\boldsymbol{I},\boldsymbol{I}^{(2)},\ldots,\ \boldsymbol{I}^{(n)}\right) = T^{-1}d_tu - \chi_e T^{-1}d_t\varrho_e \\ & - \left(\gamma^{\boldsymbol{q}}_{\boldsymbol{q},1}\boldsymbol{q} + \gamma^{\boldsymbol{I}}_{\boldsymbol{q},1}\boldsymbol{I}\right)\cdot d_t\boldsymbol{q} - \left(\gamma^{\boldsymbol{q}}_{\boldsymbol{q},2}\boldsymbol{Q}^{(2)} + \gamma^{\boldsymbol{I}}_{\boldsymbol{q},2}\boldsymbol{I}^{(2)}\right)\otimes d_t\boldsymbol{Q}^{(2)} - \cdots \\ & - \left(\gamma^{\boldsymbol{q}}_{\boldsymbol{q},n}\boldsymbol{Q}^{(n)} + \gamma^{\boldsymbol{I}}_{\boldsymbol{q},n}\boldsymbol{I}^{(n)}\right)\otimes d_t\boldsymbol{Q}^{(n)} - \left(\gamma^{\boldsymbol{q}}_{\boldsymbol{I},1}\boldsymbol{q} + \gamma^{\boldsymbol{I}}_{\boldsymbol{I},1}\boldsymbol{I}\right)\cdot d_t\boldsymbol{I} \\ & - \left(\gamma^{\boldsymbol{q}}_{\boldsymbol{I},2}\boldsymbol{Q}^{(2)} + \gamma^{\boldsymbol{I}}_{\boldsymbol{I},2}\boldsymbol{I}^{(2)}\right)\otimes d_t\boldsymbol{I}^{(2)} - \cdots - \left(\gamma^{\boldsymbol{q}}_{\boldsymbol{I},n}\boldsymbol{Q}^{(n)} + \gamma^{\boldsymbol{I}}_{\boldsymbol{I},n}\boldsymbol{I}^{(n)}\right)\otimes d_t\boldsymbol{I}^{(n)} \end{aligned} \tag{8.34}$$

where the symbol $\otimes$ denotes the inner product of the corresponding tensors. The subsequent step is the formulation of the entropy flux $\boldsymbol{J}^s$. It is natural to expect that it is not simply given by the classical expression $\left(T^{-1}\ \boldsymbol{q} - \chi_e T^{-1}\boldsymbol{I}\right)$, but that it will depend on higher order fluxes in a similar way as for the generalized Gibbs equation, so that it is assumed that

$$\begin{aligned} \boldsymbol{J}^s = & T^{-1}\boldsymbol{q} - \chi_e T^{-1}\boldsymbol{I} + \Gamma_{\boldsymbol{q},1}\boldsymbol{Q}^{(2)}\cdot\boldsymbol{q} + \cdots + \Gamma_{\boldsymbol{q},n-1}\boldsymbol{Q}^{(n)}\otimes\boldsymbol{Q}^{(n-1)} \\ & - \Gamma_{\boldsymbol{I},1}\boldsymbol{I}^{(2)}\cdot\boldsymbol{I} - \cdots - \Gamma_{\boldsymbol{I},n-1}\boldsymbol{I}^{(2)}\cdot\boldsymbol{I}, \end{aligned} \tag{8.35}$$

where the Γ's are the material coefficients in analogy to the γ's. The next step is the derivation of the rate of entropy production per unit volume η^s which, referring to Eq. (8.32), is given by

$$\eta^s = d_t\mathcal{S} + \nabla\cdot\boldsymbol{J}^s \geq 0. \tag{8.36}$$

After substituting the expressions of $d_t\mathcal{S}$ and $\boldsymbol{J}^s$ from Eqs. (8.34) and (8.35) in Eq. (8.36), respectively, and eliminating d_tu via the energy conservation law for rigid heat conductors $(d_tu = -\nabla\cdot\boldsymbol{q} + \boldsymbol{I}\cdot\boldsymbol{E})$ and $d_t\varrho_e$ via the continuity law for electric charge $(d_t\varrho_e = -\nabla\cdot\boldsymbol{I})$, one obtains

$$\begin{aligned} \eta^s = & -\left(-\nabla T^{-1} + \gamma^{\boldsymbol{q}}_{\boldsymbol{q},1}d_t\boldsymbol{q} + \gamma^{\boldsymbol{q}}_{\boldsymbol{I},1}d_t\boldsymbol{I} - \Gamma_{\boldsymbol{q},1}\nabla\cdot\boldsymbol{Q}^{(2)}\right)\cdot\boldsymbol{q} \\ & - \left(-T^{-1}\boldsymbol{E} + \nabla\chi_e T^{-1} + \gamma^{\boldsymbol{I}}_{\boldsymbol{q},1}d_t\boldsymbol{q} + \gamma^{\boldsymbol{I}}_{\boldsymbol{I},1}d_t\boldsymbol{I} - \Gamma_{\boldsymbol{I},1}\nabla\cdot\boldsymbol{Q}^{(2)}\right) \\ & \times\boldsymbol{I} - \cdots - \sum_{n=2}^{N}\boldsymbol{Q}^{(n)}\otimes\left(\gamma^{\boldsymbol{q}}_{\boldsymbol{q},n}d_t\boldsymbol{Q}^{(n)} + \gamma^{\boldsymbol{q}}_{\boldsymbol{I},n}d_t\boldsymbol{I}^{(n)} - \Gamma_{\boldsymbol{q},n}\nabla\cdot\boldsymbol{Q}^{(n+1)} - \Gamma_{\boldsymbol{q},n-1}\nabla\boldsymbol{Q}^{(n-1)}\right) \\ & - \cdots - \sum_{n=2}^{N}\boldsymbol{I}^{(n)}\otimes\left(\gamma^{\boldsymbol{I}}_{\boldsymbol{q},n}d_t\boldsymbol{Q}^{(n)} + \gamma^{\boldsymbol{I}}_{\boldsymbol{I},n}d_t\boldsymbol{I}^{(n)} - \Gamma_{\boldsymbol{I},n}\nabla\cdot\boldsymbol{I}^{(n+1)} - \Gamma_{\boldsymbol{I},n-1}\nabla\boldsymbol{I}^{(n-1)}\right) \geq 0 \end{aligned} \tag{8.37}$$

The expression for η^s is a bilinear relationship in the fluxes $\boldsymbol{q}$ and $\boldsymbol{I}$ and the subsequent higher order of fluxes $\boldsymbol{Q}^{(n)}$ and $\boldsymbol{I}^{(n)}$. The quantities represented by the terms between the parentheses are usually called the thermodynamic forces $\boldsymbol{X}_I$. The simplest way to guarantee the positiveness of the entropy production σ^s is to assume a linear flux–force relation of the forms $\boldsymbol{Q}^{(n)} = \beta_{\boldsymbol{q}}\boldsymbol{X}_{\boldsymbol{q}}$ and $\boldsymbol{I}^{(n)} = \beta_{\boldsymbol{I}}\boldsymbol{X}_{\boldsymbol{I}} (n = 1, 2, 3, \ldots, N)$, where the β_I's are phenomenological coefficients. As such, we obtain

$$\nabla T^{-1} - \gamma^{\boldsymbol{q}}_{\boldsymbol{q},1} d_t \boldsymbol{q} - \gamma^{\boldsymbol{q}}_{\boldsymbol{I},1} d_t \boldsymbol{I} + \Gamma_{\boldsymbol{q},1} \nabla \cdot \boldsymbol{Q}^{(2)} = \nu_{\boldsymbol{q},1} \boldsymbol{q}, \tag{8.38a}$$

$$T^{-1}\boldsymbol{E} - \nabla \mu_e T^{-1} - \gamma^{\boldsymbol{I}}_{\boldsymbol{q},1} d_t \boldsymbol{q} - \gamma^{\boldsymbol{I}}_{\boldsymbol{I},1} d_t \boldsymbol{I} + \Gamma_{\boldsymbol{I},1} \nabla \cdot \boldsymbol{I}^{(2)} = \nu_{\boldsymbol{I},1} \boldsymbol{I} \tag{8.38b}$$

$$\Gamma_{q,n-1} \nabla \boldsymbol{Q}^{(n-1)} - \gamma^{\boldsymbol{q}}_{\boldsymbol{q},n} d_t \boldsymbol{Q}^{(n)} - \gamma^{\boldsymbol{q}}_{\boldsymbol{I},n} d_t \boldsymbol{I}^{(n)} + \Gamma_{\boldsymbol{q},n} \nabla \cdot \boldsymbol{Q}^{(n+1)} = \nu_{\boldsymbol{q},n} \boldsymbol{Q}^{(n)}, (n = 2, 3, \ldots, N), \tag{8.39a}$$

$$\Gamma_{I,n-1} \nabla \boldsymbol{I} - \gamma^{\boldsymbol{I}}_{\boldsymbol{q},n} d_t \boldsymbol{Q}^{(n)} - \gamma^{\boldsymbol{I}}_{\boldsymbol{I},n} d_t \boldsymbol{I}^{(n)} + \Gamma_{Iq,n} \nabla \cdot \boldsymbol{I}^{(n+1)} = \nu_{\boldsymbol{I},n} \boldsymbol{I}^{(n)}, (n = 2, 3, \ldots, N), \tag{8.39b}$$

compatible with positive entropy production at the condition that $\nu_{\boldsymbol{q},1} \geq 0$, $\nu_{\boldsymbol{I},1} \geq 0$, $\nu_{\boldsymbol{q},n} \geq 0$ and $\nu_{\boldsymbol{I},n} \geq 0$. In Eqs. (8.38)–(8.39), d_t stands for the partial time derivative, if we assume the material to be at rest. Also, γ_i, Γ_i and ν_i are phenomenological coefficients related to the relaxation times, correlation lengths and transport coefficients, respectively. Equation (8.38a) reduces to the well-known Cattaneo's law [20] when the terms $\nabla \cdot \boldsymbol{Q}^{(2)}$ and $d_t\boldsymbol{I}$ are omitted. Furthermore, if the term $d_t\boldsymbol{q}$ is neglected, we obtain Fourier's law. Note that Eq. (8.38b) also reduces to a Cattaneo-like law when one omits $\nabla \cdot \boldsymbol{I}^{(2)}$ and $d_t\boldsymbol{q}$. Ohm's law is obtained when furthermore the term $d_t\boldsymbol{I}$ is neglected. In order to simplify the following procedure, we focus on only the heat flux. We now consider an infinite number of flux variables ($N \to \infty$) and apply the spatial Fourier transform

$$\hat{\boldsymbol{q}}(\boldsymbol{k}, t) = \int_{-\infty}^{+\infty} \boldsymbol{q}(\boldsymbol{r}, t) e^{-i\boldsymbol{k}\cdot\boldsymbol{r}} d\boldsymbol{r} \tag{8.40}$$

to Eqs. (8.38a) and (8.39a) neglecting for simplification the cross terms $d_t\boldsymbol{I}^{(n)} (n = 1, 2, 3, \ldots)$, with $\hat{\boldsymbol{q}}$ the Fourier transform of $\boldsymbol{q}$, $\boldsymbol{r}$ the spatial variable, t the time and $\boldsymbol{k}$ the wave number vector. This procedure results into obtaining the following time-evolution equation for the heat flux:

$$\overline{\tau}(\boldsymbol{k}) \partial_t \hat{\boldsymbol{q}}(\boldsymbol{k}, t) + \hat{\boldsymbol{q}}(\boldsymbol{k}, t) = -i\boldsymbol{k}\lambda^p_{c,i}(\boldsymbol{k}) \hat{T}(\boldsymbol{k}, t) \tag{8.41}$$

where $\overline{\tau}(\boldsymbol{k}) = \gamma_1/\mu_1$ designates a renormalized relaxation time depending generally on $\boldsymbol{k}$. $\lambda^p_{c,i}(\boldsymbol{k})$ is given [following the development of Eq. (2.38)] by the continued fraction for the $\boldsymbol{k}$-dependent effective thermal conductivity:

$$\lambda^p_{c,i}(\boldsymbol{k}) = \frac{\lambda^{p,0}_{c,i}}{1 + \frac{\boldsymbol{k}^2 l_1^2}{1 + \frac{\boldsymbol{k}^2 l_2^2}{1 + \frac{\boldsymbol{k}^2 l_3^2}{1 + \cdots}}}}, \tag{8.42}$$

where $\lambda^{p,0}_{c,i}$ is the classical bulk thermal conductivity, given by Eq. (8.27), independent of the dimension of the system. The correlation lengths, the choice of wave number and the asymptotic limit of Eq. (8.42) are explained between Eqs. (2.38) and (2.39). Therefore, we give the result here for $\lambda^p_{c,i}$:

$$\lambda^p_{c,i} = \frac{3\lambda^{p,0}_{c,i}}{4\pi^2 Kn_c^2}\left[\frac{2\pi Kn_c}{\arctan(2\pi Kn_c)} - 1\right], \tag{8.43a}$$

where Kn_c is given by Eq. (8.28). The same can be found for the electrical conductivity in analogy with the thermal conductivity. Hereby, the correlation lengths are only associated with the mean free path of the electrons. The result is

$$\sigma_{e,i}^{p} = \frac{3\sigma_{e,i}^{p,0}}{4\pi^2 Kn_e^2}\left[\frac{2\pi Kn_e}{\arctan\left(2\pi Kn_e\right)} - 1\right], \tag{8.43b}$$

with Kn_e given by Eq. (8.28) with $c = e$. We should note that the relation between Eqs. (8.43a) and (8.43b) is still valid by Eq. (8.29).

8.4 Optoelectric Model for the PV Device

8.4.1 Basic Considerations

For the PV device, we use the semiconductor equations for electron transport

$$\nabla \cdot (\varepsilon \nabla V) = -e_C\left(p - n + N_D - N_A\right) \tag{8.44}$$

$$\nabla \cdot \boldsymbol{I}_p = e_C\left(G - R\right) \tag{8.45}$$

$$\nabla \cdot \boldsymbol{I}_n = -e_C\left(G - R\right) \tag{8.46}$$

$$\boldsymbol{I}_P = -\mu_p\left(e_C p \nabla V + k_B T \nabla p\right) \tag{8.47}$$

$$\boldsymbol{I}_n = -\mu_n\left(e_C n \nabla V - k_B T \nabla n\right) \tag{8.48}$$

where ε is the dielectric constant of the semiconductor ($\varepsilon = \varepsilon_r \varepsilon_0$, with ε_r the relative permittivity with respect to the vacuum permittivity ε_0), V the electric potential, e_C the electron charge, p and n the hole and electron carrier concentrations, respectively, N_D and N_A the donor and acceptor doping concentrations, respectively, $\boldsymbol{I}_p$ and $\boldsymbol{I}_n$ the hole and electron current densities, respectively, and G and R the generation and recombination rates, respectively. The Eqs. (8.50)–(8.54) represent Poisson's equation relating the electrical potential to the space charge density, the continuity equations for holes and electrons and the hole and electron current densities, respectively. Note that in Poisson's equation $\nabla \cdot (\varepsilon \nabla V) = \nabla \cdot (\varepsilon \mathbf{E})$, where $\mathbf{E}$ is the electric field.

The generation rate G is given by

$$G = \int_{k_{\min}}^{hc/E_g} A\left(k\right) e^{-A(k)z} S_{AM1.5}\left(k\right) dk \tag{8.49}$$

where h is Planck's constant, c the speed of light, E_g the band-gap energy, A the absorption coefficient, k the photon wavelength, z the space coordinate and $S_{AM1.5}$ is defined as the standard number of incident photons per surface per wavelength: $S_{AM1.5}\left(k\right) \equiv I_{AM1.5}/E_{photon}\left(k\right)$, where E_{photon} is the photon energy. $I_{AM1.5}$ is defined as the incident photon energy per surface per wavelength, where the subscript stands for the air mass number. In Eq. (8.49), k_{min} is the minimum photon wavelength for solar irradiance, being 400 nm [21]. Note that k_{max}, needed later on, is the maximum one, being 1,050 nm [21]. The particularity of $S_{AM1.5}\left(k\right)$ is that it depends on the wavelength of the different "light spectra", i.e. blue light has a lower wavelength than for instance red light and has thus higher photon energy. Therefore, we have to integrate over the whole spectrum from k_{min} to hc/E_g in Eq. (8.49). We define $I_{AM1.5} \equiv \partial P_s/\partial k$, where P_s is the solar energy per surface for a given wavelength. There is no formulation for $I_{AM1.5}$. We, therefore, take an average integrated value over the photon wavelengths: $\langle I_{AM1.5}\rangle \equiv \int_{k_{min}}^{hc/E_g} I_{AM1.5} dk = \int_{k_{min}}^{hc/E_g} \frac{\partial P_s}{\partial k} dk \approx P_s$. Typically, P_s is of the order of magnitude of 1,000 W m^{-2}. The absorption coefficient is also averaged over the wavelength: $\langle A\rangle \equiv \int_{k_{min}}^{hc/E_g} A\left(k\right) dk$. We also take an averaged value for the thickness dependence of G,

$$\left\langle Ae^{-Az}\right\rangle = \frac{1-e^{-A\delta}}{\delta} \tag{8.50}$$

We also average the photon energy $E_{photon} = \frac{hc}{2\pi k}$

$$\left\langle E_{photon}\right\rangle = \frac{hc}{2\pi\left(\frac{hc}{E_g} - k_{min}\right)} \ln\left(\frac{\frac{hc}{E_g}}{k_{min}}\right) \tag{8.51}$$

Finally, we have

$$G \approx P_s \frac{1-e^{-A\delta}}{\delta} \frac{2\pi\left(\frac{hc}{E_g} - k_{min}\right)}{hc} \frac{1}{\ln\left(\frac{\frac{hc}{E_g}}{k_{min}}\right)} \tag{8.52}$$

This form has also been suggested in [22]. Finally, note that thusly we do not take G as a given constant value, often done in analytical developments, but dependent on the device's thickness. The recombination rate R is given by

$$R = R_{SRH} + R_{Aug} + gR_{surf} \tag{8.53}$$

where

$$R_{SRH} = \frac{pn - n_{ie}^2}{\tau_n\left(p + n_{ie}\right) + \tau_p\left(n + n_{ie}\right)} \tag{8.54a}$$

$$R_{Aug} = (B_p p + B_n n)(pn - n_{ie}^2) \tag{8.54b}$$

$$R_{surf} = \frac{p_s n_s - n_{ies}^2}{\tau_{nsf}\left(p_s + n_{ies}\right) + \tau_{psf}\left(n_s + n_{ies}\right)} \tag{8.54c}$$

and in Eqs. (8.59) and (8.60), we used

$$g \equiv \frac{N}{N_s} \tag{8.55}$$

$$n_{ie} = \sqrt{N_C N_V} e^{-\frac{E_g}{2k_B T}} \tag{8.56}$$

In Eqs. (8.53)–(8.56), R_{SRH}, R_{Aug} and R_{surf} are the so-called SRH, Auger recombination and surface recombination rates, respectively, τ_p and τ_n the minority hole and electron SRH lifetimes, W_p and W_n the p- and n-sides of the quasi-neutral layers (explained later in the chapter), S_{rp} and S_{rn} the minority hole and electron surface recombination velocities, n_{ie} the intrinsic carrier concentration in the recombination sites, B_p and B_n the hole and electron Auger recombination coefficients, respectively, τ_{psf} and τ_{nsf} the hole and electron surface recombination lifetimes, respectively, and N_C and N_V the effective density of states for the conduction and valence bands, respectively. Note that in Eqs. (8.53) and (8.55), g is a proportionality factor, being defined here as the ratio of a bulk (equilibrium) carrier concentration ($N = p, n, n_{ie}$) and a surface (equilibrium) carrier concentration ($N_s = p_s, n_s, n_{ies}$). Radiative recombination is neglected [23].

In order to solve for all these equations, we need to first make a difference between the depletion region and the quasi-neutral regions. We will start with solving for the depletion region in Section 8.4.2. In Section 8.4.3, we treat the quasi-neutral regions. The findings are synthesized in Section 8.4.4 to obtain the PV efficiency.

8.4.2 Depletion Region

We make certain assumptions in order to solve the equations analytically:

- Depletion approximation: the electric field is confined to the junction region (depletion region), and there is none in the quasi-neutral regions.
- The number of free carriers in the depletion region is small, assuming that the electric field sweeps them out of the depletion region quickly. This means that we do not consider the transport equation for the carrier concentrations and that we can neglect recombination. We consider generation in the depletion region.
- Abrupt or step-doping profile, where all dopants are ionized: the acceptor and donor doping concentrations are constant in their respective regions.
- One-dimensional system.
- We define $z = 0$ as the point of the p–n junction, $z = -z_p$ the p-side of the depletion region and $z = z_n$ the n-side of the depletion region.
- Constant electric permittivity at each side of the p–n junction.

The system to solve becomes

$$\frac{d^2V}{dz^2} = \begin{cases} \frac{e_C}{\varepsilon_p} N_A, & -z_p \leq z < 0 \\ -\frac{e_C}{\varepsilon_n} N_D, & 0 \leq z < z_n \end{cases} \tag{8.57}$$

$$\frac{dI_p}{dz} = e_C G \tag{8.58}$$

$$\frac{dI_n}{dz} = -e_C G \tag{8.59}$$

The subscripts of the relative permittivity denote that it concerns either the p- or the n-side of the depletion region. So we take into account the difference in permittivity of the two materials, which is not often done. Since we have assumed that there is no electric field outside the depletion region, we have as boundary conditions $\frac{dV}{dz} = 0$ at $z = -z_p$ and $z = z_n$. One is usually interested in the potential difference across the junction. We define here the voltage on the p-side as V_0 (applied voltage in case of a diode or induced voltage in case of a solar cell), at $z = -z_p$. At the n–p interface, the potential should be equal on both sides, so that $V_p = V_n$ at $z = 0$ (where the subscripts p and n indicate the p- and n-sides of the interface, respectively). The electric potential is then given by

$$V(z) = \begin{cases} V_0 + \frac{e_C N_A}{2\varepsilon_p} (z + z_p)^2, & -z_p \leq z < 0 \\ V_0 + \frac{e_C N_A}{2\varepsilon_p} z_p^2 + \frac{e_C N_D}{\varepsilon_n} \left(z_n - \frac{z}{2}\right) z, & 0 \leq z < z_n \end{cases} \tag{8.60}$$

It is now necessary to know z_p and z_n. At the interface $z = 0$, we assume displacement vector ($\boldsymbol{D} = \varepsilon \boldsymbol{E}$) continuity. Thus, $\varepsilon_p \frac{dV_p}{dz} = \varepsilon_n \frac{dV_n}{dz}$ at $z = 0$. This gives

$$N_A z_p = N_D z_n \tag{8.61}$$

The maximum voltage across the junction is at $z = z_n$ being

$$V(z_n) = V_0 + \frac{e_C}{2} \left(\frac{N_A z_p^2}{\varepsilon_p} + \frac{N_D z_n^2}{\varepsilon_n} \right) \tag{8.62}$$

This voltage is also equal to the built-in voltage (which can be found from the difference in Fermi levels between the p- and n-side materials) across the p–n junction, V_D, so that

$$V_D = V_0 + \frac{e_C}{2}\left(\frac{N_A z_p^2}{\varepsilon_p} + \frac{N_D z_n^2}{\varepsilon_n}\right) \tag{8.63}$$

with

$$V_D = \frac{k_B T}{e_C}\ln\left(\frac{N_A N_D}{n_{ie}^2}\right) \tag{8.64}$$

We can finally find that

$$z_p = \sqrt{\frac{2\left(V_D - V_0\right)}{e_C}\frac{N_D}{N_A\left(\frac{N_A}{\varepsilon_n} + \frac{N_D}{\varepsilon_p}\right)}} \tag{8.65}$$

$$z_n = \sqrt{\frac{2\left(V_D - V_0\right)}{e_C}\frac{N_A}{N_D\left(\frac{N_A}{\varepsilon_n} + \frac{N_D}{\varepsilon_p}\right)}} \tag{8.66}$$

The total width of the depletion region is then

$$W_d = z_p + z_n = \sqrt{\frac{2\left(V_D - V_0\right)}{e_C}\left(\frac{\varepsilon_n}{N_A} + \frac{\varepsilon_p}{N_D}\right)} \tag{8.67}$$

which shows that the depletion region depends on the material properties of both the p- and n-side ones as well as on an applied/induced voltage. Also note that

$$z_p = W_d \frac{\frac{N_D}{\varepsilon_n}}{\frac{N_A}{\varepsilon_p} + \frac{N_D}{\varepsilon_n}} \tag{8.68}$$

$$z_n = W_d \frac{\frac{N_A}{\varepsilon_p}}{\frac{N_A}{\varepsilon_p} + \frac{N_D}{\varepsilon_n}} \tag{8.69}$$

Now, under solar illumination, there is also a generation in the depletion region, as shown in Eqs. (8.58) and (8.59). Integrating between the depletion borders gives $I_n\left(z = z_n\right) - I_n\left(z = -z_p\right) = -e_C\left(C_{Gp} z_p + C_{Gn} z_n\right)$, so that we define a surplus hole current density through the depletion region $\Delta I_{p,d}$:

$$\Delta I_{p,d} = e_C\left(C_{Gp} z_p + C_{Gn} z_n\right) \tag{8.70}$$

Note that the same can be done for the electron current density $\Delta I_{n,d}$ giving

$$\Delta I_{n,d} = -e_C\left(C_{Gp} z_p + C_{Gn} z_n\right) \tag{8.71}$$

It can be easily seen that the net current density through the depletion region is constant.

8.4.3 Quasi-Neutral Regions

We continue now with the quasi-neutral regions. Having said that there is no electrical field in the quasi-neutral regions, we can omit Poisson's equation from this development (the electric potential in the p- and n-sides of the quasi-neutral regions are constant and equal to the ones at the respective depletion region borders). The drift terms in the transport equations also becomes zero. The system to solve becomes

$$\frac{dI_p}{dz} = e_C\,(G - R) \tag{8.72}$$

$$\frac{dI_n}{dz} = -e_C\,(G - R) \tag{8.73}$$

$$I_p = -\mu_p k_B T \frac{dp}{dz} \tag{8.74}$$

$$I_n = \mu_n k_B T \frac{dn}{dz} \tag{8.75}$$

We can rewrite this set of equations as

$$k_B T \frac{\mu_p}{e_C} \frac{d^2 p}{dz^2} + (G - R) = 0 \tag{8.76}$$

$$k_B T \frac{\mu_n}{e_C} \frac{d^2 n}{dz^2} + (G - R) = 0 \tag{8.77}$$

For the generation rate, we have established an averaged expression, depending now only on the total thickness of the device so that we define $G \equiv C_G$. We also define an average generation rate for the n-side

$$C_{Gn} = \frac{P_s \frac{1-\mathrm{e}^{-A\delta_n}}{\delta_n} \frac{2\pi(hc - k_{min}E_{gn})}{hcE_{gn}}}{\ln\left(\frac{hc}{k_{min}E_{gn}}\right)} \tag{8.78}$$

We can also define an average generation rate for the p-side but with the assumption that only a fraction $\mathrm{e}^{-A\delta_n}$ remains, so that

$$C_{Gp} = \frac{\mathrm{e}^{-A\delta_n} P_s \frac{1-\mathrm{e}^{-A\delta_p}}{\delta_p} \frac{2\pi(hc - k_{min}E_{gp})}{hcE_{gp}}}{\ln\left(\frac{hc}{k_{min}E_{gp}}\right)} \tag{8.79}$$

As for R, we make the assumption of low injection, where it is assumed that the majority carrier concentrations are unperturbed throughout the quasi-neutral regions. This means that $p_p \gg n_p$ and $p_p = p_{p0}$ in the p-side material (the subscript p denoting that it concerns the p-side), p_{p0} being the equilibrium majority hole concentration, which is equal to the intrinsic hole concentration plus the doping one. In case of doping, the intrinsic carrier concentration is often to be neglected. For the n-side material, we can say in analogy that $n_n \gg p_n$ and $n_n = n_{n0}$. Note that in equilibrium, the product of the majority and minority carrier concentrations is a constant, being mathematically expressed by the so-called mass action law

$$p_{i0} n_{i0} = n_{ie}^2 \tag{8.80}$$

with subscript i being either p or n, depending whether it concerns the p- or the n-side, and n_{ie} is the equilibrium carrier concentration. Therefore, at equilibrium, carrier concentrations are given by

$$p_{p0} = N_A \,\&\, n_{p0} = n_{ie,p}^2 / N_A \quad \left(\text{where}\; n_{ie,p} \equiv n_{ie}|_{E_g \equiv E_{gp}}\right) \tag{8.81}$$

for the majority and minority carriers in the p-side, respectively, and

$$n_{n0} = N_D \,\&\, p_{n0} = n_{ie,n}^2 / N_D \quad \left(\text{where}\; n_{ie,n} \equiv n_{ie}|_{E_g \equiv E_{gn}}\right) \tag{8.82}$$

for the majority and minority carriers in the n-side, respectively. Due to carrier doping, we also assume that $n_n \gg n_{ie}$ and $n_p \gg n_{ie}$. Furthermore, we assume that the lifetimes and surface recombination velocities do not vary dramatically in the p- and n-side materials, such that $\tau n_p \gg \tau n_n$ in the p-side material and $\tau n_n \gg \tau n_p$ in the n-side material. Using Eqs. (8.53)–(8.56), the recombination rates for the minority carrier concentrations $(n_p - n_{p0})$ on the p-side and $(p_n - p_{n0})$ on the n-side, respectively, become

$$R_n = (n_p - n_{p0})\left(\frac{1}{\tau_n} + \frac{1}{\tau_{ns}} + \frac{1}{\tau_{nA}}\right) = (n_p - n_{p0})\,C_{Rn} \tag{8.83}$$

$$R_p = (p_n - p_{n0})\left(\frac{1}{\tau_p} + \frac{1}{\tau_{ps}} + \frac{1}{\tau_{pA}}\right) = (p_n - p_{n0})\,C_{Rp} \tag{8.84}$$

$$\tau_{nA} \equiv \frac{1}{B_p N_A^2} \tag{8.85a}$$

$$\tau_{pA} \equiv \frac{1}{B_n N_D^2}, \tag{8.85b}$$

$$\tau_{ns} = \frac{W_p}{S_{rn}} + \frac{4}{D_n}\left(\frac{W_p}{\pi}\right)^2 \tag{8.85c}$$

$$\tau_{ps} = \frac{W_n}{S_{rp}} + \frac{4}{D_p}\left(\frac{W_n}{\pi}\right)^2 \tag{8.85d}$$

Here, $1/C_{Rn}$ and $1/C_{Rp}$ are the effective recombination lifetimes of the minority electrons and holes, respectively. Very important here is to note that the electron and hole Auger lifetimes are obtained under the assumption that the dopant concentration is much higher than the minority ones [24]. If not, the Auger lifetimes should not be calculated as in Eq. (8.85a) or (8.85b) but can be obtained elsewhere, e.g. in [25,26] for silicon. The expression for the surface recombination lifetime is given in [27], assuming identical surfaces and splitting the expression for the p- and n-sides. The set of equations to solve are now (noting that $\frac{d^2 n_p}{dz^2} = \frac{d^2(n_p - n_{p0})}{dz^2}$ and $\frac{d^2 p_n}{dz^2} = \frac{d^2(p_n - p_{n0})}{dz^2}$)

$$k_B T \frac{\mu_n}{e_C} \frac{d^2(n_p - n_{p0})}{dz^2} - C_{Rn}(n_p - n_{p0}) + C_{Gp} = 0 \tag{8.86}$$

$$k_B T \frac{\mu_p}{e_C} \frac{d^2(p_n - p_{n0})}{dz^2} - C_{Rp}(p_n - p_{n0}) + C_{Gn} = 0 \tag{8.87}$$

for the minority carrier concentrations $(n_p - n_{p0})$ on the p-side of the quasi-neutral region and $(p_n - p_{n0})$ minority carriers on the n-side of the quasi-neutral region, respectively. Let us furthermore define the characteristic diffusion lengths $L_n \equiv \sqrt{\frac{k_B T \mu_n}{C_{Rn} e_C}}$ and $L_p \equiv \sqrt{\frac{k_B T \mu_p}{C_{Rp} e_C}}$ and the characteristic diffusion coefficients $D_n \equiv \frac{k_B T \mu_n}{e_C}$ and $D_p \equiv \frac{k_B T \mu_p}{e_C}$ of the electrons (in the p-side) and the holes (in the n-side), respectively. We recall that $z = -z_p$ and $z = z_n$ at the edge of, respectively, the p- and n-side of the depletion region and $z \to -(W_p + z_p)$ and $z \to (W_n + z_n)$ for the surface of the device for the p- and n-sides, respectively, where W_p and W_n are the thickness of the p- and n-side quasi-neutral regions, respectively. On the p-side, we have the following boundary conditions:

$$D_n \frac{d(n_p - n_{p0})}{dz} = S_{rn}(n_p - n_{p0}) \,\text{at}\, z \to -(W_p + z_p) \tag{8.88}$$

$$(n_p - n_{p0}) = n_{p0} e^{\frac{e_C V_0}{k_B T}} - n_{p0} \,\text{at}\, z = -z_p \tag{8.89}$$

On the n-side, we have the following boundary conditions:

$$-D_p \frac{d\left(p_n - p_{n0}\right)}{dz} = S_{rp}\left(p_n - p_{n0}\right) \text{at } z \to \left(W_n + z_n\right) \tag{8.90}$$

$$\left(p_n - p_{n0}\right) = p_{n0} \mathrm{e}^{\frac{e_C V_0}{k_B T}} - p_{n0} \text{ at } z = z_n \tag{8.91}$$

In Eqs. (8.89) and (8.91), V_0 is the applied voltage or in the case of an illuminated solar cell, the open-circuit voltage, here denoted by $V_0 \equiv \Delta V$. Also S_{rn} and S_{rp} are the minority electron and hole surface recombination velocities, respectively. We find the following minority carrier densities in the p- and the n-sides, respectively,

$$\begin{aligned} n_p[z] = n_{p0} &+ \frac{\left(\mathrm{e}^{\frac{e_C \Delta v}{k_B T}} - 1\right) n_{p0} D_n \left(D_n - L_n S_{rn} + \left(D_n + L_n S_{rn}\right) \mathrm{e}^{2\frac{W_p + Z_p + z}{L_n}}\right)}{\mathrm{e}^{\frac{z_p + z}{L_n}} D_n \left(D_n - L_n S_{rn} + \left(D_n + L_n S_{rn}\right) \mathrm{e}^{2\frac{W_p}{L_n}}\right)} \\ &- \frac{C_{Gp}\left(\mathrm{e}^{\frac{z_p + z}{L_n}} - 1\right) L_n^2 \left(D_n\left(\mathrm{e}^{\frac{2W_p + z_p + z}{L_n}} - 1\right) + \left(\mathrm{e}^{\frac{W_p}{L_n}-1}\right)\left(\mathrm{e}^{\frac{W_p + z_p + z}{L_n}} - 1\right) L_n S_{rn}\right)}{\mathrm{e}^{\frac{z_p + z}{L_n}} D_n \left(D_n - L_n S_{rn} + \left(D_n + L_n S_{rn}\right) \mathrm{e}^{2\frac{W_p}{L_n}}\right)} \end{aligned} \tag{8.92}$$

$$\begin{aligned} p_n[z] = p_{n0} &+ \frac{\left(\mathrm{e}^{\frac{e_C \Delta V}{k_B T}} - 1\right) p_{n0} D_p \left(\left(D_p - L_p S_{rp}\right) \mathrm{e}^{\frac{2z}{L_p}} + \left(D_p + L_p S_{rp}\right) \mathrm{e}^{2\frac{W_n + z_n}{L_p}}\right)}{\mathrm{e}^{\frac{z_n + z}{L_p}} D_p \left(D_p - L_p S_{rp} + \left(D_p + L_p S_{rp}\right) \mathrm{e}^{\frac{2W_n}{L_p}}\right)} \\ &- \frac{C_{Gn}\left(\mathrm{e}^{\frac{z}{L_p}} - \mathrm{e}^{\frac{z_n}{L_p}}\right) L_p^2 \left(D_p\left(\mathrm{e}^{\frac{z}{L_p}} - \mathrm{e}^{\frac{2W_n + z_n}{L_p}}\right) + \left(\mathrm{e}^{\frac{W_n}{L_p}} - 1\right)\left(\mathrm{e}^{\frac{z}{L_p}} - \mathrm{e}^{\frac{W_n + z_n}{L_p}}\right) L_p S_{rp}\right)}{\mathrm{e}^{\frac{z_n + z}{L_p}} D_p \left(D_p - L_p S_{rp} + \left(D_p + L_p S_{rp}\right) \mathrm{e}^{\frac{2W_n}{L_p}}\right)} \end{aligned} \tag{8.93}$$

The majority carrier currents are equal to zero in each of the respective quasi-neutral regions. The minority carrier currents can be calculated using the one-dimensional Eqs. (8.74) and (8.75) giving for the p- and n-sides, respectively,

$$\begin{aligned} I_n = e_C &\frac{\left(\mathrm{e}^{\frac{e_C \Delta V}{k_B T}} - 1\right) n_{p0} D_n \left(-D_n + L_n S_{rn} + \left(D_n + L_n S_{rn}\right) \mathrm{e}^{2\frac{W_p + z_p + z}{L_n}}\right)}{\mathrm{e}^{\frac{z_p + z}{L_n}} L_n \left(D_n - L_n S_{rn} + \left(D_n + L_n S_{rn}\right) \mathrm{e}^{\frac{2W_p}{L_n}}\right)} \\ &- \frac{C_{Gp}\left(\mathrm{e}^{\frac{z_p + z}{L_n}} - 1\right) L_n^2 \left(D_n\left(\mathrm{e}^{2\frac{W_p + z_p + z}{L_n}} - 1\right) + \left(\mathrm{e}^{\frac{W_p}{L_n}} - 1\right)\left(\mathrm{e}^{\frac{W_p + 2z_p + 2z}{L_n}} - 1\right) L_n S_{rn}\right)}{\mathrm{e}^{\frac{z_p + z}{L_n}} L_n \left(D_n - L_n S_{rn} + \left(D_n + L_n S_{rn}\right) \mathrm{e}^{\frac{2W_p}{L_n}}\right)} \end{aligned} \tag{8.94}$$

$$I_p = e_C \frac{2\mathrm{e}^{\frac{W_n}{L_p}} \left(\begin{array}{c} C_{Gn} L_p^3 S_{rp} \cosh\left(\frac{z - z_n}{L_p}\right) - \left(C_{Gn} L_p^2 - \left(\mathrm{e}^{\frac{e_C \Delta V}{k_B T}} - 1\right) p_{n0} D_p\right) \\ \left(L_p S_{rp} \cosh\left(\frac{w_n - z + z_n}{L_p}\right) + D_p \sin\mathrm{h}\left(\frac{w_n - z + z_n}{L_p}\right)\right) \end{array}\right)}{L_p \left(D_p - L_p S_{rp} + \left(D_p + L_p S_{rp}\right) \mathrm{e}^{\frac{2W_n}{L_p}}\right)} \tag{8.95}$$

Having determined the minority current densities and the depletion region properties, we have enough information to proceed with the PV efficiency.

8.4.4 PV Efficiency

We make the assumption that the total current through the solar cell is constant:

$$\frac{dI}{dz} = \frac{dI_n}{dz} + \frac{dI_p}{dz} = -e_C\left(G_n - R_n\right) + e_C\left(G_p - R_p\right) = 0 \tag{8.96}$$

This assumption can be made by stating that each electron generates a hole, and each recombining electron uses one up: $G_n = G_p$ and $R_n = R_p$. This means that the total number of electrons and holes do not change in a semiconductor, and therefore, the total current also does not change. This also suggests that if one finds the total current anywhere in the device it will be the same everywhere in it. Conveniently, we can choose either sides of the depletion region. We choose to calculate the total current at the n-side, i.e. $z = z_n$. As stated earlier, both current densities across the depletion region (the edges for mathematical purposes included) are constant if it were not for the generation in the depletion region. Thus, the electron current density at the n-side of the depletion region ($z = z_n$) equals the minority electron current density at the p-side of the depletion region ($z = -z_p$) plus the one calculated due to generation in the depletion region (whether it be positive or negative): $I_n\left(z = z_n\right) = I_n\left(z = -z_p\right) + \Delta I_{n,d}$. The total current density can then be found by simply adding the hole current density at the n-side $I_P\left(z = z_n\right)$. Therefore, we can find the total current to be $I_T = I_P\left(z = z_n\right) + I_n\left(z = -z_p\right) + \Delta I_{n,d}$. Recalling that $p_{n0} = n_{ie}^2/N_D$ and $n_{p0} = n_{ie}^2/N_A$, we finally find

$$\begin{aligned} I_T = {} & \mathrm{e}\left(\frac{e_C\Delta V}{k_B T} - 1\right) \\ & \times\left(e_C\frac{D_p}{L_p}\frac{n_{ie,n}^2}{N_D}\frac{\frac{D_p}{L_p S_{rp}} + \coth\left(\frac{W_n}{L_p}\right)}{1 + \frac{D_p}{L_p S_{rp}} + \coth\left(\frac{W_n}{L_p}\right)} + e_C\frac{D_n}{L_n}\frac{n_{ie,p}^2}{N_A}\frac{\frac{D_n}{L_n S_{rn}} + \coth\left(\frac{W_p}{L_n}\right)}{1 + \frac{D_n}{L_n S_{rn}} + \coth\left(\frac{W_p}{L_n}\right)}\right) \\ & - e_C\left(C_{Gp}Z_p + C_{Gn}Z_n + C_{Gn}L_p\frac{\frac{D_p}{L_p S_{rp}} + \tanh\left(\frac{W_n}{2L_p}\right)}{1 + \frac{D_p}{L_p S_{rp}}\coth\left(\frac{W_n}{2L_p}\right)} + C_{Gp}L_n\frac{\frac{D_n}{L_n S_{rn}} + \tanh\left(\frac{W_p}{2L_n}\right)}{1 + \frac{D_n}{L_n S_{rn}}\coth\left(\frac{W_p}{2L_n}\right)}\right) \end{aligned} \tag{8.97}$$

We can typically define

$$I_0 \equiv e_C\frac{D_p}{L_p}\frac{n_{ie,n}^2}{N_D}\frac{\frac{D_p}{L_p S_{rp}} + \coth\frac{W_n}{L_p}}{1 + \frac{D_p}{L_p S_{rp}} + \coth\frac{W_n}{L_p}} + e_C\frac{D_n}{L_n}\frac{n_{ie,p}^2}{N_A}\frac{\frac{D_n}{L_n S_{rn}} + \coth\frac{W_p}{L_n}}{1 + \frac{D_n}{L_n S_{rn}} + \coth\frac{W_p}{L_n}}, \tag{8.98}$$

$$\begin{aligned} I_{SC} \equiv {} & e_C \\ & \times\left(C_{Gp}Z_p + C_{Gn}Z_n + C_{Gn}L_p\frac{\frac{D_p}{L_p S_{rp}} + \tanh\left(\frac{W_n}{2L_p}\right)}{1 + \frac{D_p}{L_p S_{rp}} + \coth\left(\frac{W_n}{2L_p}\right)} + C_{Gp}L_n\frac{\frac{D_n}{L_n S_{rn}} + \tanh\left(\frac{W_p}{2L_n}\right)}{1 + \frac{D_n}{L_n S_{rn}} + \coth\left(\frac{W_p}{2L_n}\right)}\right), \end{aligned} \tag{8.99}$$

standing for the dark saturation current (without solar illumination, corresponding to the diode current) and the short-circuit current, respectively. It should be noted that the signs are rather based on conventions. When speaking about solar cells, the expression (8.97) is often inversed in sign. Making the remark that in the case of solar cells, the short-circuit current is also called the light-generated current, i.e. $I_L \equiv I_{SC}$, we find the final expression for the total current

$$I_T = I_L - I_0\left(\mathrm{e}^{\frac{e_C\Delta V}{k_B T}} - 1\right) \tag{8.100}$$

It can be seen here that the total current density, for a specific set of p- and n-materials with a specific thickness, depends on both the built-in and the induced voltage difference, V_D and ΔV. The first is material dependent and the second depends on the operating conditions. Therefore, $I_T = I_T(\Delta V)$. The PV efficiency η_{pv} is then defined by the ratio of electric power P_{pv} and the input solar power P_s.

$$\eta_{pv} = \frac{P_{pv}}{P_s} = \frac{\Delta V \; I_T(\Delta V)}{P_s} \tag{8.101}$$

It is well known that an increasing ΔV decreases I_T and can therefore expect a certain maximum in η_{pv}. Therefore, we can define a maximum PV efficiency as

$$\eta_{pv,max} = \frac{max\left(\Delta V I_T(\Delta V)\right)}{P_s} \tag{8.102}$$

The optimal (subscript *opt*) electric potential ΔV_{opt} is the one where Eq. (8.102) holds, so that the corresponding optimal current density will be

$$I_{T,opt} = \frac{max\left(\Delta V I_T(\Delta V)\right)}{\Delta V_{opt}} \tag{8.103}$$

We can then also define

$$\eta_{pv,max} = \frac{\Delta V_{opt} I_{T,opt}}{P_s} \tag{8.104}$$

The maximum total efficiency of the photothermoelectric efficiency η_{pvte} is then given by

$$\eta_{pvte,max} = \left(1 - \eta_{pv,max}\right)\eta_{te,max} + \eta_{pv,max} \tag{8.105}$$

We have developed models for both the TE and the PV mechanism. Having said, that it is aimed to increase the PV efficiency by a cooled TE device, it is necessary to perform an analysis on the heat management of the entire cooled PVTE hybrid system in Section 8.5.

8.5 Analysis of the Heat Management of the Cooled Hybrid System

Before we can analyse the performance of the hybrid system, an analysis on the heat management should be performed. Let us recall for clarity that wherever an entity (for instance, the temperature) is used in both the PV and theTE model, it will be distinguished by subscripts: the subscript *te* denotes the TE device and *pv* the PV one. We can divide the heat flows in three parts. The first part is the net heat flow at the illuminated PV surface. The second part is the heat transfer from the PV device to the TE element. The third part is the heat removal from the TE element by the CL device (which is powered by the PV device).

8.5.1 Heat Generation in the PV Device

The heat generation rate H is given by the sum of the energy transferred to the lattice by lattice thermalization, H_{hp}, the energy of photons whose energy is lower than the band-gap energy, H_{lp} and the joule heat, H_j:

$$H = H_{hp} + H_{lp} + H_j \tag{8.106}$$

where

$$H_{hp} = G \int_{k_{min}}^{hc/E_g} \left(E_{photon} - E_g\right) dk \tag{8.107a}$$

$$H_{lp} = G \int_{hc/E_g}^{k_{max}} E_{photon} dk \tag{8.107b}$$

$$H_j = -\nabla V \cdot (\boldsymbol{I}_p + \boldsymbol{I}_n) + E_g R \tag{8.107c}$$

where G is given by Eqs. (8.78) and (8.79) for the n- and p-sides, respectively. The total heat generation rate is then expressed by

$$H = C_H + \frac{\Delta V_{opt}}{\delta} \left(I_{p,opt} + I_{n,opt}\right) \tag{8.108}$$

where the latter term is the joule heating, where we assume a linear relation for the voltage gradient with $\delta = \delta_n + \delta_p$ the total PV device thickness. The first term in Eq. (8.108) is given by

$$C_H = \frac{\delta_p}{\delta_p + \delta_n} C_{Hp} + \frac{\delta_n}{\delta_p + \delta_n} C_{Hn} \tag{8.109a}$$

$$C_{Hp} = C_{Gp} \left(E_{gp} k_{min} - hc + \frac{hc}{2\pi} \ln \left(\frac{k_{max}}{k_{min}} \right) \right) + E_{gp} R_{n,bulk} \tag{8.109b}$$

$$C_{Hn} = C_{Gn} \left(E_{gn} k_{min} - hc + \frac{hc}{2\pi} \ln \left(\frac{k_{max}}{k_{min}} \right) \right) + E_{gn} R_{p,bulk} \tag{8.109c}$$

In Eq. (8.109a), we have taken into account the fact that a thicker layer will generate more heat. Moreover, we distinguish between the heat generated in the p- and n-sides of the PV device, the differences being given by the respective thicknesses δ_p and δ_n, the respective band gap energies E_{gp} and E_{gn}, the respective generation rates C_{gp} and C_{gn} and the respective recombination rates $R_{n,bulk}$ and $R_{p,bulk}$. Note that the subscripts of the recombination rates are different now. On one hand, it concerns the recombination of electrons or holes (subscript n or p for the recombination rate) in the p- or n-side (subscript p or n for the heat generation). On the other hand, we only consider the bulk recombination rate as far as it concerns the heat generation. For this, we take Eqs. (8.83) and (8.84), where we fill in Eqs. (8.92) and (8.93) and then take the average value of the carrier concentrations over the respective quasi-neutral region. In doing so, we do not consider the contribution of the surface recombination in C_{Rn} [see Eq. (8.83)] and C_{Rp} [see Eq. (8.84)], respectively. This finally results into

$$R_{n,bulk} \approx \frac{1}{W_p} \int_{-W_p - z_p}^{-z_p} \left(n_p - n_{p0}\right) C_{Rp} dz = \left(\frac{1}{\tau_n} + \frac{1}{\tau_{nA}} \right)$$

$$\times \frac{\left(\mathrm{e}^{\frac{e_C \Delta V}{k_B T}} - 1 \right) n_{p0} D_n \left(L_n S_{rn} \left(\cosh\left(\frac{W_p}{L_n}\right) - 1 \right) + D_n \sinh\left(\frac{W_p}{L_n}\right) \right)}{\frac{D_n W_p}{2L_n} \mathrm{e}^{-\frac{W_p}{L_n}} \left(D_n - L_n S_{rn} + (D_n + L_n S_{rn}) \mathrm{e}^{\frac{2W_p}{L_n}} \right)} + \left(\frac{1}{\tau_n} + \frac{1}{\tau_{nA}} \right)$$

$$\times \frac{C_{Gp} L_n \left(\left(D_n W_p - 2L_n^2 S_{rn} \right) \cosh\left(\frac{W_p}{L_n}\right) + L_n \left(2L_n S_{rn} + \left(W_p S_{rn} - D_n\right) \sinh\left(\frac{W_p}{L_n}\right) \right) \right)}{\frac{D_n W_p}{2L_n} \mathrm{e}^{-\frac{W_p}{L_n}} \left(D_n - L_n S_{rn} + (D_n + L_n S_{rn}) \mathrm{e}^{\frac{2W_p}{L_n}} \right)} \tag{8.110}$$

$$R_{n,bulk} \approx \frac{1}{W_n} \int_{z_n}^{W_n+z_n} (p_n - p_{n0}) C_{Rn} dz = \left(\frac{1}{\tau_p} + \frac{1}{\tau_{pA}}\right)$$

$$\times \frac{\left(e^{\frac{e_C \Delta V}{k_B T}} - 1\right) p_{n0} D_p \left(L_p S_{rp} \left(\cosh\left(\frac{W_n}{L_p}\right) - 1\right) + D_p \sinh\left(\frac{W_n}{L_p}\right)\right)}{\frac{D_p W_n}{2L_p} e^{-\frac{W_n}{L_p}} \left(D_p - L_p S_{rp} + (D_p + L_p S_{rp}) e^{\frac{2W_n}{L_p}}\right)} + \left(\frac{1}{\tau_p} + \frac{1}{\tau_{pA}}\right)$$

$$\times \frac{C_{Gn} L_p \left(\left(D_p W_n - 2L_p^2 S_{rp}\right) \cosh\left(\frac{W_n}{L_p}\right) + L_p \left(2L_p S_{rp} + (W_n S_{rp} - D_p) \sinh\left(\frac{W_n}{L_p}\right)\right)\right)}{\frac{D_p W_n}{2L_p} e^{-\frac{W_n}{L_p}} \left(D_p - L_p S_{rp} + (D_p + L_p S_{rp}) e^{\frac{2W_n}{L_p}}\right)} \tag{8.111}$$

where $\Delta V = \Delta V_{pv,opt}$ corresponding to the one where Eq. (8.102) holds. The last term of Eq. (8.108) is the so-called Joule heating, obtained at optimal conditions, defined by Eqs. (8.102) and (8.103).

8.5.2 Temperature Profiles

The quasi-stationary energy balance equation in the PV device is given by

$$\nabla \cdot (\lambda_{pv} \nabla T_{pv}) + H = 0 \tag{8.112}$$

where λ_{pv} is the thermal conductivity of the PV device. We define an overall thermal conductivity of the PV device:

$$\lambda_{pv} = \frac{\delta_n + \delta_p}{\frac{\delta_n}{\lambda_{pvn}} + \frac{\delta_p}{\lambda_{pvp}}} \tag{8.113}$$

where the subscripts λ_{pvn} and λ_{pvp} denote the thermal conductivities of the n- and p-side materials of the PV device and δ_n and δ_p the n- and p-sides' total thicknesses, respectively. At the upper surface of the PV device, we assume that the only heat transfer mechanisms are solar radiation P_s (increasing the surface temperature) and heat convection (decreasing the surface temperature). The boundary condition at the upper surface is then

$$-\lambda_{pv} \left.\frac{\partial T_{pv}}{\partial y}\right|_{y=0} = P_s - h_s \left(T(y=0) - T_{amb}\right) \tag{8.114}$$

where $y = 0$ at the illuminated upper surface and $y = \delta$ at the dark side. In this section, we choose to use a different symbol for the space coordinate just for the sake of clarity and convenience. In Eq. (8.114), h_s stands for the convective heat transfer coefficient towards the ambient air with temperature T_{amb}. Since h_s is variable due to external factors, we will take a typical mean value of $h_s = 10$ W m^{-2} K^{-1} [28]. We can see from Eq. (8.5a) and (8.5b) that under the assumptions of this work, the heat flux in the TE device is considered to be constant. Assuming heat flux continuity through the PV–TE interface, we have the following boundary condition at $y = \delta$:

$$-\lambda_{pv} \left.\frac{\partial T_{pv}}{\partial y}\right|_{y=\delta} = q_{pv,opt} \tag{8.115a}$$

$$q_{pv,opt} = q_{te,opt} \tag{8.115b}$$

where $q_{te,opt}$ is the optimal heat flux from the sum of Eq. (8.5a) and (8.5b), wherein the material properties are taken at T_{ref} and the optimal TE current density $I_{te,opt}$ is used from Eq. (8.10). Using Eq. (8.12), the total heat flux in the TE element can also be written as

$$q_{te,opt} = \lambda_{te,tot} \frac{T_{te,h} - T_{te,c}}{L} \sqrt{1 + ZT} \tag{8.116}$$

Let us make some short reflections on Eq. (8.116). If the Peltier effect was neglected, Eq. (8.116) would reduce to the simple Fourier law. For bulk materials, ZT is a positive constant. For nanocomposites, ZT depends on the nanoparticle radius and volume fraction via the size-dependent material properties. Recalling that the TE device consists of a p- and an n-leg, the total effective thermal conductivity is derived by means of the Matthiessen's rule uniquely for the purposes of this section:

$$\lambda_{te,tot} = \left(\frac{1}{\lambda_{ph,p} + \lambda_{e,p}} + \frac{1}{\lambda_{ph,n} + \lambda_{e,n}} \right)^{-1} \tag{8.117}$$

Having calculated the current through the PV device, we can calculate the maximum generated heat through Eqs. (8.112) and (8.113)–(8.117). The n-side layer is often much smaller than the p-side layer, so that we can neglect the effect of the former's thickness on the temperature variation. We solve Eq. (8.112) at optimal conditions and find the following temperature profile in the PV device

$$T_{pv,h}(z) = T_{amb} + \frac{P_s - q_{te,opt} + H\delta}{h_s} - \frac{(Hz + 2q_{te,opt} - 2H\delta)\, z}{2\lambda_{pv}} \tag{8.118}$$

The temperature at the PV–TE interface is thus

$$T_{pv,h} = T(y = \delta) = T_{amb} + \frac{P_s - q_{te,opt} + H\delta}{h_s} + \frac{H\delta^2 - 2\delta q_{te,opt}}{2\lambda_{pv}} \tag{8.119}$$

8.5.3 Operating Temperatures

In order to perform all the calculations concerning the PV device, we have seen in Sections 8.2–8.4 that there is a temperature dependence. It is not the purpose here to study this temperature dependence, but in order to have realistic values, we should evaluate the calculations for the PV device at a certain realistic operating temperature, valid wherever no high (T_h) or low (T_c) temperature is explicitly specified. We define this operating temperature $T \equiv T_{op,pv}$ as an approximation of Eq. (8.119), valid for $P_s \gg q_{te,opt}$ and $P_s \gg H\delta$:

$$T_{op,pv} = T_{amb} + \frac{P_s}{h_s} \tag{8.120}$$

Concerning the TE device, we assume that it operates under an average temperature between the PV one and the low cooling temperature:

$$T_{op,te} = \frac{T_{op,pv} + T_{env}}{2} \tag{8.121}$$

As far as material properties are concerned, we assume they are evaluated at a constant ambient temperature. We assume temperature continuity through the PV–TE interface, so that we define $T_{pv,h} \equiv T_{te,h}$. Using a simple Newton's law of cooling, we can state that at quasi-stationary conditions, the heat (assumed to be constant) through the TE device is equal to

$$q_{te,opt} = h_c (T_{te,c} - T_{env}) \tag{8.122}$$

where h_c is the convective heat transfer coefficient towards the CL device (e.g. air fin and heat pipe) and T_{env} the temperature far away from the contact surface (possibly but not necessarily equal to T_{amb}). The heat transfer coefficient towards the cooling air, h_c, can be found [7] by

$$h_c = \frac{\lambda_c Nu}{L_c} \tag{8.123}$$

$$Nu = 0.664 Re^{1/2} Pr^{1/3} \tag{8.124}$$

$$Re = \frac{\rho_c v_c L_c}{\mu_c} \tag{8.125}$$

$$Pr = \frac{c_{p,c}\mu_c}{\lambda_c} \tag{8.126}$$

where N_U, Re, Pr, λ_c, ρ_c, v_c, μ_c, $c_{p,c}$ and L_c are the dimensionless Nusselt, Reynolds and Prandtl numbers, the thermal conductivity, density, velocity, dynamic viscosity, heat capacity of the cooling fluid and the characteristic length (typically the length of the hybrid system) at which the cooling takes place, respectively. The power flux P_c needed for cooling depends on the flow mean velocity v_c and the pressure drop ΔP to overcome:

$$P_c = v_c \Delta P \tag{8.127}$$

where the pressure drop is obtained via the standard Hagen–Poiseuille equation for laminar flow between two horizontal plates:

$$\Delta P = \frac{12\mu_c L_c v_c}{W_c^2} \tag{8.128}$$

where W_c is the depth of the cooling channel.

Rearranging Eqs. (8.116), (8.119) and (8.122) gives finally the expressions for $T_{te,h}$ and $T_{te,c}$:

$$T_{te,h} = \frac{h_c L \left(h_s H\delta^2 + 2\lambda_{pv}\left(P_s + h_s T_{amb} + H\delta\right)\right)}{2h_c h_s L\lambda_{pv} + 2\lambda_{te,tot}\sqrt{1+ZT}\left(h_c h_s \delta + \lambda_{pv}\left(h_c + h_s\right)\right)} + \frac{\lambda_{te,tot}\sqrt{1+ZT}\left(h_s\delta\left(2h_c T_{env} + H\delta\right) + 2\lambda_{pv}\left(P_s + h_s T_{amb} + H\delta + h_c T_{env}\right)\right)}{2h_c h_s L\lambda_{pv} + 2\lambda_{te,tot}\sqrt{1+ZT}\left(h_c h_s \delta + \lambda_{pv}\left(h_c + h_s\right)\right)} \tag{8.129}$$

$$T_{te,c} = \frac{h_c L T_{env} + T_{te,h}\lambda_{te,tot}\sqrt{1+ZT}}{h_c L + \lambda_{te,tot}\sqrt{1+ZT}} \tag{8.130}$$

It is interesting to note that the Nu number in Eq. (8.124) is linked to the Kn number via Eqs. (8.122), (8.123) and (8.130). If the cooling area would be a nano-channel, the correlation in Eq. (8.124) would not be valid and would also be dependent explicitly on the Kn number [29].

8.5.4 Total Efficiency of the Hybrid System

It is the idea to extract the power flux for the CL device directly from the electricity produced from the PV device, which causes a loss of efficiency equal to

$$\eta_c = \frac{P_c}{P_s} \tag{8.131}$$

The net efficiency of the total hybrid system is then

$$\eta_{net} = \eta_{pv,max} + \left(1 - \eta_{pv,max}\right)\eta_{te,max} - \eta_c \tag{8.132}$$

The net efficiency in this work is considered to be influenced by five parameters: the PV thickness δ, the TE thickness L, the nanoparticle radius in the TE material a_p and its volume fraction φ and the cooling velocity v_c. The model presented in this chapter will be applied in a case study in Chapter 9.

References

[1] Su, S., Wang, Y., Wang, J., Xu, Z., Chen, J. 2014. Material optimum choices and parametric design strategies of a photon-enhanced solar cell hybrid system. *Solar Energy Materials and Solar Cells* 128:112–118.

[2] Huang, M.J., Eames, P.C., Norton, B., Hewitt, N.J. 2011. Natural convection in an internally finned phase change material heat sink for the thermal management of photovoltaics. *Solar Energy Materials and Solar Cells* 95:1598–1603.

[3] Liu, W., Lucas, K., McEnaney, K., Lee, S., Zhang, Q., Opeil, C., Chen, G., Ren, Z. 2013. Studies on the Bi_2Te_3–Bi_2Se_3–Bi_2S_3 system for mid-temperature thermoelectric energy conversion. *Energy and Environmental Science* 6:552–560.

[4] Sumithra, S., Takas, N.J., Misra, D.K., Nolting, W.M., Poudeu, P.F.P., Stokes, K.L. 2011. Enhancement in thermoelectric figure of merit in nanostructured Bi_2Te_3 with semimetal nanoinclusions. *Advanced Energy Materials* 1:1141–1147.

[5] Jou, D., Sellitto, A., Cimmelli, V.A. 2014. Multi-temperature mixture of phonons and electrons and nonlocal thermoelectric transport in thin layers. *International Journal of Heat and Mass Transfer* 71:459–468.

[6] Machrafi, H. 2017. Enhancement of a photovoltaic cell performance by a coupled cooled nanocomposite thermoelectric hybrid system, using extended thermodynamics. *Current Applied Physics* 17: 890–911.

[7] Zhang, J., Xuan, Y., Yang, L. 2014. Performance of photovoltaic-thermoelectric hybrid systems. *Energy* 78:895–903.

[8] Candadai, A.A., Kumar, V.P., Barshilia, H.C. 2016. Performance evaluation of a natural convective-cooled concentration solar thermoelectric generator coupled with a spectrally selective high temperature absorber coating. *Solar Energy Materials and Solar Cells* 145:333–341.

[9] Ju, X., Wang, Z.F., Flamant, G., Li, P., Zhao, W.Y. 2012. Numerical analysis and optimization of a spectrum splitting concentration photovoltaic–thermoelectric hybrid system. *Solar Energy* 86:1941–1954.

[10] Yang, D.J., Yin, H.M. 2011. Energy conversion efficiency of a novel hybrid solar system for photovoltaic, thermoelectric, and heat utilization. *IEEE Transactions Energy Conversion* 26:662–670.

[11] Benghanem, M., Al-Mashraqi, A.A., Daffallah, K.O. 2016. Performance of solar cells using thermoelectric module in hot sites. *Renewable Energy* 89:51–59.

[12] Li, Y.L., Witharana, S., Cao, H., Lasfargues, M., Huang, Y., Ding, Y.L. 2014. Wide spectrum solar energy harvesting through an integrated photovoltaic and thermoelectric system. *Particuology* 15:39–44.

[13] Chávez-Urbiola, E.A., Vorobiev, Y.V., Bulat, L.P. 2012. Solar hybrid system with thermoelectric generators. *Solar Energy* 86:369–378.

[14] Tan, M., Deng, Y., Hao, Y. 2014. Enhancement of thermoelectric properties induced by oriented nanolayer in $Bi_2Te_{2.7}Se_{0.3}$ columnar films. *Materials Chemistry Physics* 146:153–158.

[15] Topic, M., Campa, A., Filipic, M., Berginc, M., Krasovec, U.O., Smole, F. 2010. Optical and electrical modelling and characterization of dye-sensitized solar cells. *Current Applied Physics* 10:S425–S430.

[16] Omer, S.A., Infield, D.G. 1998. Design optimization of thermoelectric devices for solar power generation. *Solar Energy Materials and Solar Cells* 53:67–82.

[17] Jou, D., Casas-Vazquez, J., Lebon, G. 2010. *Extended Irreversible Thermodynamics*, fourth ed. New York: Springer.

[18] Machrafi, H. 2016. Heat transfer at nanometric scales described by extended irreversible thermodynamics. *Communication Applied Industrial Mathematics* 7:177–195.

[19] Hua, Y.C., Cao, B.Y. 2016. Ballistic-diffusive heat conduction in multiply-constrained nanostructures. *International Journal of Thermal Sciences* 101:126–132.

[20] Cattaneo, C. 1948. Sulla conduzione del calore. *Atti del Seminario Matematico e Fisico delle Università di Modena* 3:83–101.

[21] Norton, M., Gracia, A.A.M., Galleano, R. 2015. Comparison of solar irradiance measurements using the average photon energy parameter. *Solar Energy* 120:337–344.

[22] Khriachtchev, L. 2009. *Silicon Nanophotonics, Basic Principles, Present Status and Perspectives.* Singapore: Stanford Publishing, p. 317.

[23] Palankovski, V., Quay, R. 2004. *Analysis and Simulation of Heterostructure Devices.* Vienna: Springer-Verlag.

[24] Richter, A., Glunz, S.W., Werner, F., Schmidt, J., Cuevas, A. 2012. Improved quantitative description of Auger recombination in crystalline silicon. *Physical Review B* 86:165202.

[25] Cuevas, A., Macdonald, D. 2003. Measuring and interpreting the lifetime of silicon wafers. *Solar Energy* 76:255–262

[26] Alamo, J.A., Swanson, R.M. 1987. Modelling of minority-carrier transport in heavily doped silicon emitters. *Solid-State Electronics* 30:1127–1136.

[27] Lu, M. 2008. Silicon heterojunction solar cell and crystallization of amorphous silicon, PhD thesis University of Delaware, p. 46.

[28] Salazar, A., Apiñaniz, E., Mendioroz, A., Oleaga, A. 2010. A thermal paradox: Which gets warmer? *European Journal Physics* 31:1053–1059.

[29] Balaj, M., Roohi, E., Akhlaghi, H., Myong, R.S. 2014. Investigation of convective heat transfer through constant wall heat flux micro/nano channels using DSMC. *International Journal of Heat and Mass Transfer* 71:633–638.

Part III

Advanced Applications and Perspectives

9

Optimal Enhancement of Photovoltaic Energy by Coupling to a Cooled Nanocomposite Thermoelectric Hybrid System

9.1 Case Study: Material Properties and Operating Conditions

The model developed in Chapter 8 is here applied in order to optimize the operation of a photovoltaic cell connected to a cooled thermoelectric nanocomposite.

9.1.1 Material Properties for the Photovoltaic Materials

In this work, we consider two types of photovoltaic materials. One is monocrystalline silicon (c-Si) and the other is thin-film polycrystalline silicon (p-Si, not to be confused with the p-side of a solar cell, which is considered in both the Si materials). The electron and hole properties of c-Si and p-Si are shown in Table 9.1. Typical doping concentrations are taken from [1] and presented in Table 9.2. The hole Auger lifetime is extracted from Auger lifetime studies [2,3] as $\tau_{pA} = 1/\left(B_n N_{dop}^2\right)$ and $\tau_{nA} = 1/\left(B_p N_{dop}^2\right)$, where $N_{dop} = N_D, N_A$ is the net dopant concentration in the p- and n-sides, respectively. This expression is only valid for net dopant concentrations $N_{dop} > 5 * 10^{24}\,\mathrm{m}^{-3}$, which is only the case for the n-side of both the Si solar cells (see doping concentrations in Table 9.2) here. For the p-side, we extract the value from [4,5] to be $\tau_{nA} = 8 * 10^{-5}$ s. The Auger recombination coefficients B_p and B_n are taken from [2]. The effective densities of state are taken from [6]. All these mentioned values are taken to be the same for c-Si and p-Si. The differences are in the Shockley–Read–Hall (SRH) lifetimes, the electron and hole mobilities and the absorption

TABLE 9.1
Electron and Hole Material Properties of c-Si and p-Si Thin Film at T = 300 K.

General Material Property	Hole		General Material Property	Electron	
τ_{pA} [s]	$3.6*10^{-10}$		τ_{nA} [s]	$8.5*10^{-5}$	
B_p [m^6 s^{-1}]	$9.9*10^{-44}$		B_n [m^6 s^{-1}]	$2.8*10^{-43}$	
N_V [m^{-3}]	$1.04*10^{25}$		N_C [m^{-3}]	$2.8*10^{25}$	
Specific Material Property	**c-Si**	**p-Si**	**Specific Material Property**	**c-Si**	**p-Si**
μ_p [m^2 V^{-1} s^{-1}]	$5*10^{-2}$	$56.3*10^{-4}$	μ_n [m^2 V^{-1} s^{-1}]	$1*10^{-1}$	$95.0*10^{-4}$
τ_p [s]	$1*10^{-6}$	$3*10^{-10}$	τ_n [s]	$1*10^{-6}$	$1*10^{-9}$

TABLE 9.2
Other Material Properties of c-Si and p-Si Thin Film at $T = 300$ K.

Material Property	c-Si	p-Si
E_g [eV]	1.12	1.12
ε_r [−]	11.8	11.7
N_A [m^{-3}]	1*10^{22}	1*10^{22}
N_D [m^{-3}]	1*10^{26}	1*10^{26}
A [m^{-1}]	4*10^{3}	2.5*10^{5}
λ_{pv} [W K^{-1} m^{-1}]	148	20

coefficient. The SRH lifetimes τ_p and τ_n for p-Si are obtained from the Scharfetter relation [7], which represents them as a function of the total dopant concentration:

$$\tau_{n,p} = \frac{\tau_{max,n,p}}{\left(\frac{N_D+N_A}{N_{n,p,ref}}\right)^n} \tag{9.1}$$

The parameters for this relation are given in [7]. For $\tau_{max,n} = 10^{-5}$ s, $\tau_{max,p} = 3 * 10^{-6}$ s, $N_{n,p,ref} = 10^{22}$ m^{-3} and $n = 1$, we obtain the SRH lifetimes given in Table 9.1. The SRH lifetimes for c-Si are extracted from [8]. The mobility for p-Si depend on the doping concentrations, and this can well be approximated by [9,10]

$$\mu_n = 92 * 10^{-4} + \frac{1,268 * 10^{-4}}{1 + \left(\frac{N_D+N_A}{1.3*10^{23}}\right)^{0.91}} \tag{9.2}$$

$$\mu_p = 54.3 * 10^{-4} + \frac{406.9 * 10^{-4}}{1 + \left(\frac{N_D+N_A}{2.35*10^{23}}\right)^{0.88}} \tag{9.3}$$

The mobility for c-Si is taken from [6]. The absorption coefficient for p-Si is much higher than for c-Si [11]. For the absorption coefficient, we have taken a value averaged over the whole considered wavelength spectrum following the example of a correlation given in [12].

Other photon and carrier information is given in Table 9.2. Note that the permittivity ε is given as the relative one ε_r (in Table 9.2) multiplied by the vacuum one ε_0, i.e. $\varepsilon = \varepsilon_r\varepsilon_0$. The thermal conductivity for c-Si and p-Si is taken from [13] and [14], respectively.

Finally, it should be noted that, although some properties of the considered materials in this chapter have the same values in the n- and p-sides, they are formally different in the photovoltaic model. This is done so that the model remains generally applicable to other photovoltaic materials, where the n- and p-sides are made out of distinctly different materials.

9.1.2 Material Properties for the Thermoelectric Materials

The p-leg of the thermoelectric element is composed of p-type antimony telluride (p-Sb_2Te_3) nanoparticles (NPs) dispersed in a matrix of p-type bismuth telluride (p-Bi_2Te_3). The n-leg is composed of n-type bismuth selenide (n-Bi_2Se_3) NPs dispersed in a matrix of n-type Bi_2Te_3 (n-Bi_2Te_3).

We start with p-Bi_2Te_3. The data for p-Bi_2Te_3 have already been treated and are given in Table 5.1. Given, from Table 5.1, the electron (c_e, v_e and $\ell_{e,b}$) and phonon (c_{ph}, v_{ph} and $\ell_{ph,b}$) material properties, the thermal conductivity can easily be calculated by using Eq. (8.18) and $l_{tot} \equiv l_{ph} + l_e$. Equation (8.21) then gives the electrical conductivity, and the Seebeck coefficient is obtained from Table 5.1.

As for p-Sb_2Te_3, the electric conductivity is $4.76 \times 10^5\ \Omega^{-1}\ m^{-1}$ [15]. Via Eq. (8.21), we can calculate the electron thermal conductivity to be 3.49 W K^{-1} m^{-1}. Given that the total thermal conductivity is 4.2 W K^{-1} m^{-1} [16], we can deduce that the phonon thermal conductivity is 0.71 W K^{-1} m^{-1}. The Seebeck coefficient and the phonon velocity are 79 μV K^{-1} and 2,900 m s^{-1}, respectively [16]. The total specific heat capacity is 1.34 MJ m^{-3} K^{-1} [17,18]. For the phonon mean free path, it can be found that it is 3 nm [19]. Then via Eq. (8.18), we can calculate that the phonon contribution of the specific heat is equal to 0.245 MJ m^{-3} K^{-1}. Assuming that the total specific heat is the sum of the phonon and electron contributions [20], the electron contribution of the specific heat is then equal to 1.09 MJ m^{-3} K^{-1}. The electron mean free path is 1 nm [21]. This makes the thermal electron velocity equaling 9,562 m s^{-1} via Eq. (8.18).

We assume that for the n-Bi_2Te_3 at room temperature, the phonon and electron mean free paths as well as the total specific heats remain the same as for the p-type materials. As for the n-Bi_2Te_3, the total thermal conductivity is 1.27 W K^{-1} m^{-1}, while the electric conductivity equals $0.96 \times 10^5\ \Omega^{-1}\ m^{-1}$ [22]. The latter results into an electron thermal conductivity of 0.70 W $K^{-1}m^{-1}$ via Eq. (8.21) and according to the definition under Eq. (8.8), a phonon thermal conductivity of 0.57 W $K^{-1}m^{-1}$. The phonon velocity is given to be 1,750 m s^{-1} [23]. This gives a phonon-specific heat of 0.324 MJ m^{-3} K^{-1} via Eq. (8.18) and, knowing that the total specific heat is 1.2 MJ $m^{-3}K^{-1}$ [20,24], thus, an electron-specific heat of 0.876 MJ m^{-3} K^{-1}. The thermal electron velocity is therefore 2,634 m s^{-1} via Eq. (8.18). The Seebeck coefficient is −241 μV K^{-1} [25].

It remains to evaluate the properties of n-Bi_2Se_3. The electric conductivity is $2.4 \times 10^5\ \Omega^{-1}\ m^{-1}$ [25], which gives an electron thermal conductivity of 1.76 W $K^{-1}m^{-1}$ via Eq. (8.21). Knowing from [25] that the phonon thermal conductivity is 0.66 W $K^{-1}m^{-1}$, the total thermal conductivity is thus 2.42 W $K^{-1}m^{-1}$. The phonon velocity is 2,900 m s^{-1} [26]. For the phonon-specific heat, we have obtained the value from [27,28] being 0.255 MJ m^{-3} K^{-1}. We can deduce via Eq. (8.18) that the phonon mean free path is 2.7 nm. The total heat capacity being 1.33 MJ m^{-3} K^{-1} [25], we can calculate the electron heat capacity to be 1.08 MJ m^{-3} K^{-1}. The Seebeck coefficient is −50 μV K^{-1} [25]. The electron mean free path is said to be approximately 1 nm [29]. We can then finally calculate the electron velocity to be 4,889 m s^{-1} via Eq. (8.18).

A resume of these properties is given in Table 9.3, where we indicate whether it concerns the NPs or the matrix (M).

9.1.3 Other Operating Characteristics and General Physical Properties

Table 9.4 presents the characteristics of the cooler, and Table 9.5 presents some general physical constants used throughout this chapter.

Setting $T_{\text{amb}} = T_{env} = 300$ K, we can deduce from Eqs. (8.120) and (8.121), the operating temperatures $T_{op,pv} = 400$ K and $T_{op,te} = 350$ K.

9.2 Case Study: Photovoltaic Performance

9.2.1 Optimal Thickness of the Photovoltaic Device

First of all, we investigate the influence of the thickness of the photovoltaic device and the surface recombination velocity on the electric power. Note that the photovoltaic efficiency

TABLE 9.3
Electron and Phonon Material Properties of p-Bi_2Te_3 and n-Bi_2Te_3, p-Sb_2Te_3 and n-Bi_2Se_3 at $T = 300$ K.

Material	Role	Electron				Phonon		Electrical	
		c_e [MJ m^{-3} K^{-1}]	v_e [km s^{-1}]	$\ell_{e,b}$ [nm]	c_{ph} [MJ m^{-3} K^{-1}]	v_{ph} [km s^{-1}]	$\ell_{ph,b}$ [nm]	σ_e [kΩ^{-1} m^{-1}]	S_b [μV K^{-1}]
p-Sb_2Te_3	NP	1.09	9.65	1.0	0.245	2.9	3.0	476	79
p-Bi_2Te_3	M	1.01	7.83	0.91	0.19	8.43	3.0	328	188
n-Bi_2Se_3	NP	1.08	4.89	1.0	0.255	2.9	2.7	250	–50
n-Bi_2Te_3	M	0.876	2.63	0.91	0.324	1.75	3.0	96	–241

TABLE 9.4
Cooler Characteristics [6].

Property	Value
h_s [W K^{-1} m^{-2}]	10
P_s [W m^{-2}]	1,000
λ_c [W K^{-1} m^{-1}]	0.0242
L_c [m]	0.025
ρ_c [kg m^{-3}]	1.2
μ_c [Pa s]	1.84*10^{-5}
$c_{p,c}$ [J kg^{-1} K^{-1}]	1,007
W_c [m]	0.025

is this electric power output divided by the solar power input P_s = 1,000 W m^{-2}. As for the surface recombination velocity, we assume that it is equal in n- and p-sides, although it is typically somewhat smaller in the n-side [30]. So, we take $S_{rp} = S_{rn} = S_r$. Figure 9.1 shows the photovoltaic power output as a function of the p-side thickness for four n-side thicknesses for c-Si, i.e. δ_n = 50 nm, δ_n = 250 nm, δ_n = 2 μm and δ_n = 100 μm. The same for p-Si is shown in Figure 9.2. The δ_n value of 50 nm appears to be the smallest manufacturable n-side thickness for the doping concentration used in this work [1]. The δ_n value of 100 μm is chosen for the reason that it is much larger than the characteristic diffusion lengths of both the Si solar cells.

The optimal p-side thickness for the c-Si varies from 100 to 200 μm, for recombination velocities ranging from 10^{-1} to 10^5 m s^{-1}, respectively, whatever the considered n-side thicknesses. Noting that, for this given range of surface recombination velocities, this velocity is of more importance in a thinner device than in a thicker one, it is understandable that the photovoltaic efficiency is more affected by the surface recombination velocity at smaller thicknesses. In [30], it is said that the recombination velocity for an unprepared sample is of the order of 10^2 m s^{-1}. After passivation of dangling bonds by an HF/H_2O solution, it was lowered to about ~10 m s^{-1} [30]. We can see anyway that not much increase in the electric power output is observed in Figures 9.1 and 9.2, when decreasing further the surface recombination velocity, so that we use in the following the typical value of S_r = 10 m s^{-1}. The corresponding optimal p-side thickness for c-Si is 125 μm, and for p-Si, it is around 1 μm. These values also correspond to typical industrial values found in the literature [31,32], where it is confirmed that p-Si only needs a thickness of the order of 1 μm and c-Si two orders more, due to the formers' larger absorption coefficient [11]. The maximum electric power that defines these optimal thicknesses needs some reflection. Two opposing factors [33] govern the increase or decrease of the electric output: the short-circuit current and the open-circuit voltage [see Eqs. (8.98)–(8.100)]. Physically, these factors are influenced by the photonic energy absorption and the recombination rate. Although a too thin cell will have much less bulk recombination [see Eqs. (8.83), (8.84), (8.92) and (8.93)], it will still not absorb enough photonic energy [see Eq. (8.52)], which will cause the electric output to decrease for thinner cells. At too thick cells, much larger than the characteristic diffusion length [see text under Eq. (8.87)], bulk recombination becomes predominant, even though a maximum of photonic energy is absorbed, and the electric output decreases as well. It is then understandable that a maximum can be found for intermediary thicknesses, as can be seen in Figures 9.1 and 9.2. The difference in these intermediary thicknesses for c-Si and p-Si lies in the difference of the aforementioned characteristic diffusion lengths, determined by the material properties. Figures 9.1d and 9.2d show that for large enough

TABLE 9.5
Physical Constants at $T = 300$ K.

Physical Constant	$\boldsymbol{k_B}$ $[10^{-23}\ \mathbf{J\ K^{-1}}]$	$\boldsymbol{h}$ $[10^{-34}\ \mathbf{Js}]$	$\boldsymbol{e_c}$ $[10^{-19}\ \mathbf{C}]$	$\boldsymbol{L}$ $[10^{-8}\ \mathbf{W\ \Omega\ K^{-2}}]$	**c** $[10^{8}\ \mathbf{m\ s^{-1}}]$	$\boldsymbol{\varepsilon_0}$ $[10^{-12}\ \mathbf{C\ V^{-1}\ m^{-1}}]$
Value	1.38	6.626	1.602	2.44	3.0	8.854

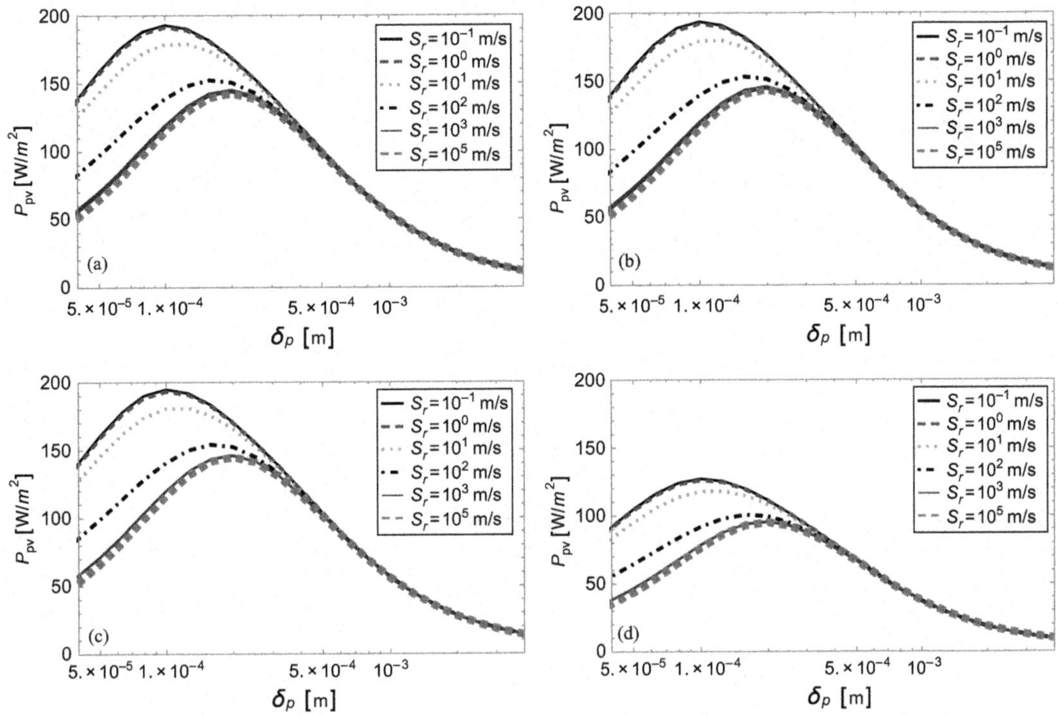

FIGURE 9.1
Electric power P_{pv} of the c-Si as a function of δ_p for a range of recombination velocities S_r ($10^{-1} - 10^5$ m s^{-1}), with P_s = 1,000 W m^{-2} and several δ_n: (a) 50 nm, (b) 250 nm, (c) 2 μm and (d) 100 μm. (From: Machrafi, H. 2017. Enhancement of a photovoltaic cell performance by a coupled cooled nanocomposite thermoelectric hybrid system, using extended thermodynamics. *Current Applied Physics* 17:890–911.)

n-side thicknesses, the electric power (even at optimal p-side thickness) decreases. Figure 9.3 shows the electric power as a function of the n-side thickness.

Figure 9.3 shows indeed that for sufficiently large n-side thicknesses, the electric power drops towards zero. The electric power starts already to drop considerably for the n-side thicknesses comparable to the p-side ones. The n-side being heavily doped, the recombination rate is much higher than in the p-side for the same thickness [1]. Therefore, the photo-generated free carriers recombine in the n-side, even before they can significantly be of any use in the p-side. This considerably reduces the photonic energy absorption in the p-side, which reduces the electric power. It is then understandable to reduce the n-side thickness as much as possible. For the intermediate n-side thicknesses, however, a small maximum in the electric power is visible in Figure 9.3. We have discussed the effect of too large n-side thicknesses. At too small n-side thicknesses, the recombination and photonic energy absorption will be insignificant with respect to those in the p-side. At intermediate n-side thicknesses, the photo-generation in the n-side is significantly large enough to generate more electricity, but small enough to limit recombination and, therefore, also insignificant enough to interfere with the photo-generation in the p-side. In the foregoing, we proceed with the optimal n- and p-side thicknesses of δ_n = 2 μm and δ_p = 125 μm for c-Si and δ_n = 250 nm and δ_p = 1 μm for p-Si, respectively.

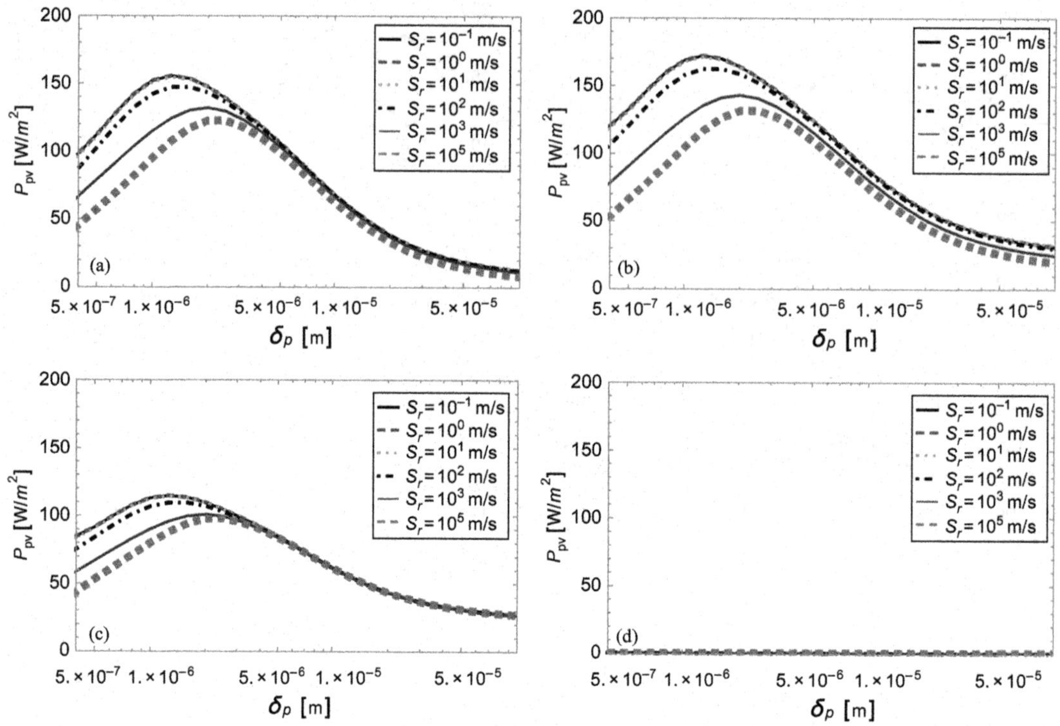

FIGURE 9.2
Electric power P_{pv} of the p-Si as a function of δ_p for a range of recombination velocities S_r ($10^{-1} - 10^5$ m s^{-1}), with P_s = 1,000 W m^{-2} and several δ_n: (a) 50 nm, (b) 250 nm, (c) 2 µm and (d) 1,000 µm. (From: Machrafi, H. 2017. Enhancement of a photovoltaic cell performance by a coupled cooled nanocomposite thermoelectric hybrid system, using extended thermodynamics. *Current Applied Physics* 17:890–911.)

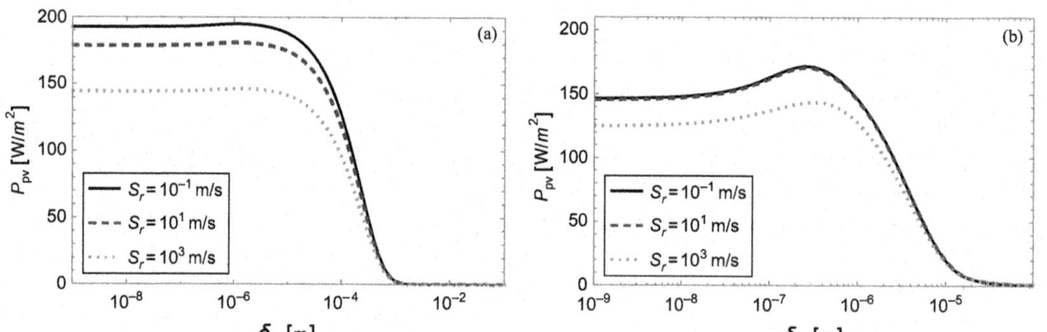

FIGURE 9.3
Electric power P_{pv} of the c-Si (a) and p-Si (b) as a function of the n-side thickness δ_n for a range of recombination velocities S_r (10^{-1}, 10 and 10^3 m s^{-1}) and corresponding optimal p-side thicknesses δ_p (100, 125 and 200 µm), for P_s = 1,000 W m^{-2}. (From: Machrafi, H. 2017. Enhancement of a photovoltaic cell performance by a coupled cooled nanocomposite thermoelectric hybrid system, using extended thermodynamics. *Current Applied Physics* 17:890–911.)

9.2.2 Influence of Nanocomposite Characteristics on Thermoelectric Efficiency

Figure 9.4 shows the thermoelectric efficiency as a function of the NP volume fraction for different NP radii and thermoelement lengths, where the heat comes from the c-Si photovoltaic cell. Figure 9.5 shows the same results but using the p-Si photovoltaic cell instead.

First of all, we can see a clear increase of the thermoelectric efficiency with the increasing volume fraction, decreasing the NP size and increasing the thermoelectric element length. A decreasing thermoelectric NP size means that, especially if the size is of the order of magnitude of the mean free paths of phonons and electrons, the phonons and electrons are scattered more within the material [24]. As such, the heat transport [due to phonon and electron transport as Eqs. (8.1a) and (8.1b) show] is slower and more electricity can be generated, increasing the figure of merit and thusly the thermoelectric efficiency. It is interesting to note that the figure of merit can be increased to almost 2.5 for the materials considered in this work. As an example, the figure of merit of the thermoelectric nanocomposite considered in this work is put against the volume fraction for various NP sizes as shown in Figure 9.6.

Obviously, a higher volume fraction will only amplify the scattering effect at smaller NP sizes. As it can be seen, the highest efficiency is obtained for a volume fraction around 0.7.

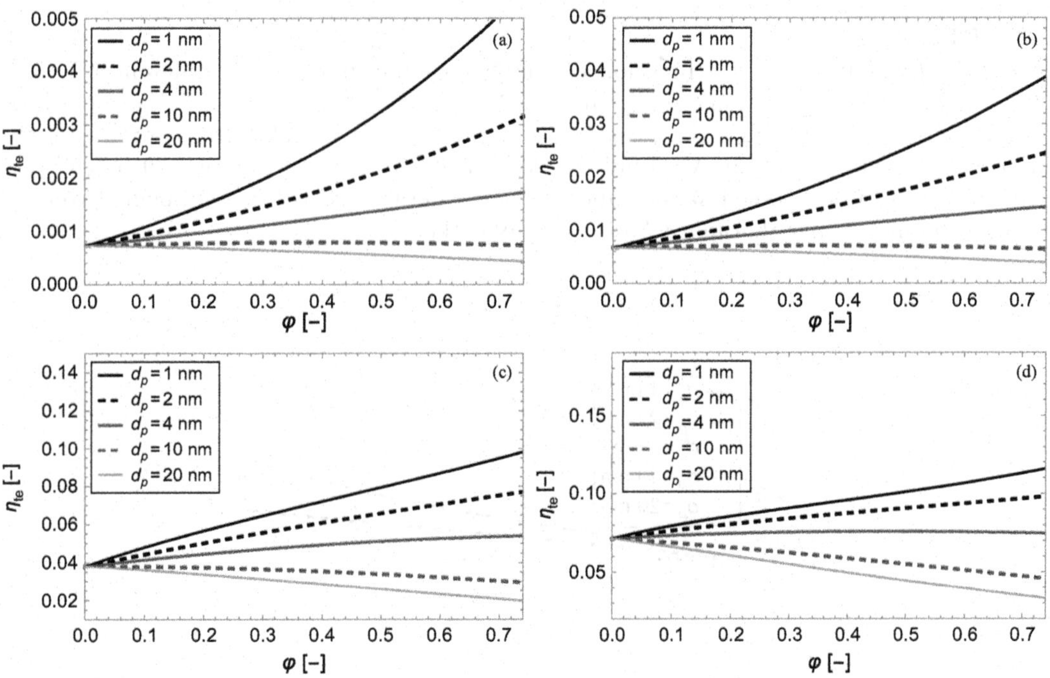

FIGURE 9.4
Thermoelectric efficiency as a function of the NP volume fraction for different NP sizes (d_p = 1, 2, 4, 10 and 20 nm), a cooling velocity of v_c = 10 m s^{-1} and thermoelectric element lengths of (a) 1 mm, (b) 1 cm, (c) 10 cm and (d) 1 m. The heat comes from the photovoltaic c-Si cell. (From: Machrafi, H. 2017. Enhancement of a photovoltaic cell performance by a coupled cooled nanocomposite thermoelectric hybrid system, using extended thermodynamics. *Current Applied Physics* 17:890–911.)

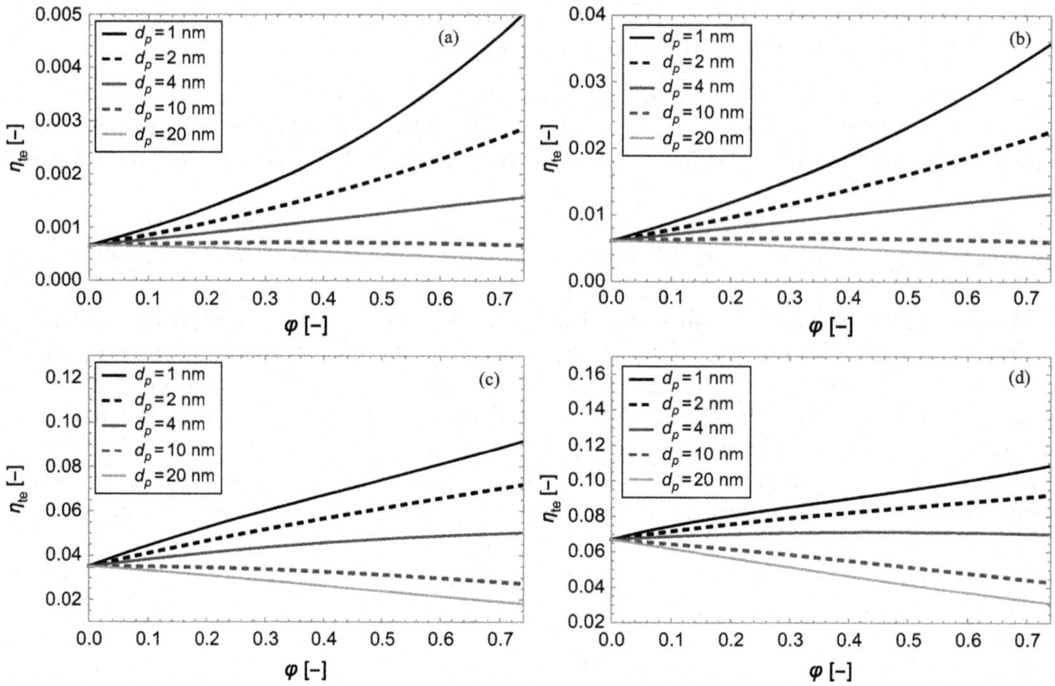

FIGURE 9.5
Thermoelectric efficiency as a function of the NP volume fraction for different NP sizes (d_p = 1, 2, 4, 10 and 20 nm), a cooling velocity of v_c = 10 m s^{-1} and thermoelectric element lengths of (a) 1 mm, (b) 1 cm, (c) 10 cm and (d) 1 m. The heat comes from the photovoltaic p-Si cell. (From: Machrafi, H. 2017. Enhancement of a photovoltaic cell performance by a coupled cooled nanocomposite thermoelectric hybrid system, using extended thermodynamics. *Current Applied Physics* 17:890–911.)

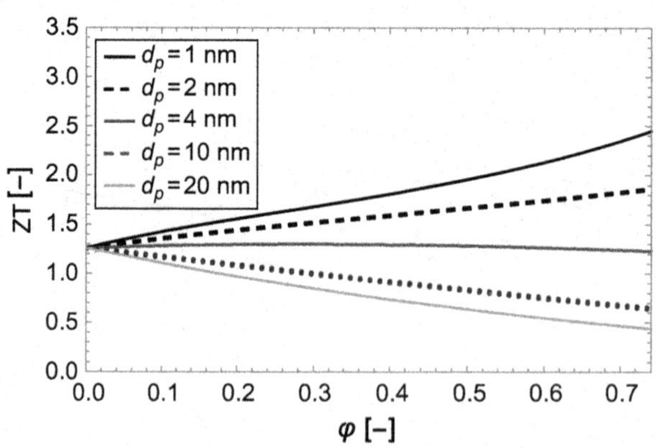

FIGURE 9.6
Figure of merit of the nanocomposite of Section 9.6.1 as a function of the volume fraction for various NP sizes (d_p = 1, 2, 4, 10 and 20 nm). (From: Machrafi, H. 2017. Enhancement of a photovoltaic cell performance by a coupled cooled nanocomposite thermoelectric hybrid system, using extended thermodynamics. *Current Applied Physics* 17:890–911.)

Note that due to geometric limitations, assuming that the NPs are spheres, the highest theoretical volume fraction is $\varphi_{max} = \pi/\sqrt{18}$. However, if the NP size is too large, the heat and electrical transport in the material is simply a function of the bulk values of the NP and matrix transports. The scattering effect will then be negligible [the Knudsen number tends to zero and Eq. (8.43) to the bulk value]. As the bulk values of the materials used for the NPs in this work exhibit figures of merit lower than that of the matrix, a higher volume fraction will only decrease the thermoelectric efficiency. Figure 9.6 shows indeed that for larger NP sizes, a higher volume fraction gives rise to a lower figure of merit. Thus, the effect of the volume fraction is not straightforward and depends on the NP size with respect to the mean free path, i.e. the Knudsen number. The length of the thermoelectric element influences the efficiency, in that a longer element can absorb more heat and the temperature difference will be larger so that more electricity can be generated. Of course, from a theoretical point of view, an element's length going mathematically to zero will reduce the temperature difference across the element to zero as well. As the element's length increases, the efficiency will attain a maximum due to heat loss to the cooling device. This is shown in Figure 9.7, where the thermoelectric efficiency is drawn against the thermoelectric element's length, with a volume fraction of 0.7 and an NP size of 2 nm. For mathematical purposes, the range of thermoelectric lengths in Figure 9.7 is extended beyond realistic limits.

To appreciate the effect of the cooling velocity on the thermoelectric efficiency, we present the results in Figure 9.7 for various cooling velocities as well. It can be seen that the cooling velocities considerably influence the thermoelectric efficiency. A higher velocity will increase the heat transfer coefficient and the temperature difference across the thermoelectric element, thereby increasing the efficiency. The results in Figure 9.7 are shown with an NP size of 2 nm. Smaller NP sizes will show even higher efficiencies. However, NPs with sizes of the order of 1 nm and below begin to exhibit quantum confinement effects [34], not taken into account in the present model. Using a size of 1 nm is at the acceptable limit and, therefore, still deemed reasonable to present in this work. Nonetheless, to be at the safe side, we chose an NP size of 2 nm in Figure 9.7 and will do so in Section 9.2.3 as well.

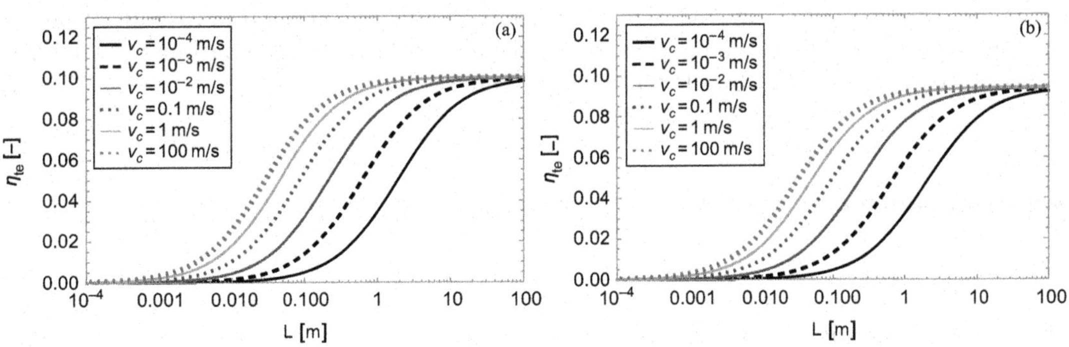

FIGURE 9.7
Thermoelectric efficiency as a function of the thermoelement length for different cooling velocities, a volume fraction $\varphi = 0.7$ and an NP size $d_p = 2$ nm, where the heat comes from the photovoltaic c-Si cell (a) and p-Si cell (b). (From: Machrafi, H. 2017. Enhancement of a photovoltaic cell performance by a coupled cooled nanocomposite thermoelectric hybrid system, using extended thermodynamics. *Current Applied Physics* 17:890–911.)

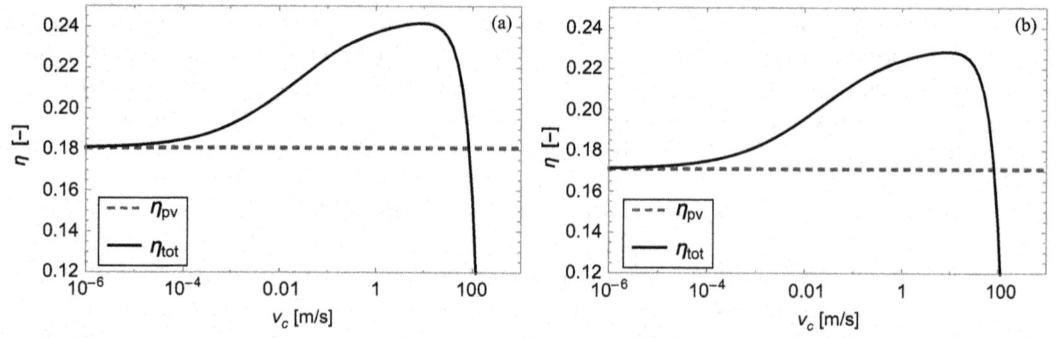

FIGURE 9.8
Total hybrid efficiency as a function of the cooling velocity compared to the photovoltaic one. Two photovoltaic cells are compared: c-Si (a) and p-Si (b). The thermoelectric element in the hybrid systems is the one from Section 9.7.2 with $\varphi = 0.7$, $d_p = 2$ nm and $L = 0.1$ m. (From: Machrafi, H. 2017. Enhancement of a photovoltaic cell performance by a coupled cooled nanocomposite thermoelectric hybrid system, using extended thermodynamics. *Current Applied Physics* 17:890–911.)

9.2.3 Optimal Hybrid Opto-Thermoelectric Efficiency

We have seen in Section 9.2.2 that an optimal total photovoltaic device thickness is around 127 μm and 1.25 μm for c-Si and p-Si, respectively, and that the highest, theoretically reasonable, thermoelectric efficiency is attained with an NP size of 2 nm and a volume fraction around 0.7. Choosing a thermoelectric element length of 10 cm (being practically reasonable [35] and close to maximum thermoelectric efficiency for higher cooling velocities), the results in the previous subsection suggest increasing the cooling velocity to a maximum. However, as the power for the cooling device is extracted directly from the electricity produced by the photovoltaic device, we can imagine that a too high cooling velocity will reduce the overall efficiency. It is therefore necessary to quantitatively assess the effect of the cooling velocity on the total efficiency. The results are shown in Figure 9.8 for both the c-Si and p-Si photovoltaic devices connected to the thermoelectric element of Section 9.2.2.

Figure 9.8 shows that indeed a too high cooling velocity makes the overall efficiency drop towards zero. An optimum cooling velocity appears to be around 10 m s^{-1} for both hybrid systems, attaining overall efficiencies of almost 25%.

9.3 Discussion

This work presents a new analytical model describing the operation of a photovoltaic-thermoelectric hybrid system assisted by a cooling device. The model is presented in three parts. The first part describes a thermoelectric model, where the material consists of NPs embedded into a matrix. The thermoelectric element is composed of an n- and a p-leg, made out of an n-type and p-Bi_2Te_3 matrix, respectively, with Sb_2Te_3 and Bi_2Se_3 NPs, respectively. The model in this work takes into account size-dependent thermoelectric properties, where non-local effects of heat transfer through phonons and electrons are important at nanoscales. These phenomena are extended to also apply for electric transfer as well as the Seebeck coefficient. The transport properties in the thermoelectric nanocomposite are derived in Chapter 5. In Chapter 8, in relation with the thermoelectric theory, a model is developed that describes the p–n junction in a photovoltaic cell, taking into account optical conversion to electricity, surface recombination effects and heat generation. Two photovoltaic materials are considered in this work: c-Si and p-Si. The last part connects the two former ones with a cooling device (powered by electricity from the photovoltaic device) and deals with the heat management analysis of the hybrid system. It should be noted, however, that the photovoltaic model is also applicable to other materials that have typically a p–n junction (such as crystalline materials or inorganic materials of the III and IV groups). The thermoelectric model is also applicable to other types of nanocomposites.

An optimal total photovoltaic device size, obtained from an analysis on the effect of both the n- and p-side thicknesses on the electric power output, has been found being around 127 μm and 1.25 μm for c-Si and p-Si, respectively, corresponding to typical values from the literature [31,32]. It has been shown that the thermoelectric efficiency is increased considerably for increasing volume fraction at the condition that the NP size is rather small, i.e. of the order smaller than 4 nm, which corresponds to Knudsen numbers of the order of 1 or higher. For larger NP sizes, the opposite effect is obtained. The enhancement of the thermoelectric efficiency is explained on one side by an increase in the figure of merit, due to scattering effects that decrease the heat and electric transport properties and on the other side by an increase of the temperature difference across the thermoelectric device. The latter is attained by a higher cooling velocity (a higher heat transfer coefficient between the thermoelectric element and the cooling device), the results of this work showing that the choice of the photovoltaic device hardly influences the thermoelectric efficiency. A study of the overall efficiency shows that a cooling velocity around 10 m s^{-1} presents the highest overall efficiency, i.e. almost 25%. Theoretically, if the NP sizes and the surface recombination velocities could be lowered even further, overall efficiencies of almost 30% could be attained.

Finally, it can be said that it in this work that a rather comprehensive, but analytical model, is able to describe electricity generation by both a photovoltaic and a thermoelectric device. It is also useful for understanding the underlying mechanisms that can increase the photovoltaic and/or photovoltaic-thermoelectric efficiencies, showing the interest of using nanotechnology for solar energy harvesting [36]. The model, presented in this chapter, has also shown how the photovoltaic efficiency can be enhanced by a cooled thermoelectric device, what the limitations are and how further enhancement could be achieved [37].

References

[1] Wang, F., Yu, H., Li, J., Wong, S., Sun, X.W., Wang, X., Zheng, H. 2011. Design guideline of high efficiency crystalline Si thin film solar cell with nanohole array textured surface. *Journal of Applied Physics* 109:084306.

[2] Richter, A., Glunz, S.W., Werner, F., Schmidt, J., Cuevas, A. 2012. Improved quantitative description of Auger recombination in crystalline silicon. *Physical Review B* 86:165202.

[3] Schmidt, J., Kerr, M., Altermatt, P.P. 2000. Coulomb-enhanced Auger recombination in crystalline silicon at intermediate and high-injection densities. *Journal of Applied Physics* 88:1494.

[4] Cuevas, A., Macdonald, D. 2003. Measuring and interpreting the lifetime of silicon wafers. *Solar Energy* 76:255–262.

[5] Alamo, J.A., Swanson, R.M. 1987. Modelling of minority-carrier transport in heavily doped silicon emitters. *Solid-State Electronics* 30:1127–1136.

[6] Zhang, J., Xuan, Y., Yang, L. 2014. Performance of photovoltaic-thermoelectric hybrid systems. *Energy* 78:895–903.

[7] Palankovski, V., Quay, R. 2004. *Analysis and Simulation of Heterostructure Devices.* Vienna: Springer-Verlag.

[8] Rein, S., Glunz, S.W. 2003. Advanced lifetime spectroscopy. *13th Workshop on Crystalline Silicon Solar Cell Materials and Processes, NREL*, pp. 18–25.

[9] Da, Y., Xuan, Y. 2013. Role of surface recombination in affecting the efficiency of nanostructured thin-film solar cells. *Optics Express* 21:A1065–A1077.

[10] Luque, A., Hegedus, S. 2004. *Handbook of Photovoltaic Science and Engineering*, second ed. Hoboken, NJ: John Wiley & Sons.

[11] Gutierrez, E.A., Deen, M.J., Claeys, C.L. 2001. *Low Temperature Electronics: Physics, Devices, Circuits, and Applications.* Amsterdam: Elsevier.

[12] Serra, J.M., Gamboa, R., Vallera, A.M. 1996. Optical absorption coefficient of polycrystalline silicon with very high oxygen content. *Materials Science and Engineering B* 36:73–76.

[13] Hopkins, P.E., Reinke, C.M., Su, M.F., Olsson III, R.H., Shaner, E.A., Leseman, Z.C., Serrano, J.R., Phinney, L.M., El-Kady, I. 2011. Reduction in the thermal conductivity of single crystalline silicon by phononic crystal patterning. *Nano Letters* 11:107–112.

[14] Uma, S., McConnell, A.D., Asheghi, M., Kurabayashi, K., Goodson, K.E. 2001. Temperature-dependent thermal conductivity of undoped polycrystalline silicon layers. *International Journal Thermophysics* 22:605–616.

[15] Lee, W.Y., Park, N.W., Hong, J.E., Yoon, S.G., Koh, J.H., Lee, S.K. 2015. Effect of electronic contribution on temperature-dependent thermal transport of antimony telluride thin film. *Journal Alloys Compounds* 620:120–124.

[16] Lošt'ák, P., Drašar, C., Horák, J., Zhou, Z., Dyck, J.S., Uher, C. 2006. Transport coefficients and defect structure of $Sb_{2-x}Ag_xTe_3$ single crystals. *Journal of Physics and Chemistry of Solids* 67:1457–1463.

[17] Cheng, C.H., Huang, S.Y., Cheng, T.C. 2010. A three-dimensional theoretical model for predicting transient thermal behavior of thermoelectric coolers. *International Journal of Heat and Mass Transfer* 53:2001–2011.

[18] Saci, A., Battaglia, J.L., Kusiak, A., Fallica, R., Longo, M. 2014. Thermal conductivity measurement of a Sb_2Te_3 phase charge nanowire. *Applied Physics Letters* 104:263103.

[19] Venkatasubramanian, R. 2000. Lattice thermal conductivity reduction and phonon localization-like behaviour in superlattice structures. *Physical Review B* 61:3091–3097.

[20] Jou, D., Sellitto, A., Cimmelli, V.A. 2014. Multi-temperature mixture of phonons and electrons and nonlocal thermoelectric transport in thin layers. *International Journal of Heat and Mass Transfer* 71:459–468.

[21] Herdt, A. 2012. *Exploring the Electronic Properties of Novel Spintronic Materials by Photoelectron Spectroscopy.* Duisburg: Forschungszentrum Jülich GmbH, p. 77.

[22] Yang, J.Y., Aizawa, T., Yamamoto, A., Ohta, T. 2000. Thermoelectric properties of n-type $(Bi_2Se_3)_x(Bi_2Te_3)_{1-x}$ prepared by bulk mechanical alloying and hot pressing. *Journal Alloys Compounds* 312:326–330.

[23] Bessas, D., Töllner, W., Aabdin, Z., Peranio, N., Sergueev, I., Wille, H.C., Eibl, O., Nielsch, K., Hermann, R.P. 2013. Phonon spectroscopy in a Bi_2Te_3 nanowire array. *Nanoscale* 5:10629–10635.

[24] Machrafi, H. 2016. An extended thermodynamic model for size-dependent thermoelectric properties at nanometric scales: Application to nanofilms, nanocomposites and thin nanocomposite films. *Applied Mathematical Modelling* 40:2143–2160.

[25] Liu, W., Lucas, K., McEnaney, K., Lee, S., Zhang, Q., Opeil, C., Chen, G., Ren, Z. 2013. Studies on the Bi_2Te_3–Bi_2Se_3–Bi_2S_3 system for mid-temperature thermoelectric energy conversion. *Energy Environmental Science* 6:552–560.

[26] Sarma, S.D., Li, Q. 2013. Many-body effects and possible superconductivity in the two-dimensional metallic surface states of three-dimensional topological insulators. *Physical Review B* 88:081404R.

[27] De Wette, F.W., Kulkarni, A.D. 1992. Phonon dispersion, phonon specific heat, and Debye temperature of high-temperature superconductors. *Physical Review B* 46:14922–14925.

[28] Lai, Y.P., Chen, H.J., Wu, K.H., Liu, J.M. 2014. Temperature-dependent carrier-phonon coupling in topological insulator Bi_2Se_3. *Applied Physics Letters* 105:232110.

[29] Cho, S., Butch, N.P., Paglione, J., Fuhrer, M.S. 2011. Insulating behaviour in ultrathin bismuth selenide field effect transistors. *Nanoletters* 11:1925–1927.

[30] Baek, D., Rouvimov, S., Kim, B., Jo, T.C., Schroder, D.K. 2005. Surface recombination velocity of silicon wafers by photoluminescence. *Applied Physics Letters* 86:112110.

[31] Mishima, T., Taguchi, M., Sakata, H., Maruyama, E. 2011. Development status of high-efficiency HIT solar cells. *Solar Energy Material Solar Cells* 95:18–21.

[32] Yamamoto, K., Yoshimi, M., Suzuki, T., Okamoto, Y., Tawada, Y., Nakajima, A. 1997. Thin film poly-Si solar cell with "Star Structure" on glass substrate fabricated at low temperature. *Conference Record 26th. IEEE Photovoltaic Specialists Conference*, pp. 575–580, Anaheim, USA.

[33] Sah, R.L.Y., Yamakawa, K.A., Lutwack, R. 1982. Effect of thickness on silicon solar cell efficiency. *Electron Devices, IEEE Transactions* 29:903–908.

[34] Lebon, G., Machrafi, H. 2015. Thermal conductivity of tubular nanowire composites based on a thermodynamical model. *Physica E* 71:117–122.

[35] Hashim, H., Bomphrey, J.J., Min, G. 2016. Model for geometry optimization of thermoelectric devices in a hybrid PV/TE system. *Renewable Energy* 87:458–463.

[36] Abdin, Z., Alim, M.A., Saidur, R., Islam, M.R., Rashmi, W., Mekhilef, S., Wadi, A. 2013. Solar energy harvesting with the application of nanotechnology. *Renewable Sustainable Energy Review* 26:837–852.

[37] Machrafi, H. 2017. Enhancement of a photovoltaic cell performance by a coupled cooled nanocomposite thermoelectric hybrid system, using extended thermodynamics. *Current Applied Physics* 17:890–911.

10

Nanomedicine: Permeation of Drug Delivery Through Cell Membrane

10.1 Transporters of Drugs

10.1.1 Background

Combinatorial chemistry and high-throughput screening allowed the generation of medical drug candidates. However, it also gave a number of medications that are poorly soluble or absorbable in the human body. Therefore, there is a need to improve the permeation of medical drugs through cell membranes, i.e. to enhance membrane permeability [1,2]. It is important to note that the absorption of certain complex molecules, such as oligonucleotides vaccines and small peptides, is quite different from standard synthetic small organic molecules. Nonetheless, in all these cases, one can define their solubility or permeation in relation to their absorption profile. As such, four types of classification can be brought forward following the Biopharmaceutical Classification System (BCS), depending on the four combinations of high/low solubility and high/low permeability. Due to the inauspicious physicochemical and chemical properties of many drug molecules, they show a poor permeability, which often requires an excipient to be added externally in order to enhance permeation transiently [3]. In the past decades, important medical advances are made improving new dosage forms and techniques for drug delivery. Other than the oral route, which offers more advantages of therapeutic effectiveness and patient compliance, other routes exist for drug delivery, such as by injection, transdermal or pulmonary routes [4]. Drug delivery by oral route is preferred because it finds a great convenience with the patients. Indeed, tablets and capsules can be prepared at low price and at large quantities. The good functioning depends on intestinal permeability, solubility during gastrointestinal transit, liberation from dosage form and metabolism, of which, the importance is highlighted in the BCS, as a scientific basis assuring in vivo bioavailability and bioequivalence studies [5]. Although several experimental systems, backed up by animal models, improved the understanding of permeability enhancement, these models, being inherently complex, render difficult to define experiments for the determination of biochemical mechanisms. In this, it is incumbent to have a better understanding of permeability enhancement [6].

10.1.2 Challenges

In order to know how to enhance permeability, a non-exhaustive discussion is following on a few barriers. To be more concrete, the examples that follow address the intestinal permeation of drugs. The barriers for permeation can be the mucous layer, the apical and basal cell membranes and the capillaries.

10.1.2.1 Mucous Layer

The mucous layer consists of water glycoproteins, electrolytes, proteins and nucleic acids. It covers the epithelial cells of the entire intestine. It is bonded to the apical surface by a 500 nm thick glycoprotein structure that is covalently linked to the lipids and proteins of the border membrane. The mucous layer itself is around 50–100 μm controlling the pH around 6 by acting as a buffer.

10.1.2.2 Apical Cell Membrane

The apical cell membrane is a 1 μm thick brush border containing a 10 nm thick double layer of polar lipid molecules that have both a hydrophobic and a hydrophilic part. The transport of molecules through the lipid bilayer is often quantified by a lipid–water coefficient. The absorption of strongly hydrophilic substances is limited by the lipid bilayer, which is the case for antibiotics and peptides. Other mechanisms are necessary to make drug permeation successful, such as diffusion through pores or carrier-mediated transport.

10.1.2.3 Basal Cell Membrane

The basal cell membrane consists of a 9 nm phospholipid bilayer containing proteins. The fluidity of lipids in the basolateral membrane is higher than the apical membrane. Therefore, the barrier function of the basal membrane is probably less important than that of the apical membrane.

10.1.2.4 Capillary Wall

The capillary wall is 500 nm under the basal membrane. The endothelial cell membrane contains small perforations (equivalent to pores) of around 0.4–1 nm, and the blood capillary wall has capillaries of radii of 20–30 nm. It is interesting to note that the lymphatic capillaries have radii up to 300 nm. Due to the relatively large pores, the intestinal blood and lymph capillaries should not comprise too much of a barrier, if it were not for the strong interaction hydrophilic drugs can have with the capillary walls, thereby limiting permeation [7,8].

10.2 Permeability Enhancement

10.2.1 Nano-emulsions

Nano-emulsions can be efficient vehicles for oral drug delivery, since they can be produced from components that have solubilizing and permeation enhancement properties. Nano-emulsions of which the droplet size is less than 150 nm are often of the type oil/water. They have the capability to enhance gastrointestinal absorption, have large interfacial area and show efficient drug release properties. There are also nano-emulsions that present the characteristic to self-emulsify in aqueous media. They can be administered as dry pre-concentrates that form in situ nano-emulsions, once in contact with the gastrointestinal tract fluids. Formulations of such pre-concentrates have an influence on the intestinal permeation of the transcellularly transported drugs.

10.2.2 Spray Freeze-Drying

Another technique to enhance cell permeability is spray freeze-drying. In this procedure, one applies the freeze-drying to the drug together with a wetting agent and a permeation enhancer, making it kinetically stable, amorphous solid dispersions with higher dissolution performance.

10.2.3 Chitosan Derivatives

Chitosan is a non-toxic polymer that is biocompatible and has a number of applications in drug delivery. When the pH is kept at 6.5, chitosan has the capability to increase the permeability of peptide drugs across mucous epithelia. Trimethyl chitosan chloride (synthesized from chitosan at different degrees of quarterization) has shown to considerably increase the permeation and absorption of peptides across intestinal epithelia. The mechanism of enhancing permeation is by interacting reversibly with components that are present in tight junctions (junctions that seal adjacent epithelial cells in a narrow band just below the apical surface, consisting out of a network of proteins) in order to widen the paracellular routes [1].

10.2.4 Straight Chain Fatty Acids

Another way to loosen the tight junctions is by medium-chain fatty acids, such as lauric acid and capric acid, as well as by long-chain fatty acids, such as oleic acids. They have the ability, in this way, to allow hydrophilic drugs to pass in between the junctions, by dilating them. The advantage here is that they procure a certain comfort in that they can be incorporated in oral form without the need of complex formulation techniques [9].

10.2.5 Self-Micro-Emulsifying Drug Delivery Systems

It concerns systems that are isotropic mixtures of oils, surfactants and solvents. When administered, due to the gastric motility providing for local agitation, it results in a spontaneous formation of fine oil-in-water micro-emulsions of droplets less than 50 nm. As such, they can permeate more easily through the intestinal capillaries.

10.3 Modeling Considerations for Drug Delivery Permeation

10.3.1 Context

Having discussed the different ways of permeability enhancement, we can see that very often it concerns the passage dimensions and the wettability of the passage walls (albeit capillaries, membranes, and so on). This allows bringing these observations to a physical and thermodynamic level and relating them to the work presented in this book. We will in the following propose how such systems in the field of nano-medicine could be analysed theoretically and how the model can be useful for the improvement of drug delivery.

10.3.2 Solubility and Permeability

From typical clinical observations that lead to data in the literature (as such discussed before), it appears that one of the most important issues to be studied for improving drug delivery is through enhancing the permeability. In Chapter 7, we have seen that it depends on various parameters and also on the size of the capillaries or membrane pores. The basis for that model was based on mass diffusion given by Eq. (7.15). It is important to note, however, that for the present case, there is a difference from that given in Eq. (7.15). In the latter, the concentration gradient was negligible with respect to the flux, which is often the case for nanoporous flow that are pressure-driven by some external pressure. There are some cases in the human body that are comparable from a conceptual point of view, such as oxygen and carbon dioxide transfer from the lungs to the veins within the alveoles. Nonetheless, porous flow within the cell membrane pores often presents the difference that, due to the absence of externally imposed pressure, the concentration gradient can be quite important with respect to the flux. Moreover, the concentration gradient is not simply a gradient of the concentrations from both sides of the membrane, since concentration jumps can exist at the entry and exit of those membranes due to lipid bilayer interaction or wettability effects. The gradient that counts is the one within the membrane. Therefore, it is important to consider the fact that the flux by diffusion across a membrane (even if it were uniform) depends on the solubility. As there are different solubilities at both sides of the membrane and within the membrane as well, we can define partition coefficients ϑ_{in} and ϑ_{out} for the entry and exit of the membrane, respectively:

$$\vartheta_{in} = \frac{N_{mb,in}}{N_{in}} = \frac{c_{mb,in}(1 - c_{in})}{c_{in}(1 - c_{mb,in})} \tag{10.1}$$

$$\vartheta_{out} = \frac{N_{mb,out}}{N_{out}} = \frac{c_{mb,out}(1 - c_{out})}{c_{out}(1 - c_{mb,out})} \tag{10.2}$$

where $N_{mb,in}$ and $N_{mb,out}$ are the concentrations within the membrane from the entry and exit sides, respectively. Also, N_{in} and N_{out} stand for the same concentrations, but at the medium sides (outside the membrane), respectively. The equivalent expressions in the mass fraction c are also indicated in Eqs. (10.1) and (10.2). For the evoluation of the mass flux, Equation (7.15) can be used:

$$\tau \partial_t \boldsymbol{J} + \boldsymbol{J} = -\rho D \nabla c + \ell^2 \nabla^2 \boldsymbol{J}, \tag{10.3}$$

We can follow the same procedure as in Chapter 7 for finding the permeability, with the difference expressed by Eq. (10.3). Also, there is an absence of an external pressure gradient, but there is an osmotic pressure caused by the concentration gradient within the membrane.

10.4 Example: Cell Permeation

10.4.1 Permeation as a Set of Barrier Resistances

When speaking about cell permeation and the body uptake efficiency of drugs into the body that it entails, many works take the apparent cell permeability, P_{app}, through monolayers of human intestinal epithelial cells as a widely accepted standard [10–12]. The values of P_{app} (often presented in medicine as a volumetric flux per unity of area) of such cells are also often used for the estimation of the blood–brain barrier [13,14]. It is useful to understand the apparent cell permeation as a series of parallel and serial resistances (from layer to another). This vision is based on the concept of solubility diffusion [15]. The permeability is seen here, therefore, not as an imposed force, but is rather passive in a way that only spontaneous diffusion governs the permeation process. This diffusion process is determined by a mass-fraction gradient. The mass-fraction gradient is described by Fick's law or, in case of pore sizes (within the membranes separating the monolayers) at nanoscale, by the evolution equation (7.15), describing the extended diffusion law. Active transport, as could be caused by transporter proteins, is not considered here. Although, in reality, chemicals/drugs could be transported by proteins (and thus have an effect on P_{app}), it is not taken into account here. The model is limited by passive permeability, since it is still an important aspect of cell permeation and needs more investigation, being an inherent process. Such an investigation will be done by considering growing cells on a filter, proposing a mechanical model of resistances (defined by the reciprocal of a permeability). The pathway that is proposed here consists of a chemical going through a series of barriers before attaining their end goal. The chemicals go first through an unstirred apical water monolayer (with permeability P_{wla}), then the epithelial cell membrane (with permeability P_{cm}), then the cell itself (with permeability P_c), then again the cell membrane and finally the unstirred basal water monolayer (with permeability P_{wlb}).

It is well admitted that generally three pathways are possible through a cellular monolayer, i.e. in between the cells (the paracellular way), through the cell (considered here) and along the membrane solely (the lateral way) [16,17]. In this case, parallel resistances of the three pathways would apply. Here, we only consider the pathway through the cell. As we define $P_{app} = \frac{1}{R_{app}}$, where R_{app} is the apparent diffusion resistance, we can write

$$R_{app} = R_{wla} + R_{cm} + R_c + R_{cm} + R_{wlb}$$

Thus, the overall permeability can be seen as a series of resistances for the different barriers. All the resistances are estimated at a pH of 7.4. Following the pH partition hypothesis [18], it can be assumed that only the neutral form of the drug chemicals can pass the membrane resistances. Therefore, the chemicals that are ionized should undergo an acid–base reaction in between the water layer and the membrane before they are able to pass the membrane as a neutral species. According to the aforementioned pH partition hypothesis, such a reaction can be considered to occur instantaneously, meaning that no kinetic hinderness (and therefore no additional resistance) needs to be considered.

10.4.2 Solubility Diffusion Theory

All the resistances, which will be discussed afterwards, are based on the solubility diffusion model. Therefore, a short discussion on it is presented here. Let us imagine the existence of two compartments of well-mixed water that are separated by barrier. This barrier could well be a real barrier, such as a membrane, but it could also be a stagnant water layer of a certain thickness. Another option would be a porous medium of a length that allows limited diffusion. Let us then introduce a solute in the donor compartment (the analogy of the apical water layer). At quasi-stationary state, also assuming homogenous mass-fractions in all the compartments, the resulting mass-fractiongradients will be constant and, in approximation, linear. A linear Fick's law for the barrier can then be applied here:

$$\boldsymbol{J} = -\rho D \frac{c_a - c_d}{\delta_b} \tag{10.4}$$

where c_a is the mass fraction of the acceptor, c_d the mass fraction of the donor and δ_b the thickness of the barrier. Imagine, if the barrier contains pores (which could be the case for a membrane) of sizes that are of the order of magnitude of the mean free path of the chemicals, then the non-local effects become important and the flux (at quasi-stationarity) is then given by

$$\boldsymbol{J} = -\rho D \frac{c_a - c_d}{\delta_b} + \ell^2 \frac{\partial^2 \boldsymbol{J}}{\partial r^2}, \tag{10.5}$$

where we neglect spatial variations of the mass flux in the flow direction and assume a constant mass-fraction gradient. Thus, the non-local effects are only considered in the direction perpendicular to the flow direction (here represented by a derivative with respect to the radial coordinate, r, assuming the pores are cylindrical), which is often observed in porous systems [19]. For more information, one is referred to Chapter 7. Assuming maximum flux in the middle of the pore ($r = 0$), i.e. $\frac{\partial \boldsymbol{J}}{\partial r}\big|_{r=0} = 0$, and first-order slip (with a slip length ℓ_s) at the wall ($r = R$, with R the pore radius), i.e. $\boldsymbol{J} = -\, C_1 \ell_s \frac{\partial \boldsymbol{J}}{\partial r}\big|_{r=R}$, the flux is then given by

$$\boldsymbol{J} = -\rho D \frac{c_a - c_d}{\delta_b} \left(1 - \frac{2Kn^2 \ \mathcal{B}\left[1, \frac{1}{Kn}\right]}{Kn \ \mathcal{B}\left[0, \frac{1}{Kn}\right] + B_s \ \mathcal{B}\left[1, \frac{1}{Kn}\right]} \right) \tag{10.6}$$

with $Kn = \frac{\ell}{R}$, $B_s = \frac{\ell_s}{R}$ and $\mathcal{B}$ the Bessel-I function. It should be noted that Eq. (10.6) depends on how slip is observed and how the mass flux is defined. Other effects can influence the mass flux, but here it is the purpose to show how physical models can be used in the field of nanomedicine.

In an effort to give tools (which is the purpose of this chapter) for the use of such models, it is interesting to relate the mass flux with the permeability in a porous system as follows

$$P = \frac{\boldsymbol{J}}{\rho \varepsilon} \tag{10.7}$$

where ε is the porosity. The permeability can then be found using Eq. (10.4) or (10.6). This is the basis for all the permeability, although the final form might be different. In liquid or watery layers, the permeability, or rather solubility, would be given by simple diffusion. Finally, Eq. (10.7) gives the permeability through a barrier using Eq. (10.6).

References

[1] Fuyuki, M., Koji, T., Masateru, M., Ken-ichi, O., Masaaki, O., Kazutaka, H., Toshikiro, K. 2009. Novel oral absorption system containing polyamines and bile salts enhances drug transport via both transcellular and paracellular pathways across Caco-2 cell monolayers. *International Journal of Pharmaceutics* 367:103–108.

[2] Shaikh, M.S.I., Derle, N.D., Bhamber, R. 2012. Permeability enhancement techniques for poorly permeable drugs: A review. *Journal of Applied Pharmaceutical Science* 2:34–39.

[3] Pradeep, S., Manthena, V.S.V., Harmander, P.S.C., Ramesh, P. 2005. Absorption enhancement, mechanistic and toxicity studies of medium chain fatty acids, cyclodextrins and bile saltsas peroral absorption enhancers. *Farmaco* 60:884–893.

[4] Seulki, L., Sang, K.K., Dong, Y.L., Kyeongsoon, P., Tadiparthi, S., Su, Y.C., Youngro, B. 2005. Cationic analog of deoxycholate as an oral delivery carrier for Ceftriaxone. *Journal of Pharmaceutical Sciences* 94:2541–2548.

[5] Carsten, B., Adrian, F., Urte, K., Torsten, W. 2006. Comparison of permeation enhancing strategies for an oral factor Xa inhibitor using the Caco-2 cell monolayer model. *European Journal of Pharmaceutics and Biopharmaceutics* 64:229–237.

[6] Edward, L.L., Steven, C.S. 1997. In vitro models for selection of development candidates. Permeability studies to define mechanisms of absorption enhancement. *Advanced Drug Delivery Reviews* 23:163–183.

[7] Ewoud, J.V.H., Albertus, G.D.B, Douwe, D.B. 1989. Intestinal drug absorption enhancement: An overview. *Pharmacology & Therapeutics* 44:407–443.

[8] Machrafi, H., Lebon, G. 2018. Fluid flow through porous and nanoporous media within the prisme of extended thermodynamics: Emphasis on the notion of permeability. *Microfluidics and Nanofluidics* 22:65.

[9] Dimple, P., Fatemeh, A., Hossein, Z. 2010. Intestinal permeability enhancement of levothyroxine sodium by straight chain fatty acids studied in MDCK epithelial cell line. *European Journal of Pharmaceutical Sciences* 40:466–472.

[10] Alsenz, J., Haenel, E. 2003. Development of a 7-day, 96-well Caco-2 permeability assay with high-throughput direct UV compound analysis. *Pharmaceutical Research* 20:1961–1969.

[11] Lee, K.J., Johnson, N., Castelo, J., Sinko, P.J., Grass, G., Holme, K., Lee, Y.H. 2005. Effect of experimental pH on the in vitro permeability in intact rabbit intestines and Caco-2 monolayer. *European Journal of Pharmaceutical Sciences* 25:193–200.

[12] Thiel-Demby, V.E., Humphreys, J.E., Williams, S.L.A., Ellens, H.M., Shah, N., Ayrton, A.D., Polli, J.W. 2009. Biopharmaceutics classification system: Validation and learnings of an in vitro permeability assay. *Molecular Pharmaceutics* 6:11–18.

[13] Doan, K.M.M. 2002. Passive permeability and P-glycoprotein-mediated efflux differentiate central nervous system (CNS) and non-CNS marketed drugs. *Journal of Pharmacology and Experimental Therapeutics* 303:1029–1037.

[14] Garberg, P., Ball, M., Borg, N., Cecchelli, R., Fenart, L., Hurst, R.D., Lindmark, T., Mabondzo, A., Nilsson, J.E., Raub, T.J., Stanimirovic, D., Terasaki, T., Oberg, J.O., Osterberg, T. 2005. In vitro models for the blood-brain barrier. *Toxicology in Vitro* 19:299–334.

[15] Diamond, J.M., Katz, Y. 1974. Interpretation of nonelectrolyte partition coefficients between dimyristoyl lecithin and water. *Journal of Membrane Biology* 17:121–154.

[16] Bittermann, K., Goss, K.U. 2017. Predicting apparent passive permeability of Caco-2 and MDCK cell-monolayers: A mechanistic model. *PLoS One* 12:e0190319.

[17] Heikkinen, A.T., Mönkkönen, J., Korjamo, T. 2010. Determination of permeation resistance distribution in in vitro cell monolayer permeation experiments. *European Journal of Pharmaceutical Sciences* 40:132–142.

[18] Avdeef, A. 2012. *Absorption and Drug Development: Solubility, Permeability, and Charge State*. Hoboken, NJ: John Wiley & Sons.

[19] Machrafi, H., Lebon, G. 2018. Self-assembly of carbon nanotube-based composites by means of evaporation-assisted depositions: Importance of drop-by-drop self-assembly on material properties. *Materials Chemistry and Physics* 218:1–9.

11

Self-Assembled Nanostructures as Building Blocks for Nanomedicine Carriers: Thermal and Electrical Conductance

11.1 Context

Self-assembly as a procedure is applied in many fields. Some examples can be found in the medical sector [1], in the energy storage domain [2], in the thin film industry for the development of transistors [3] or even in the technology of membrane fabrication [4]. For a proper self-assembly procedure, it is important to understand its mechanism and how the type of material and the way of preparation can influence the characteristics of the deposited material. This can be done by building nanoporous structures or by surface functionalization [5]. The goal of the procedure can be quite different: it can serve for depositing a suspension containing nanoparticles on a surface for the preparation of different kinds of coatings [6] or even for improving the initial adhesion of osteoblast-like cells [7]. Other examples include cross reactive molecular markers recognition [8], liquid crystalline pattern formation of DNA [9], humidity sensors [10] and gas sensors [11], to mention a few. Many of them rely on making depositions on substrates. Different types of deposition methods exist, such as dip coating, sedimentation, spray coating and electrostatic assembly, while convective deposition, and more particularly drop evaporation, is a convenient way to deposit microparticles and nanoparticles [12]. The amount of fluid used is minimized, possibly inducing economic advantages, and the outcome can easily be controlled, by choosing initial parameters.

The interest lies in creating patterned structures out of evaporating drops, which can be of use for energetic and medical applications. The deposited patterns that are left by the evaporated colloidal drops can present a multiplicity of structures, such as the ring structure, a central bump, a uniform deposit, or more complex structures such as multiple rings and hexagonal arrays [13–16]. This variety of patterns is a reflection of the multi-scale attractive forces and transport phenomena taking place during the droplet evaporation and the effect they can have on the structure deposition of the substrate. This can have a large effect on the wettability of the substrate, as well as on the thermal and electrical properties. The effect of self-assembly on the wettability has been studied in a previous work [17]. Here, we focus on characterizing the obtained structures for their morphology, electrical and thermal properties.

The morphological, thermal and electrical properties have been the subject of many studies preparing self-assembled nanocomposites, each method having its advantage and issues to be solved. For instance, solution combustion synthesis is a technique that is adopted worldwide for the fabrication of nanomaterials [18,19] due to its simplicity and time-effectiveness. However, this method still copes with the difficulty of controlling the final morphology and phase of the product [18]. Coordinative layer-by-layer-assembled films have been studied [20], which are prepared by combining electrostatic and coordinative interactions between organic and inorganic building blocks. In [21], it was shown that alternating the aforementioned interactions resulted into a controlled formation of multilayered films with a well-controlled nanometric thickness range of the layers. However, such a technique is not systematically applicable. In general, a variety of deposition techniques exist for the layer-by-layer assembling method, the dip coating being the most widely used [22,23]. As an advantage, it is very simple to use, but it can be time-consuming. Other techniques rely on the inherent properties of the components that are to be assembled, such as hydrogen-bonded self-assembled composites [24]. In general, these techniques offer great possibilities, but the versatility and controllability of the methods remain to be improved. A rather new technique is to deposit subsequent evaporating drops containing the building blocks of the nanomaterial [17]. The advantage with respect to dip coating is that an increase of temperature can considerably increase the speed of the procedure due to the controllable size of the droplet and a short injection time. Another advantage is the possibility to combine this method with other existing methods, such as layer-by-layer deposition or functionalized/charged components. Finally, it is also possible to coat substrates with more complex geometries, without additional complications. Self-assembly in evaporating mono-component nanofluid droplets has already been investigated and reasonably understood [25–29]. However, self-assembly by evaporating multiple nanofluid droplets at the same place, one after the other, is still an open field, especially when it concerns composites. Since such an investigation has not yet been studied much, it is the purpose of this chapter to focus on the morphological, thermal and electrical properties as a function of the number of deposited droplets. The effect of the initial nanoparticle concentration on electrical properties as well as on the presence of the coffee ring has already been studied [17,30]. It appeared that too high concentrations resulted into the formation of coffee rings. In order to avoid the formation of a coffee ring, it would be better to make multiple depositions at lower initial nanoparticle concentration. Therefore, in order to understand the behaviour of these properties, influenced by the content of the droplet, the study in this chapter is limited to the effect of the number of deposited droplets and the droplets' content. This understanding can then be used later for tailoring (in combination with other techniques or not) controlled thermal and electrical properties of nanocomposites. Besides, while often the thermal and electrical properties show the same behaviour (which is linked in many cases), the method in this chapter provides an easy way to propose nanocomposites of which such behaviour can be surprising.

By using drop-by-drop evaporation, containing carbon nanotubes (CNTs), we can deposit multiple self-assembling layers of CNTs on a substrate. The way of deposition depends on the type of substrate, type of the nanoparticles and the evaporation process. This can lead to various forms and structures, depending on various physical phenomena, such as buoyancy, temperature effects, surface tension changes, colloidal forces and substrate–particle interactions.

11.2 Theory

Based on Eq. (3.38), we can define the thermal conductivity of a CNT porous nanostructure, $\lambda_{p,CNT}$. Since pure CNTs have a very high thermal and electrical conductivity, that of air can be largely neglected. This results, not considering nanostructures with porosities close to one (which would not make any sense in this context), into

$$\lambda_{p,CNT} = \lambda_{CNT}\frac{2-2\varepsilon}{2+\varepsilon}. \tag{11.1}$$

In this equation, ε is the porosity and λ_{CNT} the thermal conductivity of a CNT. If the porous material in question has a relatively low thermal conductivity, then Eq. (3.38) and the equations that are in relation with Eq. (3.38) should be used. For the electrical conductivity of a CNT porous nanostructure, $\sigma_{p,CNT}$ (see Chapter 5 for explanation on the analogy), we can write the same type of equation (with the same conditions):

$$\sigma_{p,CNT} = \sigma_{CNT}\frac{2-2\varepsilon}{2+\varepsilon}, \tag{11.2}$$

where σ_{CNT} denotes the electrical conductivity of a CNT. Equations (11.1) and (11.2) are to be used for a porous matrix. If nanoparticles are homogeneously inserted in such a porous matrix (for instance, by depositing a droplet containing a mixture of CNTs and nanoparticles) then, for the thermal conductivity, Eq. (11.1) is to be used as the matrix thermal conductivity in Eq. (3.1). Again, the thermal conductivity of the nanoparticle is much smaller than that of the matrix, so that we can neglect the former. For the effective thermal conductivity, Equation (3.1) then becomes

$$\lambda_{CNT}^{eff} = \lambda_{p,CNT}\frac{2-2\varphi}{2+\varphi}. \tag{11.3}$$

If the nanocomposite material in question has a relatively low thermal conductivity, Eq. (3.1) and the equations that are in relation with Eq. (3.1) should be used. In the same way, we obtain the effective electrical conductivity (with the same conditions)

$$\sigma_{CNT}^{eff} = \sigma_{p,CNT}\frac{2-2\varphi}{2+\varphi} \tag{11.4}$$

Since Eqs. (11.1) and (11.2) can be obtained through Eqs. (11.3) and (11.4), respectively, we continue with the latter two, which will be used to explain the behaviour of porous nanocomposites and their thermal and electrical properties. For practical purposes (applications as biosensors, batteries, rather than insulators, and semi-conductors), the interest in this chapter lies more into the conductivities in the longitudinal direction of the CNTs. For a preferred substrate direction, even though the CNTs are not perfectly aligned, the longitudinal axis of the majority of the CNTs will still be in that preferred direction. The experimental data used for this purpose are taken from [31]. Nevertheless, for the sake of completeness, important details are represented in the following sections.

11.3 Experimental

11.3.1 Self-Assembly Process

It is the purpose of our drop-deposition experiment to deposit droplets that contain nanomaterial and let them evaporate at room temperature at ambient humidity (60% humidity) for 7 h. This duration was needed not so much to evaporate the droplet as such, but, since the deposited nanomaterial creates a nanoporous network, to allow the water held back in the pores by the capillary forces to be evacuated. The nanomaterial considered is composed of a mixture of CNTs and SiO_2 nanoparticles. For interpretation purposes, pure CNT nanoporous networks are also obtained. The 3 g L^{-1} aqueous 5 nm multi-walled CNT dispersion has been supplied by Nanocyl, and the 0.3 g L^{-1} aqueous 175 nm SiO_2 has been supplied by Bangs Laboratories. The CNT solutions are kept in homogeneous dispersion by the presence of anionic surfactants, which guarantee a long-lasting stable homogeneous aqueous dispersion of the CNTs. As for the SiO_2 solutions, they are found to remain in a stable homogeneous aqueous dispersion due to Si-OH surface groups. As recommended by the fabricants, the aqueous solutions are sonicated before depositing the droplets. The deposited droplet that commence to evaporate triggers convection. The nanomaterial will move along the flow patterns and, after evaporation, settle on the substrate, which creates a covered substrate. Drop-by-drop deposition has been studied previously [17] but using only CNTs. The depositions were performed in a motorized way. The position of the droplet is controlled by a motor with a precision of 0.01 mm. Each droplet is deposited by a syringe on a spot that is delimited by a groove, which creates pinning conditions. This results into 40 μL droplets with a diameter of 12 mm. The pinning behaviour created by the grooves allow avoiding uncontrolled spreading of the nanofluid droplets, guaranteeing approximately the same spherical form (so that we can be sure that the obtained results are caused by the difference in the number of deposited drops) and assuring the ability to concentrate the nanomaterial on a small controlled surface to facilitate self-assembly. A camera, equipped with a detection software, is used to check the constant droplet size for all the measurements (with a maximum deviation of 2 μL). The drop-deposition setup has been developed in the laboratory and is made of a bi-dimensional translation stage (Moons STM17S-1AE) and a home-made double syringe pump using the same motorized stages. The software automatically drives the setup and acquires images of the drop after each deposition (camera JAI BM-500GE) to control the volume of the drop. The drying time was set to 7 h.

Schematically, the procedure is as follows:

a. We deposit a certain number of droplets next to each other.

b. After evaporation, the nanomaterial sticks to the polycarbonate substrate; we add another droplet, except for the first spot.

c. After evaporation, a thicker deposition is obtained on the second and subsequent spots; we again add another droplet, except for the first two spots.

d. After evaporation, the third spot from the left shows an even thicker deposition.

e. This can then be repeated as much as wanted, which results into a series of incremental number of deposited drops from the first to the last spot.

11.3.2 Measurements

We measure the bulk electrical and thermal conductivities of the deposited nanomaterials, where the preferred direction is the one parallel to the substrate. The interpretation of these results needs the measurement of the thickness of the deposited materials.

11.3.2.1 Electrical Conductivity

The electrical conductivity is measured via the sheet resistance. By passing a current, I, through the two probes, the voltage, V, is measured across those probes that are in contact with the sample to be tested. The electrical resistance that is measured is called the sheet resistance,

$$R_s = \left.\frac{V}{I}\right|_{\parallel}, \tag{11.5}$$

where the $\parallel$ sign indicates that the measurement is performed in the parallel direction with respect to the substrate. From the sheet resistance, we can calculate the sheet resistivity by multiplying by the surface-to-length ratio along the electrical current path [32,33]. It should be noted that the surface changes along the current path (as it concerns a circular spot). Therefore, an average surface (along the radial angle of the circle) is calculated. The electrical conductivity is simply the reciprocate of the resistivity and, if the thickness of the nanomaterial test sample is known, the electrical conductivity is then given by [33]

$$\sigma_{\parallel} = \frac{\pi}{2R_s\delta_m} \tag{11.6}$$

where δ_m is the thickness of the nanomaterial sample. Here, we should note that Eq. (11.6) is defined for ideal situations where the current would pass directly from one electrode to another, without being hindered by any obstacles. In reality, this is not the case and the real values of the electrical conductivities would be somewhat different. Nonetheless, Eq. (11.6) still gives the correct order of magnitude of the electrical conductivity.

11.3.2.2 Thermal Conductivity

The thermal conductivity is measured along the parallel direction. Using Fourier's law, the thermal conductivity can be calculated as

$$\lambda = -\frac{\boldsymbol{q}}{\partial T/\partial x} \tag{11.7}$$

where $\boldsymbol{q}$ is an imposed, known, heat flux (per unit surface) and $\partial T/\partial x$ is the temperature gradient in the direction of the imposed heat flux. The heat flux is approximated by $\Delta T/X$. Here, $X = 2r_m$ is the diameter of the deposited nanomaterial for the thermal conductivity. The temperature difference is measured across the corresponding direction, X, by means of thermocouples that are put on the same places as the electrodes in section 11.3.2.1. The thermocouples are connected to an Agilent Data Logger. The same comment can be made as for the electrical conductivity. In reality, the gradient in Eq. (11.7) is not a straightforward overall vector but rather a local vector that can be different as one goes from one thermocouple to the other. Nonetheless, approximating this local gradient as increments of local temperature differences, one may extend this approximation and express it as an overall temperature difference across the distance between thermocouples. This would give a correct order of magnitude for the thermal conductivity.

11.3.2.3 Thickness of Deposited Layer

As has been mentioned earlier, the thickness of the deposited nanomaterial is crucial in order to calculate the electrical and thermal conductivity. For this purpose, we used a one-dimensional confocal probe [30], with which we measured the thickness on several places across the deposition. The optical confocal system for the measurement of the film thickness has been developed in [34]. The thickness is measured locally on three spots where the electrical and thermal conductivities are measured.

11.4 Results and Discussion

11.4.1 Electrical Conductivity

The electrical conductivity is measured by placing the outer electrodes at equidistance from the border of the deposition spots, and the measurements are repeated twice. An average

TABLE 11.1
Measured Experimental Electrical Conductivity.

N_d	$\sigma_{\parallel}$ [kS m^{-1}]
1	1.98
2	2.59
3	2.80
4	3.76
5	4.00
6	3.10
7	3.42

value is calculated with an error around 15%. Table 11.1 shows the measured electrical conductivity.

Table 11.1 shows that as the number of deposited drops increases, the electrical conductivity increases. As the number of deposited drops increase, the alignment of the CNTs around the SiO_2 nanoparticles cause the CNTs to penetrate into the underlying layers as well as parallel to the layer [31]. Therefore, from the results we can argue that the structure becomes denser, caused by a locally higher degree of alignment. A lower porosity can also play a role, which will be discussed in Section 11.4.3. This kind of behaviour was also observed for the alignment of graphene in bulk copper [35]. Moreover, a recent study showed that CNTs could be aligned by adding ZnO nanoparticles, which formed a chemical bond with the CNTs, enforcing the mechanical properties as well [36]. Although the SiO_2 nanoparticles have the tendency to reduce the electrical conductivity, this seems to be sufficiently compensated by the increasing CNT network density and efficiency due to alignment. Such behaviour was also observed in [36]. The difference with the results in [31] is that in [36], a chemical bond was created by heating up to 480°C, while in [31], a rather strong hydrogen bonding is the cause for the alignment. The result is that the electrical conductivity increases for the nanocomposite. We can imagine that after depositing more drops, the density attains a maximum value and the electrical conductivity stops increasing.

It is known that the axial electrical conductivity of CNT depositions (corresponding to the one considered in this chapter) is $\sim 2{*}10^3$–10^6 S m^{-1} [37]. The values obtained for the composite (with a maximum value around 4 kS m^{-1}) are around the lower limits of what is found in the literature, which is explained by the non-uniform overall alignment and probably by a relatively high porosity (which will be the subject of section 11.4.3). Nevertheless, the order of magnitude stays reasonably within the recorded range.

11.4.2 Thermal Conductivity

Table 11.2 shows the values of the thermal conductivity.

Although values up to 6,600 W K^{-1}m^{-1} have been reported [38], the typical values reported extensively in the literature are lower, e.g. 1,300 W K^{-1}m^{-1} [39] or anywhere in between 300 and 3,000 W K^{-1}m^{-1} [40]. We can say that the values in Table 11.2, with a maximum around 1,000 W K^{-1}m^{-1}, are nicely within the range of the reported values. We discussed earlier that the nanocomposite has shown to render the whole layer denser by increasing the number of deposited drops. This assures a better contact between the CNT and the SiO_2 nanoparticles, which results into a higher thermal conductivity. A lower porosity can also play a role, which is the subject of section 11.4.3.

TABLE 11.2
Measured Experimental Thermal Conductivity.

N_d	$\lambda_{\parallel}$ [kW km^{-1}]
1	0.88
2	0.89
3	0.93
4	0.89
5	0.96
6	1.00
7	0.98

11.4.3 Porosity

The electrical and thermal conductivities of the samples investigated in this work will mainly depend on two observations, i.e. the density of the structure and the alignment of the CNTs. The density of the structure reflects on the contact of the CNTs between themselves and/or their interaction with SiO_2, which influences the "easiness" of conduction. One way of assessing the density of the structure is to measure the thickness of the depositions as a function of the number of deposited drops. Since each deposited layer is of equal mass, and the diameter of the depositions is fixed, the thickness will tell something about the density or porosity for that matter (knowing the density of CNTs). Note that this method is more reliable if no or nearly no coffee ring would be formed and if the dispersion is homogeneous. It is observed that the coffee ring had little importance, which justifies even more using a multiple deposition method at lower concentrations instead of a single deposition at a higher concentration. Table 11.3 shows the measured thickness for the porous CNT-SiO_2 nanocomposites as a function of the number of deposited droplets, N_d. The deviation from the average value is around 15%. It is now the idea to calculate the porosity from the thickness data. We define the porosity of the CNT matrix as

$$\varepsilon = \frac{V_a}{V_t - V_{Silica}} = 1 - \frac{V_{CNT}}{V_t - V_{Silica}} = 1 - \frac{(N_d + 1)\frac{C_{CNT} V_d}{\rho_{CNT}}}{\pi r_m^2 \delta_m - N_d \frac{C_{Silica} V_d}{\rho_{Silica}}}. \tag{11.8}$$

where V_a, V_{CNT} and V_{Silica} are the pore (air) volume, the volume occupied by the CNTs and the volume occupied by silica, respectively (note that the total volume is $V_t = V_a + V_{CNT} + V_{Silica}$). Moreover, C_{CNT} is the CNT concentration in the deposited droplet ($C_{CNT} = 3$ g L^{-1}), C_{Silica} is the silica concentration in the deposited droplet ($C_{Silica} = 0.3$ g L^{-1}), ρ_{CNT} the CNT "tap" density ($\rho_{CNT} = 1.6$ kg L^{-1}), ρ_{Silica} the silica

TABLE 11.3
Measured Experimental Properties and the Calculated Porosity.

N_d	$\lambda_{\parallel}$ [kW km^{-1}]	$\sigma_{\parallel}$ [kS m^{-1}]	δ_m [μm]	ε
1	0.88	1.98	8.0	0.83
2	0.89	2.59	11.9	0.83
3	0.93	2.80	18.1	0.85
4	0.89	3.76	20.9	0.84
5	0.96	4.00	22.3	0.82
6	1.00	3.10	24.0	0.80
7	0.98	3.42	28.0	0.81

density ($\rho_{Silica} = 2.65$ kg L^{-1}) and V_d the volume of each deposited droplet ($V_d = 40$ µL). The term $(N_d + 1)$ in Eq. (11.8) stands for the fact that before depositing the CNT-SiO_2 mixtures, one layer of pure CNT is deposited [31]. The calculated porosity presents an error of less than 1%.

Table 11.3 shows that, overall, the thermal and electrical conductivities increase, with an overall decreasing porosity, which suggests indeed that a denser structure is obtained, which can be explained by a better alignment of CNTs. In general, a composite structure that becomes denser than the CNT structure would lead to a relative increase of the electrical and thermal conductivities with respect to its initial value. As for the alignment of the CNTs, this is confirmed by scanning electron microscope images in [31]. In general, as the degree of alignment of CNTs is higher, the electrical and thermal conductivities should increase in the axial direction of the nanotubes.

11.5 Linking Theory to Experiment

With the calculated porosity from Table 11.3, we are now able to calculate the effective thermal and electrical conductivities from combining Eqs. (11.1) and (11.2) with Eqs. (11.3) and (11.4), respectively, which gives

$$\lambda_{CNT}^{eff} = \lambda_{CNT} \frac{2 - 2\varepsilon}{2 + \varepsilon} \frac{2 - 2\varphi}{2 + \varphi}. \tag{11.9}$$

$$\sigma_{CNT}^{eff} = \sigma_{CNT} \frac{2 - 2\varepsilon}{2 + \varepsilon} \frac{2 - 2\varphi}{2 + \varphi} \tag{11.10}$$

Since it is quite difficult to find the exact values for λ_{CNT} and σ_{CNT}, they will be taken from experimental findings from [31] of pure CNT depositions. In [31], it is found that for a pure monolayer nanoporous CNT, $\lambda_{p,CNT} = 0.9$ kW K^{-1}m^{-1} and $\sigma_{p,CNT} = 3$ kS m^{-1}. With a porosity (calculated from data from [31] and, not surprisingly, corresponding approximately to the case $N_d = 1$ in Tale 11.3) of $\varepsilon = 0.83$, we find easily as approximation that $\lambda_{CNT} =$ 7.3 kW K^{-1}m^{-1} (which is, taking into account experimental uncertainty, not far from the 6,600 W K^{-1}m^{-1} found in the literature [38]) and $\sigma_{CNT} = 26$ kS m^{-1}. Before calculating the conductivities as a function of the porosity, we should find the volume fraction of the silica nanoparticles, being defined as

$$\varphi = \frac{V_{Silica}}{V_{CNT} + V_{Silica}} \tag{11.11}$$

Note that in Eq. (11.11), the volume fraction of silica nanoparticles is taken with respect to the total volume of the nanomaterial without the air pores, since the latter is taken into account via the porosity. This would finally lead to

$$\varphi = \frac{C_{Silica}}{\frac{\rho_{Silica}}{\rho_{CNT}} C_{CNT} + C_{Silica}} \tag{11.12}$$

Using the values mentioned earlier, we find $\varphi = 0.057$. From Eqs. (11.9), (11.10) and (11.12), the effective thermal and electrical conductivities are calculated and the experimental values from Table 11.3 are added for comparison. The results are presented in Figure 11.1.

Figure 11.1 shows a good agreement between the model and the experiments, showing that a lower porosity, corresponding to higher conductivities, indeed confirms a better alignment of the CNTs.

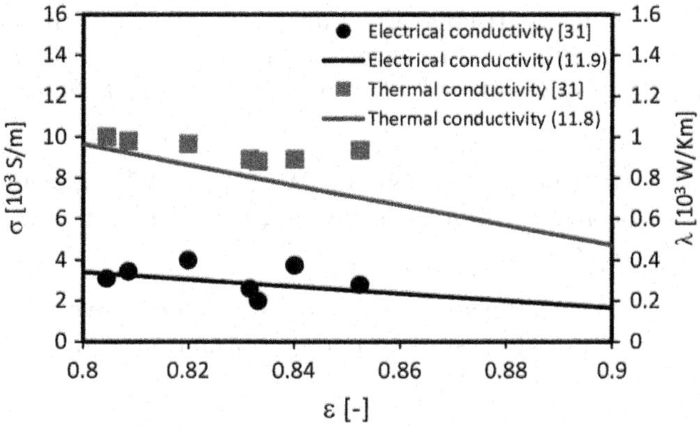

FIGURE 11.1
Effective thermal and electrical conductivities of the porous SiO_2–CNT nanocomposite as a function of porosity. Dots are the experimental results from Table 11.3 [31] and the lines represent the model from Eqs. (11.9) and (11.10).

It is interesting to note that the self-assembled structures can not only be used as scaffolds or in medical dressings but in many other applications as well. These applications are gaining a growing interest in the energetic sector, especially in the field of renewable energy, such as the valorization of CO_2 [41,42].

References

[1] Ke, L.J., Gao, G.Z., Shen, Y., Zhou, J.W., Rao, P.F. 2015. Encapsulation of aconitine in self-assembled licorice protein nanoparticles reduces the toxicity in vivo. *Nanoscale Research Letters* 10:449.

[2] Bufon, C.C.B., González, J.D.C., Thurmer, D.J., Grimm, D., Bauer, M., Schmidt, O.G. 2010. Self-assembled ultra-compact energy storage elements based on hybrid nanomembranes. *Nano Letters* 10:2506–2510.

[3] Hsieh, G.W., Beecher, P., Li, F.M., Servati, P., Colli, A., Fasoli, A., Chu, D., Nathan, A., Ong, B., Robertson, J., Ferrari, A.C., Milne, W.I. 2008. Formation of composite organic thin film transistors with nanotubes and nanowires. *Physica E* 40:2406–2413.

[4] Ding, J., Li, X., Wang, X., Zhang, J., Yu, D., Qiu, B. 2014. Pressure-assisted self-assembly technique for fabricating composite membranes consisting of highly ordered selective laminate layers of amphiphilic graphene oxide. *Carbon* 68:670–677.

[5] Rangharajan, K.K., Kwak, K.J., Conlisk, A.T., Wu, Y., Prakash, S. 2015. Effect of surface modification on interfacial nanobubble morphology and contact line tension. *Soft Matter* 11:5214–5223.

[6] Prevo, B.G., Kuncicky, D.M., Velev, O.D. 2007. Engineered deposition of coatings from nano- and microparticles: A brief review of convective assembly at high volume fraction. *Colloids and Surfaces A: Physicochemical and Engineering Aspects* 311:2–10.

[7] Zhang, R., Elkhooly, T.A., Huang, Q., Liu, X., Yang, X., Yan, H., Xiong, Z., Ma, J., Feng, Q., Shen, Z. 2017. A dual-layer macro/mesoporous structured TiO_2 surface improves the initial adhesion of osteoblast-like cells. *Materials Science and Engineering C* 78:443–451.

[8] Wang, D., Liu, S., Trummer, B.J., Deng, C., Wang, A. 2002. Carbohydrate microarrays for the recognition of cross reactive molecular markers of microbes and host cells. *Nature Biotechnology* 20:275–281.

[9] Smalyukh, I.I., Zribi, O.V., Butler, J.C., Lavrentovich, O.D., Wong, G.C.L. 2006. Structure and dynamics of liquid crystalline pattern formation in drying droplets of DNA. *Physical Review Letters* 96:177801.

[10] Zhang, D., Jiang, C., Sun, Y., Zhou, Q. 2017. Layer-by-layer self-assembly of tricobalt tetroxide-polymer nanocomposite toward high-performance humidity-sensing. *Journal of Alloys and Compounds* 711:652–658.

[11] Zhang, D., Liu, J., Chang, H., Liu, A., Xia, B. 2015. Characterization of a hybrid composite of SnO_2 nanocrystal-decorated reduced graphene oxide for ppm-level ethanol gas sensing application. *RSC Advances* 5:18666–18672.

[12] Ray, D., Sain, S. 2016. In situ processing of cellulose nanocomposites. *Composites A: Applied Science and Manufacturing* 83:19–37.

[13] Deegan, R.D., Bakajin, O., Dupont, T.F., Huber, G., Nagel, S.R., Witten, T.A. 1997. Capillary flow as the cause of ring stains from dried liquid drops. *Nature* 389:827–829.

[14] Bhardwaj, R., Fang, X., Attinger, D. 2009. Pattern formation during the evaporation of a colloidal nanoliter drop: A numerical and experimental study. *New Journal of Physics* 11:075020.

[15] Sommer, A.P., Franke, R. 2003. Biomimicry patterns with nanosphere suspensions. *Nano Letters* 3:573.

[16] Truskett, V.N., Stebe, K.J. 2003. Influence of surfactants on an evaporating drop: Fluorescence images and particle deposition patterns. *Langmuir* 19:8271–8279.

[17] Machrafi, H., Minetti, C., Dauby, P.C., Iorio, C.S. 2017. Self-assembly by multi-drop evaporation of carbon-nanotube droplets on a polycarbonate substrate. *Physica E* 85:206–213.

[18] Wen, W., Wu, J.M. 2014. Nanomaterials via solution combustion synthesis: A step nearer to controllability. *RSC Advances* 4:58090–58100.

[19] Aruna, S.T., Mukasyan, A.S. 2008. Combustion synthesis and nanomaterials. *Current Opinion in Solid State and Materials Science* 12:44–50.

[20] Tieke, B. 2011. Coordinative supramolecular assembly of electrochromic thin films. *Current Opinion in Solid State and Materials Science* 16:499–507.

[21] Decher, G., Hong, J.D. 1991. Build up of ultrathin multilayer films by a self-assembly process: I. Consecutive adsorption of anionic and cationic bipolar amphiphiles on charged surfaces. *Makromolekulare Chemie, Macromolecular Symposia* 46:321–327.

[22] Borges, J., Mano, J.F. 2014. Molecular interactions driving the layer-by-layer assembly of multilayers. *Chemical Reviews* 114:8883–8942.

[23] Hong, J., Park, H. 2011. Fabrication and characterization of block copolymer micelle multilayer films prepared using dip-, spin- and spray-assisted layer-by-layer assembly deposition. *Colloids and Surfaces A: Physicochemical and Engineering Aspects* 381:7–12.

[24] Kharlampieva, E., Sukhishvili, S.A. 2006. Hydrogen-bonded layer-by-layer polymer films. *Journal of Macromolecular Science, Part C Polymer Reviews* 46:377–395.

[25] Park, J., Moon, J. 2006. Control of colloidal particle deposit patterns within picoliter droplets ejected by inkjet printing. *Langmuir* 22:3506–3513.

[26] Sommer, A.P., Cehreli, M., Akca, K., Sirin, T., Piskin, E. 2005. Superadhesion: Attachment of nanobacteria to tissues–model simulation. *Crystal Growth and Design* 5:21–23.

[27] Andreeva, L.V., Koshkin, A.V., Lebedev-Stepanov, P.V., Petrov, A.N., Alfimov, M.V. 2007. Driving forces of the solute self-organization in an evaporating liquid droplet. *Colloids and Surfaces A: Physicochemical and Engineering Aspects* 300:300–306.

[28] Onoda, G., Somasundaran, P. 1987. Two- and one-dimensional flocculation of silica spheres on substrates. *Journal of Colloid and Interface Science* 118:169–175.

[29] Min, Y., Moon, G.D., Kim, C.E., Lee, J.H., Yang, H., Soon, A., Jeong, U. 2014. Solution-based synthesis of anisotropic metal chalcogenide nanocrystals and their applications. *Journal of Materials Chemistry C* 2:6222–6248.

[30] Gençer, A., Schütz, C., Thielemans, W. 2016. Influence of the particle concentration and marangoni flow on the formation of cellulose nanocrystal films. *Langmuir* 33:228–234.

[31] Machrafi, H., Minetti, C., Miskovic, V., Dauby, P., Dubois, F., Iorio, C. 2018. Self-assembly of carbon nanotube XE "carbon nanotube"-based composites by means of evaporation-assisted depositions: Importance of drop-by-drop self-assembly on material properties. *Materials Chemistry and Physics* 218:1–9.

[32] Smits, F.M. 1958. Measurement of sheet resistivities with the four-point probe. *Bell System Technical Journal* 34:711–718.

[33] Valdes, L.B. 1954. Resistivity measurements on germanium on transistors. *Proceedings of the Institute of Radio Engineers* 42:420–427.

[34] Glushchuk, A., Minetti, C., Machrafi, H., Iorio, C.S. 2017. Experimental investigation of force balance at vapour condensation on a cylindrical fin. *International Journal of Heat and Mass Transfer* 108:2130–2142.

[35] Roussel, F., Brun, J.F., Allart, A., Huang, L., O'Brien, S. 2012. Horizontally-aligned carbon nanotubes arrays and their interactions with liquid crystal molecules: Physical characteristics and display applications. *AIP Advances* 2:012110.

[36] Cao, M., Xiong, D.B., Tan, Z., Ji, G., Amin-Ahmadi, B., Guo, Q., Fan, G., Guo, C., Li, Z., Zhang, D. 2017. Aligning graphene in bulk copper: Nacre-inspired nanolaminated architecture coupled with in-situ processing for enhanced mechanical properties and high electrical conductivity. *Carbon* 117:65–74.

[37] Hossain, M.M., Islam, M.A., Shima, H., Hasan, M., Lee, M. 2017. Alignment of carbon nanotubes in carbon nanotube XE "carbon nanotube" fibers through nanoparticles: A route for controlling mechanical and electrical properties. *ACS Applied Materials & Interfaces* 9:5530–5542.

[38] Sinha, S., Barjami, S., Iannacchione, G., Schwab, A., Muench, G. 2005. O-axis thermal properties of carbon nanotube XE "carbon nanotube" films. *Journal of Nanoparticle Research* 7:651–657.

[39] Chu, K., Yun, D.J., Kim, D., Park, H., Park, S.H. 2014. Study of electric heating effects on carbon nanotube XE "carbon nanotube" polymer composites. *Organic Electronics* 15:2734–2741.

[40] Koziol, K.K., Janas, D., Brown, E., Hao, L. 2017. Thermal properties of continuously spun carbon nanotube XE "carbon nanotube" fibres. *Physica E* 88:104–108.

[41] Machrafi, H., Cavadias, S., Amouroux, J. 2011. CO_2 valorization by means of dielectric barrier discharge. *Journal of Physics: Conference Series* 275:012016.

[42] Amouroux, J., Cavadias, S. 2017. Electrocatalytic reduction of carbon dioxide under plasma DBD process. *Journal of Physics D: Applied Physics* 50:465501.

Epilogue

The principles of extended non-equilibrium thermodynamics have been presented, accompanied by a new methodology of applying its principles to real experimental situations. The methods proposed allow combining our theory with other fields in order to provide for comprehensive, yet easy-to-use, models that are able to describe physical phenomena at nanoscale. We studied heat transfer, electrical conduction, thermoelectric processes, nanofluids, nanoporous flow (permeability) and photovoltaic energy. The proposed models therein are in some cases extensions of existing models and in other cases stem from new developments. Finally, these models were used to study some real-life cases, such as solar panels, nanomedicine drug carriers and self-assembled structures.

The developed theory has been shown to be useful and compared well with experimental results. Nevertheless, the use of the models and the underlying theory has its limits and conditions, which are discussed at their appropriate places. Also, the themes studied are not exhaustive, but rather representative examples of topics that are of great interest nowadays. Moreover, there are other topics that are of interest, and it is the aim that this book has laid an introductory set of tools to meet the demands addressing those topics that are, of course, open for further developments.

Unit Conversions

The table hereunder shows conversion values between SI and imperial units for quantities used in this book, classified in different categories. The **SI** units are in bold. Sometimes, different units can be used for describing the same quantity. In that case, not all of them are used, but each reader familiar with the corresponding unit system can make an easy conversion (e.g. Btu h^{-1} ft^{-2} is the same as hp ft^{-2}). Also, conversions are not given for all decimal multiples and sub-multiples (e.g. kg and g). The same goes for the time (e.g. h and s).

Category	To Convert From	To	Multiply By
Pressure	lb_f in.$^{-2}$ (psi)	**Pa**	6,895
Pressure	**Pa**	lb_f in.$^{-2}$ (psi)	0.000145
Density	**g cm^{-3}**	lb ft^{-3}	62.428
Density	lb ft^{-3}	**kg m^{-3}**	16.02
Density	lb in.$^{-3}$	**kg m^{-3}**	27679.9
Density	lb ft^{-3}	**g cm^{-3}**	0.016
Electric field	volts mil^{-1}	**kV mm^{-1}**	0.0394
Distance	mil (0.001 in.)	**cm**	0.00254
Distance	**cm**	mil	393.7
Heat capacity	**J (gK)$^{-1}$**	Btu (lb °F)$^{-1}$	0.239
Heat capacity	Btu (lb °F)$^{-1}$	**J (gK)$^{-1}$**	4.184
Energy	**J**	cal	0.239
Energy	cal	**J**	4.184
Energy	**J**	Btu	0.0009485
Energy	Btu	**J**	1054.35
Work	**J**	ft lb	0.738
Work	ft lb	**J**	1.356
Power	**W**	Btu h^{-1}	3.413
Power	Btu h^{-1}	**W**	0.293
Power	**W**	hp	0.00134
Power	hp	**W**	746
Heat flux	**W m^{-2}**	Btu (h ft^2)$^{-1}$	0.317
Heat flux	Btu (h ft^2)$^{-1}$	**W m^{-2}**	3.153
Thermal conductivity	**W (mK)$^{-1}$**	Btu in. (h ft^2 °F)$^{-1}$	6.9335
Thermal conductivity	Btu in. (h ft^2 °F)$^{-1}$	**W (mK)$^{-1}$**	0.1441
Thermal conductivity	**W (mK)$^{-1}$**	J m (min m^2 K)$^{-1}$	60
Thermal conductivity	J m (min m^2 K)$^{-1}$	**W (mK)$^{-1}$**	0.0167
Heat transfer coefficient	**W m^{-2} K^{-1}**	Btu (h ft^2 °F)$^{-1}$	0.176
Heat transfer coefficient	Btu (h ft^2 °F)$^{-1}$	**W m^{-2} K^{-1}**	5.675
Permeability	**m^2**	in.^{2}	1550
Permeability	in.^{2}	**m^2**	0.0006452
Dynamic viscosity	**Pa s**	lb (ft s)$^{-1}$	0.672
Dynamic viscosity	lb (ft s)$^{-1}$	**Pa s**	1.488

(*Continued*)

Category	To Convert From	To	Multiply By
Velocity	$\mathbf{m\ s^{-1}}$	ft s^{-1}	3.281
Velocity	ft s^{-1}	$\mathbf{m\ s^{-1}}$	0.3408
Force	**N**	lb_f	0.2248
Force	lb_f	**N**	4.448
Mass	**kg**	lb	2.205
Mass	lb	**kg**	0.454
Electric conductivity	$\mathbf{S\ m^{-1}}$ $(S = 1/\Omega)$	ft $(\Omega$ circ-mil$)^{-1}$ (1 circ-mil = $\frac{\pi}{4}$ mil^2)	$1.66 * 10^{-9}$
Electric conductivity	ft $(\Omega$ circ-mil$)^{-1}$	ft $(\Omega$ circ-mil$)^{-1}$	$6.02 * 10^{8}$

Index

I

K

L

M

N

O

T